INDUSTRIAL ELECTRONICS

Devices, Systems and Applications

INDUSTRIAL ELECTRONICS

Devices, Systems and Applications

Terry Bartelt

Fox Valley Technical College

Delmar Publishers

an International Thomson Publishing company I(T)P®

Albany · Bonn · Boston · Cincinnati · Detroit · London · Madrid
Melbourne · Mexico City · New York · Pacific Grove · Paris · San Francisco
Singapore · Tokyo · Toronto · Washington

NOTICE TO THE READER

Cover Design: Anne Pompeo

Delmar Staff

Publisher: Robert D. Lynch
Administrative Editor: Paul Shepardson
Developmental Editor: Michelle Ruelos Cannistraci

Project Editor: Christopher B. Chien
Production Manager: Larry Main
Art/Design Coordinator: Mary Beth Vought

COPYRIGHT © 1997
by Delmar Publishers
a division of International Thomson Publishing Inc.
The ITP logo is a trademark under license.

Printed in the United States of America

For more information, contact:

Delmar Publishers
3 Columbia Circle, Box 15015
Albany, New York 12212-5015

International Thomson Publishing–Europe
Berkshire House 168-173
High Holborn
London WC1V 7AA
England

Thomas Nelson Australia
102 Dodds Street
South Melbourne, 3205
Victoria, Australia

Nelson Canada
1120 Birchmount Road
Scarborough, Ontario
Canada M1K 5G4

International Thomson Editores
Campos Eliseos 385, Piso 7
Col Polanco
11560 Mexico D F Mexico

International Thomson Publishing GmbH
Königswinterer Strasse 418
53227 Bonn
Germany

International Thomson Publishing Asia
221 Henderson Road
#05 - 10 Henderson Building
Singapore 0315

International Thomson Publishing–Japan
Hirakawacho Kyowa Building, 3F
2-2-1 Hirakawacho
Chiyoda-ku, Tokyo 102
Japan

2 3 4 5 6 7 8 9 10   XXX   01 00 99 98

Library of Congress Cataloging-in-Publication Data

Bartelt, Terry L. M.
    Industrial control electronics : devices, systems and applications / Terry Bartelt.
      p. cm.
    Includes index.
    ISBN 0-8273-6104-1
    1. Industrial electronics. 2. Electronic control. I. Title.
TK7881.B37   1997
629.8'9—dc20
                             96-35775
                              CIP

Contents

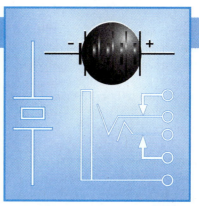

SECTION II ELECTRIC MOTORS

CHAPTER 4
DC MOTORS   101

CHAPTER 5
AC MOTORS   129

CHAPTER 6
SERVO MOTORS   159

SECTION III ELECTRONIC MOTOR DRIVE SYSTEMS

SECTION IV PROCESS CONTROL

SECTION V
PROGRAMMABLE CONTROLLERS

CHAPTER 14
INTRODUCTION TO PROGRAMMABLE CONTROLLERS 329

CHAPTER 15
FUNDAMENTAL PLC PROGRAMMING 355

CHAPTER 16
ADVANCED PROGRAMMING AND PLC APPLICATIONS 387

SECTION VI MOTION CONTROL

CHAPTER 17
MOTION CONTROL FEEDBACK DEVICES   419

CHAPTER 18
ELEMENTS OF MOTION CONTROL   443

CHAPTER 19
FUNDAMENTALS OF SERVOMECHANISMS   463

CHAPTER 20
FUNCTIONAL INDUSTRIAL SYSTEMS   485

DEDICATION

I dedicate this book to Carol, Wendy, and Holly.

Preface

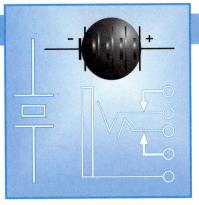

INTENDED USE AND LEVEL

This book is intended for use in electronics-based industrial courses. It is designed for two-year technology programs such as Electronics Technology, Electromechanical Technology, Automated Manufacturing Systems, and Instrumentation Technology, as well as four-year university industrial technology programs. Portions of the book also are appropriate for industrial apprenticeship programs, continuing education classes, and self-study pursuits.

This book is written at the midrange level of difficulty. It is designed to be used in courses that teach industrial motion control (servomechanisms) and process control (instrumentation) systems. It provides the concepts, then describes the operation of electronic devices, circuits, systems, and applications used in industry. It does not include any information on design concepts. To understand the material in the text the student should have completed courses in DC and AC, solid state devices, and digital electronics.

SUBJECTS AND ORGANIZATION

The book is organized in a logical sequence that first examines a block diagram of an industrial control system and then expands on the function of each block in detail. The amount of material contained in the text exceeds what can be covered in a single course. Therefore, to help identify topic areas that may be appropriate for particular courses, the book is divided into sections that contain related chapters.

SECTION I: INDUSTRIAL CONTROL OVERVIEW

Chapter 1 introduces the basic concepts of open- and closed-loop systems common to motion and process control.

Chapter 2 describes the operation of solid-state components and ICs used in circuits throughout the text. These components are utilized within each block of a control system for transmitting signals from one block to another or for providing power to an actuator that drives the manufacturing process.

Chapter 3 describes On/Off, PID, and Fuzzy Logic operations performed by the control block of a closed loop system.

SECTION II: ELECTRIC MOTORS

Chapters 4, 5, and 6 cover the most common devices used in the actuator block of a closed-loop system: DC and AC electrical motors. Several types of specialty motors used in motion control operations are also discussed.

SECTION III: ELECTRONIC MOTOR DRIVE SYSTEMS

Chapters 7 and 8 describe the operation of DC drives, VVI, pulse width modulation, and vector AC drives.

SECTION IV: PROCESS CONTROL

Chapters 9 through 12 explain the fundamental concepts of the four major categories of instrumentation: pressure, temperature, flow, and level. Various measurement methods are described, and the operation of mechanical and electronic measurement sensors is explained.

Chapter 13 describes the operation of sensors that detect the presence or absence of an object. These measurement devices are used in both motion and process control applications. The sensors include mechanical switches, proximity detectors, and photoelectronic devices. Sensor interfacing concepts are discussed.

SECTION V: PROGRAMMABLE CONTROLLERS

Chapters 14, 15, and 16 cover the generic operation of the PLC. The material includes ladder logic fundamentals, PLC hardware and wiring, PLC programming, advanced programming operations, and application examples.

SECTION VI: MOTION CONTROL

Chapters 17, 18, and 19 describe the fundamental concepts of motion control servomechanisms. Velocity and position measurement detectors are included.

Chapter 20 puts all of the foregoing concepts together and provides practical examples of motion and process control systems commonly used in industrial applications.

FEATURES

- ○ A writing style that is easy to read and understand
- ○ Over 500 two-color diagrams that illustrate the operation of devices and systems
- ○ Coverage of both motion control (servomechanisms) and process control (instrumentation) systems
- ○ Application examples throughout the book
- ○ A complete glossary of terms following the answer section in the back of the text
- ○ An instructor's resource manual which contains the answers to all chapter problems; a test book (including worksheets for selected chapters) with over 450 questions requiring critical thinking and experiments; and a list of vendors who provide laboratory trainers for some of the sophisticated circuits and equipment described in the book.

ACKNOWLEDGMENTS

The author and Delmar Publishers gratefully acknowledge the contributions of the following reviewers:

James Ahneman
Chippewa Valley Technical College
Eau Claire, WI

James L. Brubley
Athens Area Technical Institute
Athens, GA

Richard Anthony
Cuyahoga Community College
Cleveland, OH

Jacob Miller
RETS Institute
Nutley, NJ

Don Arney
IVY Tech State College
Terre Haute, IN

Roy Powell
Chattanooga State Tech College
Chattanooga, TN

Tim Baker
John A. Logan College
Marion, IL

Le Tang
California State University,
 Los Angeles
Los Angeles, CA

I would like to express my appreciation to Terry Fleischman, colleagues, students, and reviewers who provided many valuable suggestions for the manuscript. I also thank Sharon Green of Graphics West, Inc. and copy editor Mark Arnest for their conscientious work during the final stages of production.

The following companies have graciously contributed materials for this book:

Kollmorgen Motion Technologies Group

Digital Electronics, T. Bartelt, 1991, Prentice-Hall

Omron Electronics, Inc.

Data Instruments, Inc.

Copyright of Motorola, used by permission.

Honeywell's MICRO SWITCH division

Turck Inc.

Allegro Micro Systems, Inc.

Banner Engineering Corporation

Efector, Inc.

Sunx Sensors

Rockwell International—Allen-Bradley

BEI, Industrial Encoder Division

MTS Systems Corporation Consulting, Inc.

Plant Engineering Publications

Sure Controls

Terry Bartelt
Fox Valley Technical College

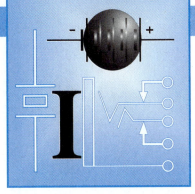
INDUSTRIAL CONTROL OVERVIEW

Section one introduces key concepts in industrial control: control system classifications, interfacing devices, and controllers. The first chapter introduces the student to the ways in which industrial control systems are classified. It then provides an introductory overview of the elements that make up an industrial control loop. The remaining sections describe each of these elements in detail so that the entire spectrum of industrial control is addressed.

Chapter two describes the operation of discrete components and integrated circuits that are used throughout the book.

Chapter three describes the operation of the "brain" of the industrial control loop: the controller element. It addresses the operational techniques performed by controllers that use discrete components or computer software to perform their functions. Various control modes such as On-Off, PID (proportional-integral-derivative), and fuzzy logic are described.

CHAPTER 1

Introduction to Industrial Control Systems

CHAPTER OBJECTIVES

At the completion of this chapter, you should be able to:

○ List the classifications of industrial control systems.

○ Describe the differences between the different industrial control systems and provide examples of each type.

○ Define the following terms associated with industrial control systems:

Servos	Continuous
Servomechanisms	Instrumentation
Batch	

○ Describe the differences between open-loop and closed-loop systems.

○ Define the following terms associated with open- and closed-loop systems:

Negative Feedback	Error Signal
Controlled Variable	Controller
Measurement Device	Actuator
Feedback Signal	Manufacturing Process
Set Point	Disturbance
Error Detection	

○ List the factors that cause process lag time.

○ Describe the operation of Feed-Forward Control.

INTRODUCTION

The industrial revolution began in England during the mid-1700s when it was discovered that productivity of spinning wheels and weaving machines could be dramatically increased by fitting them with steam-powered engines. Further inventions and new ideas in plant layouts during the 1850s enabled the United States to surpass England as the manufacturing leader of the world. Around the turn of the twentieth century, the electric motor replaced steam and water wheels as a power source. Factories became larger, machines were improved to allow closer tolerances, and the assembly line method of mass production was created.

Between World Wars I and II, the feedback control system was developed, enabling manually operated machines to be replaced by automated equipment. The feedback control system is a key element in today's manufacturing operations. The term **industrial controls** is used to define this type of system, which automatically monitors manufacturing processes being executed and takes appropriate corrective action if the operation is not performing properly.

During World War II, significant advances in feedback technology occurred due to the sophisticated control systems required by military weapons. After the war, the techniques used in military equipment were applied to industrial controls to further improve the quality of products and to increase productivity.

Because many modern factory machines are automated, the technicians who install, troubleshoot, and repair them need to be highly trained. To perform effectively, these individuals must understand the elements, operational theory, and terminology associated with industrial control systems.

Industrial control theory encompasses many fields, but uses the same basic principles whether controlling the position or speed of a motor, or the temperature and pressure of a manufacturing process.

In this chapter, the various types of industrial control systems, their characteristics, and important terminology will be studied.

1-1   INDUSTRIAL CONTROL CLASSIFICATIONS

MOTION AND PROCESS CONTROLS

Industrial control systems are often classified by *what* they control: either motion or process.

Motion Control

A **motion control** system is an automatic control system that controls the physical motion or position of an object. One example is the industrial robot arm which performs welding operations and assembly procedures.

There are three characteristics that are common to all motion control systems. First, motion control devices control the position, speed, or acceleration of an object. Second, the motion or position of the object being controlled is measured. Third, motion devices typically respond to input commands within fractions of a second, rather than seconds or minutes, as in process control.

Motion control systems are also referred to as *servos,* or *servomechanisms.*

Process Control

The other type of industrial control system is **process control**. In process control, one or more variables are regulated during the manufacturing of a product. These variables may include temperature, pressure, flow rate, liquid and solid level, pH, or humidity.

This regulated process must compensate for any outside disturbance that changes the variable. The response time of a process control system is slow, and can vary from a few seconds to several minutes. Process control is the type of industrial control system most often used in manufacturing. Process control systems are divided into two categories, *batch* and *continuous*.

Batch Process   **Batch processing** is a sequence of timed operations executed on the product being manufactured. An example is an industrial machine that produces various types of cookies, as shown in Figure 1-1. Suppose that chocolate-chip cookies are made in the first production run. First, the oven is turned on to the desired temperature. Next, the required ingredients in proper quantities are dispensed into the sealed mixing chamber. A large blender then begins to mix the contents.

After a few minutes, vanilla is added, and the mixing process continues. After a prescribed period of time, the dough is the proper consistency, the blender stops turning, and the compressor turns on to force air into the mixing chamber. When the air pressure reaches a certain point, the conveyer belt turns on. The pressurized air forces the dough through outlet jets onto the belt. The dough balls become fully baked as they pass through the oven. The cookies cool as the belt carries them to the packaging machine.

After the packaging step is completed, the mixing vat, blender, and conveyer belt are washed before a batch of raisin-oatmeal cookies is made. Products from foods to

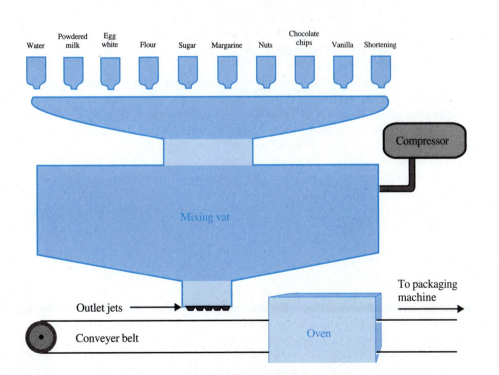

FIGURE 1-1 Batch processing cookie machine

petroleum to soap to medicines are made from a mixture of ingredients that undergo a similar batch process operation.

Batch process is also known as *sequence* (or *sequential*) *process.*

Continuous Process   In the **continuous process** category, one or more operations are being performed as the product is being passed through a process. Raw materials are continuously entering and leaving each process step. Producing paper, as shown in Figure 1-2, is an example of continuous process. Water, temperature, and speed are constantly monitored and regulated as the pulp is placed on screens, fed through rollers, and gradually transformed into a finished paper product. The continuous process can last for hours, days, or even weeks without interruption. Everything from wire to textiles to plastic bags is manufactured by using a continuous manufacturing process similar to the paper machine.

Another term commonly used to describe process control is *instrumentation.*

OPEN- AND CLOSED-LOOP SYSTEMS

Industrial control systems are also classified by *how* they control, either manually in an *open-loop* system, or automatically in a *closed-loop* system.

Open-Loop Systems

An open-loop configuration is the simplest way to control a system. In an open-loop (or a *manual control*) system, a command signal is formed to inform the output device what to do. This signal usually needs to be processed or modified to be usable by an output actuator device. In response to the command, the actuator will either cause a machine movement to occur or a process to be performed.

The problem with this configuration is that the output signal will not be appropriate in all circumstances. The signal becomes ineffective if some external disturbance changes the load on a machine or process being performed. For example, variations of the load on a motor system will cause the RPM to fluctuate. To make any necessary modifications of the command signal when the load changes occur, some degree of physical effort on the part of a human operator is required.

An example of a manually operated open-loop system is a home heated by a fireplace. A person throws on more logs when the temperature is cool than when the temperature is moderate. Another application is the speed of a car being controlled by the driver. The driver adjusts the throttle to maintain a highway speed when going uphill, downhill, or on level terrain.

Closed-Loop Systems

There are many situations in industry where the open-loop system is adequate. However, some manufacturing applications require continuous monitoring and self-correcting action of the operation for long periods of time without interruption. The automatic closed-loop configuration performs the self-correcting function. This automatic system employs a feedback loop to keep track of how well the output actuator is doing the job it was commanded to do. A feedback signal is produced by a sensing component that measures the status of the output. This signal is then fed back to the input of the system and compared to the input command signal. If a difference between them is detected, an error signal forms that represents the difference between the desired output and the actual output. The error signal is usually amplified and then used to initiate an output action that will cause the discrepancy to approach zero. The term *closed-loop* is derived from the fact that once the command signal is entered, it travels around the loop until equilibrium is restored.

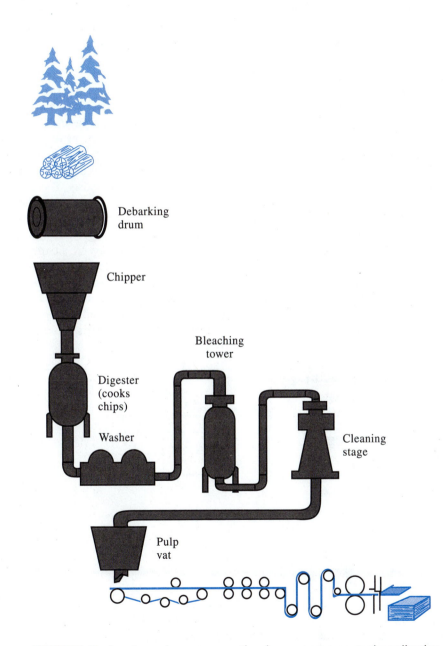

FIGURE 1-2 A pulp and paper operation is a process control application

Feedback signals may be either positive or negative. If the feedback signal's polarity aids a command input signal, it is said to be positive or regenerative feedback. Positive feedback is used in radios. If the radio signal is weak, an Automatic Gain Control (AGC) circuit is activated. Its output is a feedback signal that boosts the radio signal's overall strength.

However, when positive feedback is used in industrial closed-loop systems, the input usually loses control over the output. If the feedback signal opposes the input signal, the system is said to use negative or degenerative feedback. By combining negative feedback values from the command signal, a closed-loop system works properly.

An example of closed-loop control that uses negative feedback is the central heating system in the house. The thermostat in Figure 1-3 monitors the temperature in the house and compares it to the desired reference setting. Suppose the room temperature drops to 66 degrees from the reference setting of 72 degrees. The measured feedback value is subtracted from the setpoint command and causes a six degree discrepancy. The thermostat contacts will close and cause the furnace to turn on. The furnace supplies heat until the temperature is back to the reference setting. When the negative feedback is sufficient to cancel the command, the error no longer exists. The thermostat then opens and switches the furnace off until the house cools down below the reference. As this cycle repeats, the temperature in the house is automatically maintained without human intervention.

The speed of an automobile can also be controlled automatically by a closed-loop system called a cruise control. The desired speed is set by an electronic mechanism usually placed on the steering wheel assembly. A Hall effect speed sensor connected to the front axle generates a signal proportional to actual speed. An electronic error detector compares the actual speed to the desired speed, and then sends a signal representing the difference between them to a controller. The controller sends a demand

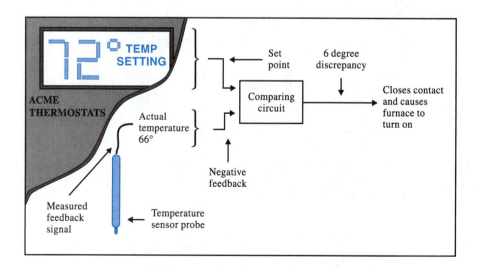

FIGURE 1-3 A thermostat uses a negative feedback signal to control the temperature of a house

signal to a vacuum device called an actuator. A part of the vacuum mechanism is a rod connected to the throttle, which varies the fuel flow to the engine. If a car that is traveling on a level road suddenly encounters an uphill grade, it begins to slow down. Because the actual speed is lower than the desired speed, the error detector sends a signal to the actuator. A vacuum is varied which causes the rod to move the throttle so that more fuel flows to the engine. The additional fuel causes the car to accelerate until it reaches the desired speed.

1-2   OPEN- AND CLOSED-LOOP LEVEL CONTROL

A reservoir system used to store water for an irrigation system is a good example of an application that may be either open-loop or closed-loop.

Figure 1-4 shows a representation of an open-loop configuration. A natural spring supplies water to the reservoir. The reservoir is used as a back-up water supply if the pipe that supplies water to the irrigation system becomes plugged or some other malfunction occurs. The diagram shows that the reservoir system is composed of a storage tank, an inlet pipe with a manual control valve, and an outlet pipe. Ideally, the control valve setting and the size of the outlet pipe are exactly the same. When this occurs, the water level in the tank remains the same. Therefore the process reaches a steady state condition, or is said to be *balanced*. The problem with this design is that any change or disturbance will upset the condition of balance. For example, a substantial rainfall may occur. Since there is more water entering the tank than exiting, the level will rise. If this situation is not corrected, the tank will eventually overflow. Excessive evaporation would also upset the balance. If it occurs over a prolonged period of time, the water level in the tank may become unacceptably low.

A human operator who periodically inspects the tank can change the control valve setting to compensate for these disturbances.

To eliminate the need for human operator corrections, the reservoir system can be modified into a closed-loop system to allow for automatic control of the process. The

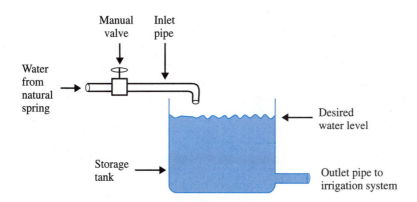

FIGURE 1-4 An open-loop reservoir system which stores water for an irrigation system

change in the system is the replacement of the manually controlled valve with an automatic valve controlled by a float, as shown in Figure 1-5.

If the level of the water in the tank goes up, the float is pushed upward; if the level goes down, the float moves downward. The float is connected to the inlet valve by a mechanical linkage. As the water level rises, the float moves upward, pushing on the lever and closing the valve, thus reducing the water flow into the tank. If the water level lowers, the float moves downward, pulling on the lever and opening the valve, thus admitting more water into the tank. To adjust for a desired level of water in the tank, the float is moved up or down on the float rod A.

Most output variables produced in automated manufacturing processes use closed-loop control. These systems that have a self-regulation capability are designed to produce continuous balance.

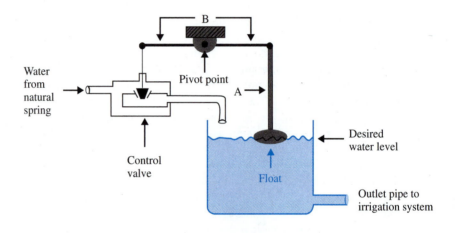

FIGURE 1-5 A closed-loop system that uses a linkage mechanism as a feedback device to provide self-correcting capabilities

1-3   ELEMENTS OF OPEN- AND CLOSED-LOOP SYSTEMS

A block diagram of a closed-loop control system is shown in Figure 1-6. Each block shows an element of the system that performs a significant function in the operation. The lines between the blocks show the input and output signals, of each element, and the arrowheads indicate the direction in which they flow.

This section describes the functions of the blocks, their signals, and common terminology used in a typical closed-loop network:

Controlled Variable.   The controlled variable is the actual output variable being manipulated in the manufacturing process. Examples in a process control system may include temperature, pressure, and flow rate. Examples in a motion control system may be position or velocity. In the water reservoir system (Figure 1-5), the water level is the controlled variable.

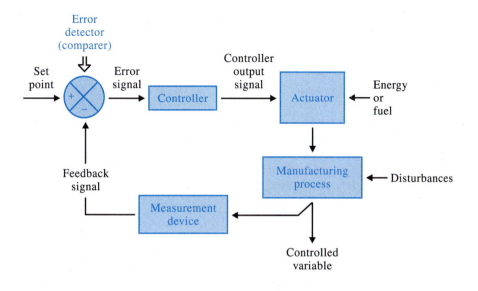

FIGURE 1-6 Closed-loop block diagram that shows elements, input/output signals, and signal direction

Measurement Device.   The measurement device is the "eye" of the system. It detects the controlled variable and produces an output signal that represents its status. Examples in a process control system may include a thermistor to measure temperature or a humidity detector to measure moisture. Examples in a motion control system may be an optical device to measure position or a transducer to measure speed. In the water reservoir system, the float is the measurement device. Other terms used are *detector, transducer,* and *sensor.*

Feedback Signal.   The feedback signal is the output of the measurement device. In the water reservoir system, the feedback signal is the angular position of member A in the linkage mechanism. Other terms used are *measured value, position feedback* if in a position loop, or *velocity feedback* if in a velocity loop.

Set Point.   The set point is the prescribed input value that indicates the desired operating point of the manufacturing process. The setting is provided either by a human operator or an electronic device such as a computer. In the water reservoir system, the set point is the location of the float above the bottom of the tank. Other terms used are *command* and *reference.*

Error Detector.   The error detector compares the set point to the feedback signal. It then produces an output signal that is proportional to the difference between them. In the water reservoir system, the error detector is the entire linkage mechanism. Other terms used are *comparator* or *comparer* and *summing junction.*

Error Signal.   The error signal is the output of the error detector. If the set point and the feedback signal are not equal, an error signal proportional to their difference develops. When the feedback and set point signals are equal, the error signal goes to zero. In the reservoir system, the error signal is the angular position of member B of the linkage mechanism. Other terms used are *difference signal* and *deviation.*

Controller.   The controller is the "brain" of the system. It receives the error signal (for closed-loop control) as its input and develops a conditioned signal to properly manipulate the controlled variable output of the system. This operation is performed by hardwired circuitry or computer software. The controller usually includes an amplifier at its output to provide the signal with enough energy to drive the device that processes the controlled variable. For open-loop control, the controller receives the setpoint signal as its input.

Actuator.   The actuator is the "muscle" of the system. It is a device that uses some type of energy or fuel to directly affect the process. The actuator alters some manufacturing process or movement to change the controlled variable so that it matches the desired setpoint value. In the reservoir system, the actuator is the control valve. Another term used is *final correcting device.*

Manufacturing Process.   The manufacturing process is the operation performed by the actuator to control a physical variable, such as the motion of a machine, or the processing of a liquid.

Disturbance.   A disturbance is a factor that upsets the manufacturing process being performed, causing a change in the controlled variable. In the reservoir system, the disturbances are the rainfall and evaporation that alter the water level.

PROCESS LAG TIME

Most of the processes or movements performed during manufacturing operations are easily controlled if the amounts they are permitted to vary fall within certain limits. However, when a process is subjected to change due to disturbances or setpoint alterations, it takes a certain amount of time for a closed-loop system to correct itself and reestablish a balanced condition. The term *process lag time* is used to describe this delay. The factors that cause process lag time include inertia, capacitance, resistance, and dead-time.

Inertia refers to the ability of a body to *continue* in a certain energy state after a change has occurred. For example, an object in motion such as a car will resist a change that would slow it down.

Capacity refers to the ability of a body to *store* energy. For example, a furnace will retain heat after it has been turned off.

Resistance refers to the ability of a body to *oppose* a transfer of energy when a process change occurs. For example, the heat held by the walls of the furnace must be dissipated before the change occurs.

Dead-time refers to the *time* required for a body to change from one value to another after the process has been altered. For example, after a furnace has been turned off, there is a duration of time when no change occurs. Capacity, resistance, and process inertia all have some influence on dead-time delay.

The four causes of process lag time all must be overcome before any significant system change can occur.

FEED-FORWARD CONTROL

Two conditions can make feedback control ineffective. The first is the occurrence of large magnitude disturbances. The second is a large process time lag. To compensate for these limitations of feedback control, feed-forward control can be used.

The operation of feed-forward control is very different than feedback control. Feedback control takes corrective action after an error develops. The object of feed-forward control is to prevent errors from occurring.

The reservoir tank in Figure 1-7 demonstrates the operation of a feed-forward control system. There are two inlet ports (P_1 and P_2) that provide water for the tank and one outlet port (P_o). The flow rates at all three ports are measured by flowmeters. Only one port, P_1, is regulated by a flow valve.

To maintain a steady level, the water must flow into the tank at the same rate it flows out of the tank, as stated by the following equation:

$$\text{(Flow In)} \qquad \text{(Flow Out)}$$
$$P_1 + P_2 \quad = \quad P_o$$

If a disturbance occurs that changes P_2 or P_o, the balanced condition will be upset and the water will deviate from the desired level. By modifying the formula, a new equation is derived to determine the corrective action needed to cancel the effect of the disturbance:

$$P_1 = P_o - P_2$$

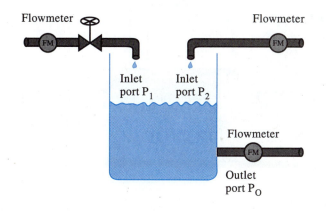

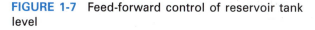

FIGURE 1-7 Feed-forward control of reservoir tank level

By subtracting the inlet flow rate of P_2 from the outlet flow rate of P_o, the flow rate necessary at P_1 can be determined to maintain the desired level. The calculation is made by a feed-forward controller. The number it produces is converted into a proportional electrical signal. The signal causes the valve at P_1 to allow greater or lesser flow before the water level has a chance to deviate significantly.

The feed-forward control system does not operate perfectly. There are always unmeasurable disturbances which cannot be detected, such as a worn flow valve, a sensor out of tolerance, or inexact mathematical calculations. Over a period of time, these unmeasurable disturbances affect operation and eventually the water in the tank would reach an unacceptable level. Due to the inaccuracy of feed-forward control, it is seldom used by itself. By adding feedback control to the system, corrections to unmeasurable disturbances can be made.

A block diagram of a closed-loop feed-forward system with feedback control is shown in Figure 1-8.

Feed-forward control adjusts the operation of the actuator to prevent changes in the controlled variable. Feed-forward controllers must make very sophisticated calculations to compute the changes of the actuator needed to compensate for variations in disturbances. Since they require highly skilled engineers, they typically are used only in critical applications within the plant.

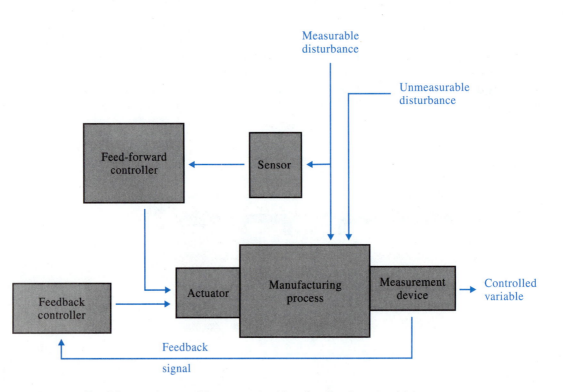

FIGURE 1-8 Feed-forward control loop used with a feedback control loop

CHAPTER PROBLEMS

1. The two classifications of industrial control systems are _____

   control and _____ control.

2. List another name for the following terms.

   Motion Control

   Process Control

   Batch Process

3. A closed-loop industrial system typically uses _____ (negative, positive) feedback.

4. List two examples of controlled variables for motion control applications and two examples for process control applications.

   Motion Control

   Process Control

5. List one example of a measurement device for a motion control application and one example for a process control application.

   Motion Control

   Process Control

6. The output of the measurement device is called the _____ _____ .

7. Define set point.

8. The difference between the set point and feedback signal is referred to as the

   _____ signal, and is produced by the _____ detector.

9. _ T _ F   The controller can be considered the brain of a closed-loop system.

10. The device that provides the muscle to perform work in the closed-loop system is referred to as the _____.

11. The factor that upsets the manufacturing process is referred to as the _____

    _____.

12. List four factors that cause process lag time.

13. Which of the following conditions are compensated for by using feed-forward control? _____
    a. Excessive lag time   c. An error signal
    b. Large disturbances   d. Feedback signal

14. Feed-forward control makes corrections for _____ disturbances,

    and feedback control makes corrections for _____ disturbances.
    a. measurable   b. unmeasurable

CHAPTER 2

Interfacing Devices

CHAPTER OBJECTIVES

When you complete this chapter, you will be able to:

○ Identify the schematic diagrams, describe the operations, and calculate the outputs of inverting, summing, and non-inverting operational amplifiers (op amps), and bipolar transistor amplifier devices.

○ Identify the schematic diagrams of the Integrator and Differentiator op amps and draw the output waveforms they produce when various input signals are applied.

○ Given applied input signals, draw the resulting outputs of the open-loop op amp, difference op amp, and digital comparator devices.

○ Describe the wave-shaping capability and operating characteristics of a Schmitt trigger.

○ Explain how to turn transistor and thyristor semiconductor switching devices on and off, and draw the output signals they produce.

○ Determine how optoelectronic devices are switched and explain the isolation function they perform.

○ Explain the operation of analog-to-digital and digital-to-analog converters, determine their resolution, and make the proper wiring connections to their integrated circuit packages.

○ Assemble monostable and astable multivibrators using a 555 monolithic integrated circuit and use calculations to determine their output.

INTRODUCTION

In Chapter 1, a block diagram was used to describe the operation of the elements of a closed-loop system. Each element plays a significant role in the operation of the system. One of the requirements of any system is the successful interfacing or connecting together of the various blocks. In a block diagram, an interface is represented by the lines between two blocks, indicating that some type of signal passes from one to another.

17

There are many components and circuits that perform the functions of each element. Sometimes the signals processed in one element are incompatible with those that can be used in the next element. To make the elements compatible, various types of conversion components are used to interface them together.

To help the reader understand the material covered in later chapters, this chapter describes the basic operation of discrete components and integrated circuits that are used within and between the elements. Readers who have learned about these components in a previous course may use this material as a review. The components covered here have been selected based on the types of circuits that are covered in the remainder of the book.

2-1 AMPLIFIERS

Amplification is a control function employed by many types of industrial equipment. Amplification involves converting a weak signal into a higher power signal. For example, the output of a controller, such as a microprocessor or computer, may be used to drive a servo valve that requires a larger signal to operate. Amplification is achieved by using several types of solid state devices. Several amplifier devices will be described, including the bipolar transistor and various operational amplifier circuits.

TRANSISTORS

The transistor is constructed by sandwiching a thin layer of one type of semiconductor material between two layers of another type of material. For example, the NPN transistor in Figure 2-1(a) is formed by placing a layer of P-type (positive) material between two layers of N-type (negative) material. The PNP transistor in Figure 2-1(b) is formed the opposite way. The three layers are called regions and are identified as the **emitter** (E), **base** (B), and **collector** (C). Leads that extend from each region are given the same names. Figure 2-1(c) shows the schematic symbols for the NPN and PNP transistors. The only difference is the direction in which the arrow at the emitter

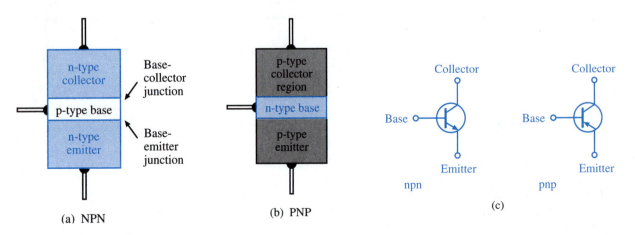

FIGURE 2-1 The bipolar junction transistor

points: The arrow on the emitter of the NPN transistor points away from the base, while the arrow on the PNP transistor points toward the base. The transistor has two PN junctions, which gives it the name "bipolar." One junction is referred to as the base-emitter, and the other is the base-collector. To operate as an amplifier, the two PN junctions must be properly biased with external DC voltages.

Figure 2-2 shows an NPN transistor with the base-emitter junction forward biased and the base-collector reverse biased. Current flows through the base-emitter junction

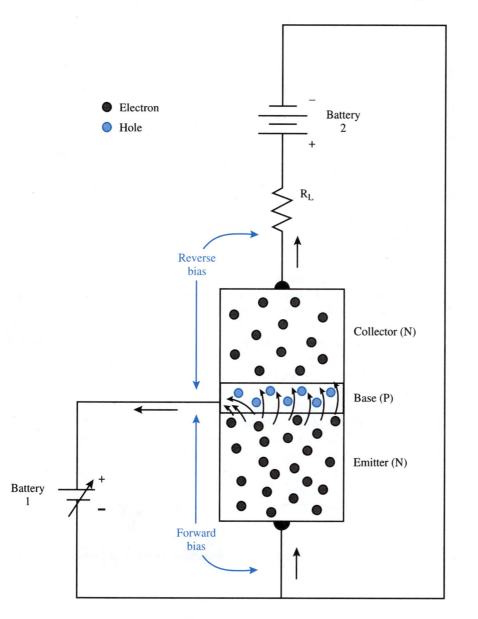

FIGURE 2-2 The biasing of an NPN transistor

the same way as a forward-biased diode, from the negative to the positive lead of battery 1. However, since the base region is lightly doped and very thin, it has a limited number of holes. Therefore, only a small percentage of the total number of electrons from the emitter combine with the holes and flow out of the base lead. The remaining electrons have no place to go except across the base-collector junction. They continue through the collector region to the positive lead of battery 2. As the voltage of battery 1 is changed, the current through the base is varied. The amount of base current determines the resistance between the emitter and collector. The higher the voltage on the base of the transistor, the more base current flows, and the lower the emitter-collector resistance becomes.

The transistor operates like the faucet in Figure 2-3. The emitter is the input and the collector is the output. The base is the valve that controls the flow. The base-emitter current controls the major current path between the emitter and collector. A few milliamps of base current can control several hundred milliamps of collector current.

Instead of using batteries to bias the transistor junction, a network of resistors and one DC source are used, as shown in Figure 2-4(a). Resistors R_1 and R_2 form a voltage divider that provides the base voltage. Resistor R_L is in series with the conducting transistor. The input signal V_{IN} is applied to the base. The output of the amplifier is located between the collector lead and the ground. The variation in collector current produces a changing voltage across the transistor, resulting in a variation in collector voltage.

When V_{IN} goes more positive, as shown between T_1 and T_2 of the waveform in Figure 2-4(b), base current increases. As collector current rises, the $I_C R_C$ drop increases, which produces a decrease in collector voltage (because $V_{OUT} = V_{CC} - I_C R_C$). Likewise, as the input voltage decreases, base current lowers, and the collector current drops. The result is that $I_C R_C$ decreases, which causes the collector voltage to increase. The waveform shows that there is a 180 degree phase inversion between the input voltage and the output signal produced. The waveform also shows amplification, since a small voltage change applied to the input causes a larger voltage change at the output. A higher positive voltage applied to an NPN input causes the transistor to turn on harder. When the voltage reaches a high enough level, the transistor is in the saturation mode because it is unable to conduct more current. A potential near zero volts will be read at the output. Likewise, when the input drops to a given voltage, the base-emitter is no longer forward biased and collector current stops. The resistance across the emitter collector rises to infinity. This condition, called the cutoff mode, causes the transistor to drop the supply voltage, similar to an open switch.

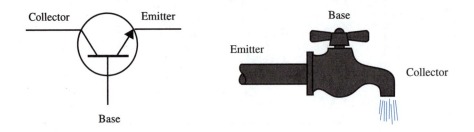

FIGURE 2-3 Transistors operate like faucets

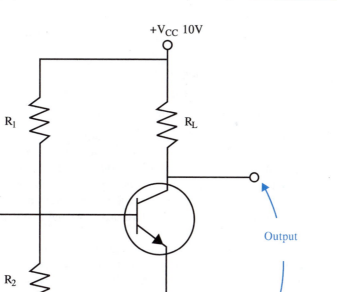

(a)

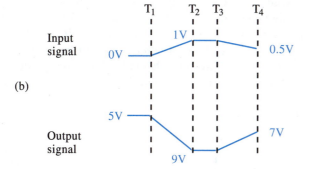

(b)

FIGURE 2-4 Transistor (NPN) amplifier

A PNP transistor operates the opposite way. A negative voltage applied to the base input causes the transistor to turn on harder, and a positive voltage will make it turn on less.

OPERATIONAL AMPLIFIERS

A very versatile amplifier device is the **operational amplifier** (op amp). The most popular op amp is the uA741, which is fabricated inside an 8-pin integrated circuit package. There are three important characteristics of op amps that make them ideal amplifiers:

1. High input impedance
2. High voltage gain
3. Low output impedance

Figure 2-5 shows the standard schematic symbol of the uA741 op amp. Represented by a triangle, the op amp has two input terminals located at the base on the left and a single output located at the apex of the triangle. There are also two separate power-supply lines. The one located at the top base is connected to a positive potential, and the other located at the bottom base is connected to a negative potential. These two power supplies allow the output voltage to swing to either a positive or negative voltage with respect to ground.

One of the inputs has a minus sign. This is called the inverting input, because any DC or AC signal applied to its input is 180 degrees out of phase at the output. The other input has a plus sign and is called the noninverting input. Any DC or AC signal applied at this input is in phase at the output.

When external components are connected to the input and output leads, the op amp is capable of performing several functions. How the components are connected determines which function the op amp performs.

Inverting Op Amp

A typical op amp can have a voltage gain of approximately 200,000. However, the output voltage level cannot exceed approximately 80 percent of the supply voltage. For example, the maximum output voltages of the op amp in Figure 2-5 are +5 volts and −5 volts because the power-supply potentials are +6.25 volts and −6.25 volts. Therefore, it only takes a 25-uV input to result in a positive or negative 5-volt output voltage, depending on the input-signal polarity and the terminal to which it is applied.

However, the op amp is used for many applications that require a voltage gain less than 200,000. A technique called *feedback* is used to control the gain of this device,

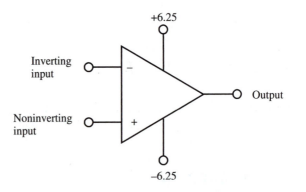

FIGURE 2-5 Standard symbol of an operational amplifier

and it is accomplished by connecting a resistor from the output terminal to an input lead. A negative-feedback circuit is shown in Figure 2-6. Its operation is as follows:

- ○ Both input terminals have high impedances; therefore, they do not allow current to flow into or out of them.

- ○ The potential at the negative input lead is called 0-volt virtual ground (that is, it acts like a 0-volt ground). The positive input lead is connected to an actual 0-volt ground potential.

- ○ Because point VG is 0 volts, there is a voltage drop of 2 volts across the 2-kilohm resistor, R_{IN}. 1 mA flows through it.

- ○ The 1 mA cannot flow into the op amp. Therefore, it flows up through the 10-kilohm feedback resistor R_F, developing a 10-volt drop across it.

- ○ Because V_{OUT} is measured with respect to the virtual ground, its voltage is −10 volts.

The voltage *gain* of the op amp is determined by:

$$Vgain = \frac{V_{OUT}}{V_{IN}}$$

The gain of the inverting op amp in Figure 2-6 is 5 because a 2-volt signal is applied to the input and an inverted −10-volt signal is at the output. A negative input voltage applied to this amplifier produces a positive output. The gain is influenced by the resistance ratio of R_F compared to R_{IN}. The larger R_F becomes compared to R_{IN}, the larger the gain.

The output voltage can also be determined by:

$$V_{OUT} = - \frac{V_{IN}}{R_{IN}} \times R_F$$

Table 2-1 provides examples of how the inverting amplifier with a gain of 10 responds to several input voltages.

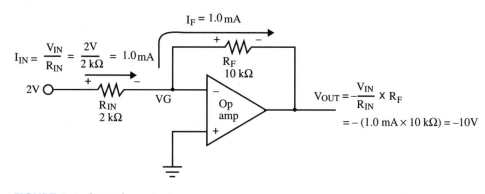

FIGURE 2-6 Inverting op amp

TABLE 2-1 **INPUT AND OUTPUT VOLTAGES OF AN INVERTING OP AMP WITH A GAIN OF 10**

V_{IN}	V_{OUT} (Volts)
+0.2	−2
−0.4	+4
0	0
+0.32	−3.2

Summing Amplifier

When two or more inputs are tied together and then applied to an input lead of an op amp, a summing amplifier is developed. This type of amplifier is capable of adding the algebraic sum of DC or AC signals. The circuit in Figure 2-7 is that of an inverting summing amplifier. It consists of a 20-kilohm feedback resistor R_F, three parallel 20-kilohm summing resistors tied together and connected to the inverting input lead, and +2-volt, +1-volt, and +3-volt signals applied to the inputs. The calculations above the diagram show how to determine the voltage at the output terminal.

The gain of each input is calculated and then summed to obtain the resulting current flow through R_F. Next the output voltage is determined by multiplying I_{RF} times R_F.

Table 2-2 provides examples of how the summing amplifier in Figure 2-7 responds to several input voltages.

Non-Inverting Amplifier

Some applications require that an amplified output signal be in phase with the input. Using an operational amplifier, this is accomplished by applying the input signal to the non-inverting input, while the feedback to control gain is still provided by connecting the output terminal to the inverting input through a resistor (R_F). One lead of resistor R_{IN} is also connected to the inverting input. The other lead of R_{IN} is connected to a 0-volt ground potential.

Figure 2-8 shows the schematic diagram of a non-inverting amplifier. The gain of the circuit is influenced by resistors R_{IN} and R_F. The equation used to determine the gain of the noninverting amplifier is derived by adding 1 to the resistance ratio of R_F and R_{IN}. Thus:

$$\text{Gain} = 1 + \frac{R_F}{R_{IN}}$$

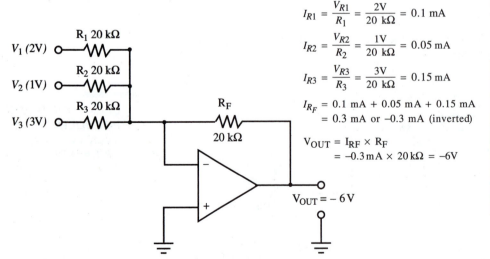

$$I_{R1} = \frac{V_{R1}}{R_1} = \frac{2V}{20\ k\Omega} = 0.1\ mA$$

$$I_{R2} = \frac{V_{R2}}{R_2} = \frac{1V}{20\ k\Omega} = 0.05\ mA$$

$$I_{R3} = \frac{V_{R3}}{R_3} = \frac{3V}{20\ k\Omega} = 0.15\ mA$$

$$I_{R_F} = 0.1\ mA + 0.05\ mA + 0.15\ mA$$
$$= 0.3\ mA\ or\ -0.3\ mA\ (inverted)$$

$$V_{OUT} = I_{RF} \times R_F$$
$$= -0.3\ mA \times 20\ k\Omega = -6V$$

$V_{OUT} = -6\ V$

TABLE 2-2   OPERATION OF AN INVERTING SUMMING AMPLIFIER

Input Voltages			Algebraic Sum of Output Voltages
V_1	V_2	V_3	
+1	+1	+1	−3
+1	−1	−1	+1
+2	−1	−1	0
−3	−1	+3	+1
+1	+2	−1	−2

FIGURE 2-7   Inverting summing amplifier

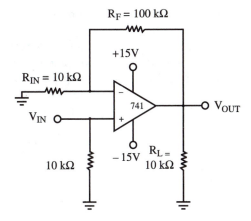

TABLE 2-3 OPERATION OF A NON-INVERTING OP AMP	
V_{IN}	V_{OUT}
+0.3	+3.3
−1.0	−11
+.75	+8.25
−.52	−5.72

FIGURE 2-8   Non-inverting operational amplifier

The output voltage is determined by:

$$V_{OUT} = 1 + \frac{R_F}{R_{IN}} (V_{IN})$$

The gain will always be greater than 1.

Table 2-3 provides examples of how the non-inverting amplifier shown in Figure 2-8 responds to several input voltages.

2-2   SIGNAL PROCESSORS

Signal processors are special devices that change or modify signals applied to their inputs. The output signals of these devices can then be used to perform specific functions. Three signal processor devices will be described: the integrator, the differentiator, and the Schmitt trigger.

INTEGRATOR OPERATIONAL AMPLIFIER

An integrator is an amplifier circuit that continuously increases its gain over a period of time. The magnitude of the output is proportional to the period of time that the input signal is present. Figure 2-9(a) shows the schematic diagram of the op amp integrator. The circuit resembles that of an inverting op amp. The difference is that a capacitor replaces the resistor as the feedback element. The waveform diagrams in Figure 2-9(b) illustrate the operation of the circuit when different DC voltages are applied to the input.

When the input voltage changes from zero to +5-volts, at T_1 of the waveform, the capacitor initially has a low impedance because it is discharged. The gain of the op amp is zero because the ratio of the feedback resistance to the input resistance is zero. This action is expressed by the formula for the inverting op amp: $V_{OUT} = R_{FB}/R_{IN}$.

As the capacitor begins to charge, the impedance path to current flow increases. Because the feedback resistance rises, the C_{FB}/R_{IN} ratio increases. The result is that the output of the op amp increases in a linear fashion. Since the inverting input is used, the output will be a negative-going waveform. Eventually the waveform levels off because the op amp reaches saturation, as shown at T_2 of the diagram.

At T_3, the input voltage changes from +5V to zero volts. The capacitor discharges and causes the output to return to zero volts. If a negative voltage is applied to the input, a positive-going signal develops at the output. If a square wave is applied to the input, a sawtooth waveform will develop at the output, as shown in Figure 2-9(c). The rate at which the output changes is determined by the capacitor and resistor values.

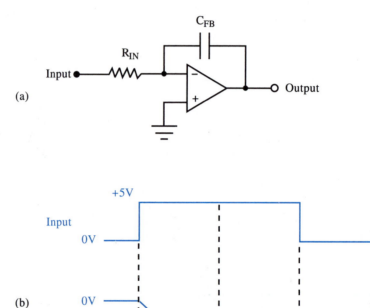

(a)

(b)

(c)

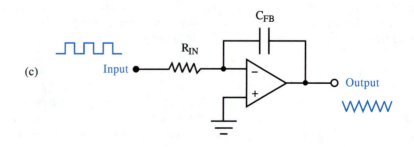

FIGURE 2-9 Integrated operational amplifier

DIFFERENTIATOR OPERATIONAL AMPLIFIER

A differentiator is an amplifier circuit that produces an output proportional to the rate of change of the input signal. Figure 2-10(a) shows the schematic diagram of the op amp differentiator. Its configuration is opposite to that of the integrator because the capacitor replaces the input resistor instead of the feedback resistor. The waveform diagrams in Figure 2-10(b) illustrate how the differentiator responds to different input signals. Since the inverting lead is used, the output signal that develops will be in the opposite direction as the signal applied to the input.

When the input voltage is DC and remains constant, as shown from T_1 to T_2 on the waveform, the output of the differentiator is 0 volts. If the voltage changes at a slow, steady rate, the output will be a small constant DC voltage, as shown from T_2 to T_3. If the voltage changes at a fast, steady rate, as shown from T_3 to T_4, the output will be a high constant DC voltage. When a sawtooth is applied to the input, a square wave signal is produced, as shown in Figure 2-10(c). As the sawtooth goes in the positive direction, the square wave alternation is negative. A negative-going sawtooth produces a positive alternation of the square wave. Figure 2-10(d) shows that when a square wave is applied to the input, a series of spikes is produced at the output. The

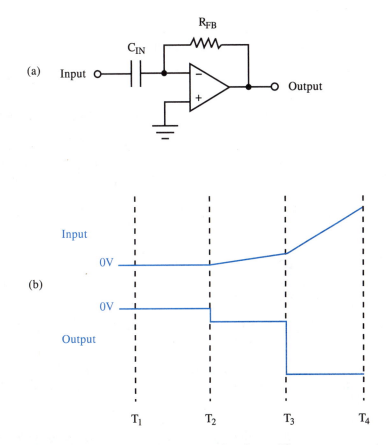

FIGURE 2-10 Differentiator operational amplifier

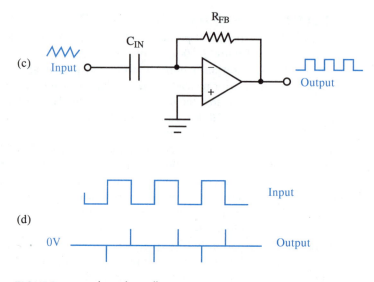

FIGURE 2-10 *(continued)*

polarity of each spike is determined by the positive or negative-going transition of the square wave.

Integrators and differentiators are used to control output actuators in closed-loop automated systems.

WAVE-SHAPING SCHMITT TRIGGER

The Schmitt trigger is a device that produces rectangular wave signals. It is often used to convert sine waves or arbitrary waveforms into crisp square-shaped signals. It is also used to restore square waves, which sometimes become distorted during transmission, back to their required square-shaped waveforms. The Schmitt trigger uses positive feedback internally to speed up level transitions. It also utilizes an effect called *hysteresis,* which means that the switching threshold on a positive-going input signal is at a higher voltage level than the switching threshold on a negative-going signal. Schmitt triggers can also be used to transform the following waveforms into rectangular-shaped signals:

○ A low-voltage AC wave.
○ Signals with slow rise times, such as those produced from charging and discharging capacitors, and temperature-sensing transducers.

Figure 2-11(a) illustrates the switching action of a Schmitt trigger inverter and also shows how the hysteresis characteristics recondition a distorted square wave.

Operation

Time Period 1.   A logic 0 is recognized at the input, and a 1 state is generated at the inverting output.

Time Period 2.   A logic 1 at the input is recognized if the input voltage exceeds the 1.7-volt positive-going threshold level that causes the output to snap to a logic 0 value. Note that the ragged spike on the input signal caused from

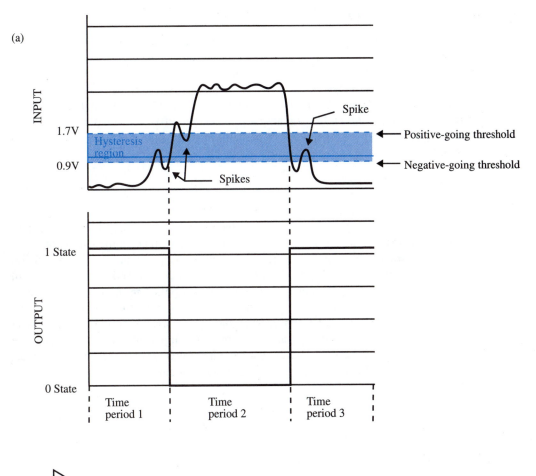

(a)

(b)

FIGURE 2-11 Schmitt trigger

noise drops below 1.7 volts into the hysteresis region during period 2. The output does not change unless the input drops below the 0.9-volt negative-going threshold level.

Time Period 3. A logic 0 at the input is recognized if the voltage drops below the 0.9-volt negative-going threshold level, which causes the output to snap to a logic-1 value. Note that a spike on the input rises above 0.9 volts into the hysteresis region during time period 3. The output does not change unless the input reaches the 1.7-volt positive-going threshold level.

The logic symbol for a Schmitt trigger inverter is shown in Figure 2-11(b). It includes a miniature hysteresis waveform inside the symbol to indicate that it is a Schmitt trigger instead of a regular inverter.

2-3 COMPARATOR DEVICES

The comparer element of a closed-loop system shown in Figure 1-6 has two inputs and one output. The command signal is applied to one input lead and the feedback signal is applied to the other input lead. The function of the comparer (also referred to as a **comparator**), is to produce an output error signal that is determined by the difference between the two inputs. The input and output signals can be either analog or digital. The op amp comparator and the op amp difference amplifier are capable of comparing analog signals, and the magnitude comparator compares digital signals. They are discussed below.

OPERATIONAL AMPLIFIER COMPARATOR

Figure 2-12 shows an op amp configuration that operates as a voltage comparator. This device compares the voltage applied to one input to the voltage applied at the other input. Any difference between the voltages drives the op amp output into either a +5 volt or a −5 volt saturation. The polarity of the output is determined by the polarity of the voltages applied at the inputs. When the voltage applied to the inverting input is more positive than the voltage at the noninverting terminals, the output swings to a −5 volt saturation potential. Likewise, when the voltage applied to the inverting input is more negative than the voltage at the noninverting input, the output swings to the +5 volt saturation potential. However, when the input voltages are the same amplitude, the output is zero. The following equations provide a summary of the operation for the voltage comparator:

Inverting input voltage < noninverting input voltage = positive output voltage

Inverting input voltage > noninverting input voltage = negative output voltage

Inverting input voltage = noninverting input voltage = zero output voltage

Table 2-4 provides examples of how the op amp, operating as a comparator, responds to several input voltages.

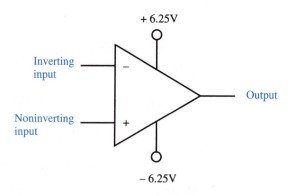

FIGURE 2-12 Op amp comparator

TABLE 2-4 **OPERATION OF AN OP-AMP COMPARATOR**

Inverting Input Terminal (Volts)	Noninverting Input Terminal (Volts)	Output Saturation Voltage (Volts)
+1	−1	−5
+1	+2	+5
+2	+1	−5
0	0	0
−1	+1	+5
0	−1	−5
0	+1	+5
+3	+3	0

DIFFERENCE OPERATIONAL AMPLIFIER

The difference operational amplifier (shown in Figure 2-13) finds the algebraic difference between two input voltages. Neither the inverting input nor the non-inverting input is grounded. Instead, signals are applied to both inputs at the same time, and the difference between them is amplified. If the signals are the same, the output voltage is zero.

Note that the circuit utilizes the closed-loop feedback configuration, which results in a controlled amplified output voltage. If all the external resistors are equal, no amplification takes place. Instead, the voltage difference op amp performs the arithmetic operation of subtraction. For example, suppose that 3 volts are applied to the inverting input 1, and 6 volts to the non-inverting input 2. The voltage difference between these inputs is 3 volts, which is developed at the op amp output.

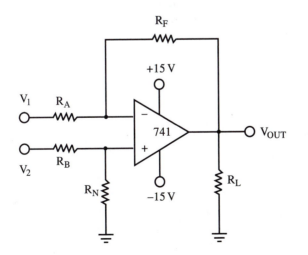

All resistors are 10 kΩ.

FIGURE 2-13 Difference op amp

The output can be calculated using the following formula:

$$V_{OUT} = \frac{R_F}{R_{IN}} (V_2 - V_1)$$

$$= \frac{10K}{10K} (6V - 3V)$$

$$= 1 \ (3) = +3V$$

TABLE 2-5   OPERATION OF A DIFFERENCE OPERATIONAL AMPLIFIER

Input Voltage		Output Voltage Algebraic Difference (Inverted)
V_1	V_2	
+2	+4	+2
+4	+2	−2
+4	−2	−6
−2	+4	+6
−4	−2	+2
−2	−4	−2

If the voltage at the inverting input is more negative than the voltage at the noninverting input, the polarity of the output will be positive, and vice versa. Table 2-5 provides examples of various input conditions and the resulting output voltages for the circuit in Figure 2-13.

If the ratios of the resistor values in the circuit are changed, the difference op amp provides amplification. The output voltages can be determined by using the formula above.

DIGITAL MAGNITUDE COMPARATOR

The magnitude comparator is capable of comparing two binary numbers and indicating whether one number is greater than, less than, or equal to the other. Figure 2-14 shows the block diagram of a 4-bit magnitude comparator. It has four lines for input A, four lines for input B, and three logic state output lines. The A>B output will go high if input A is larger than B; the A<B output will go high if input B is larger than A; and the A=B output will go high if A is equal to B.

A TTL 7485 magnitude comparator integrated circuit (IC) is shown in Figure 2-15(a). Its operation is identical to the block diagram circuit in Figure 2-14. It also has three inputs called the expansion input lines, labeled A<B, A=B, and A>B. If only 4-bit words are being compared, the A=B input should be wired to a high and the A<B and A>B inputs should be wired to a low. Figures 2-15(b) and 2-15(c) show the pin diagram and the truth table of the 7485 IC.

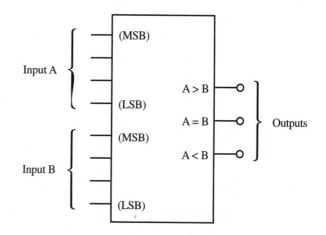

FIGURE 2-14   Block diagram of a magnitude comparator

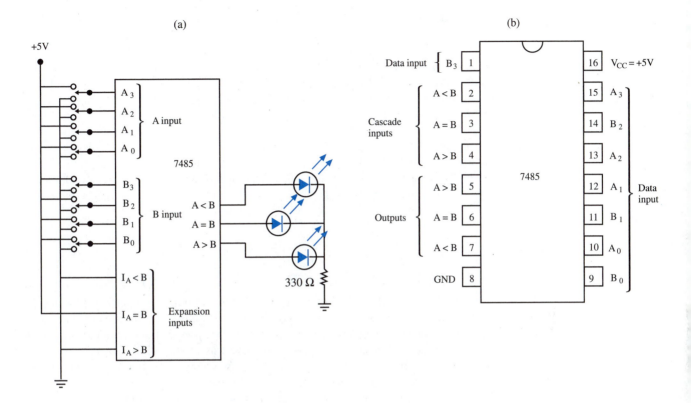

FIGURE 2-15 The 7485 magnitude comparator IC

Several 7485 ICs can be connected together to compare binary numbers larger than 4 bits. The block diagram of Figure 2-16 shows how two 7485 ICs are cascaded to make an 8-bit comparator. The four least significant bits of each 8-bit word are connected to inputs A_0–A_3 and B_0–B_3 of the comparator on the left. The four most significant bits of each 8-bit word are connected to inputs A_0–A_3 and B_0–B_3 of the comparator on the right. The A>B, A=B, and A<B outputs of the least significant comparator are connected to the expansion inputs of the most significant comparator. The expansion lines of the least significant comparator should be wired as if it were comparing only two 4-bit words. The comparison results of the two 8-bit words are generated at the three output lines of the most significant comparator.

The cascaded inputs resulting from the comparison of the low-order numbers are always overridden by the high-order numbers. The only time the cascaded input affects the output is when the two high-order numbers are equal.

2-4   SEMICONDUCTOR CONTROL DEVICES

Numerous industrial operations require controlled amounts of electrical power supplied to loads. Components that govern this power are referred to as *control devices*. The load may be a motor, a heater system, an electric power circuit, or an actuator that drives an industrial machine. The most effective way of controlling power is by using solid-state electronic switches. These convert low voltage signals from a controller to a regulated high voltage, high current supply for the load.

Since semiconductors are an integral part of control applications, their operation will be explained in this section. The coverage may provide a review for those readers who have completed a solid state course.

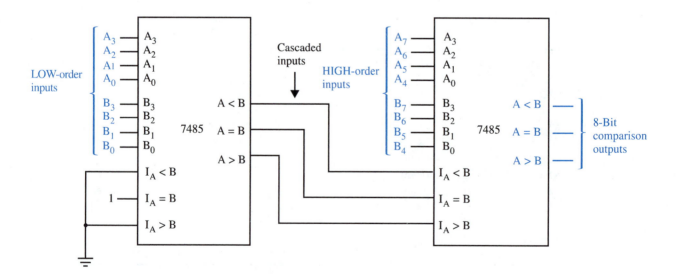

FIGURE 2-16 Magnitude comparison of two 8-bit binary strings (or binary words)

TRANSISTOR SWITCH

Transistors are commonly used as amplifiers. However, they also make effective switches. To perform as a switch, a transistor is controlled by I_b (base current), which causes it to operate alternately in the cutoff and saturation conditions. Figure 2-17(a) shows the base shorted to ground. With no forward bias on the base-emitter junction, there is no voltage to develop a base current. The result is that the transistor is cut off and there is, ideally, no connection between the collector and the emitter. It can be determined whether the transistor is not conducting by measuring the voltage from the collector to the ground. The reading should show a value equal to the supply voltage. An open switch will also have the full voltage present across its terminals in the equivalent circuit.

Figure 2-17(b) shows the base-emitter junction forward biased when the base is connected to R_A. If the resulting base current is large enough, the collector current will reach its maximum, or saturated, value. In this condition there is, ideally, a short

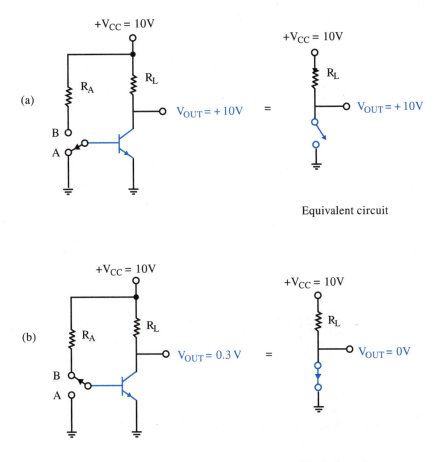

FIGURE 2-17 Bipolar transistor used as a switch

between the collector and emitter. A value barely above zero volts from collector to ground indicates that the transistor is fully on. A closed switch in the equivalent circuit will have zero volts present across its terminals.

Instead of using a switch at the base, the controlled signal may be provided by a microprocessor, a computer, a programmable controller, or an electronic circuit, as shown in Figure 2-18. Here, the input to the transistor is a 5-volt signal from the controller, which causes the transistor to saturate. A zero volt input will drive the transistor into the cutoff condition. Resistor R_b limits the base current to a safe value. Resistor R_c represents the load, such as a relay coil or motor.

THYRISTORS

Transistors have several limitations. For example, they can drive only low-current, low voltage devices such as relays, coils, low-voltage DC motors, and stepper motors. Another disadvantage is that they consume a high and continuous base current to remain on.

A more efficient method of controlling power delivered to a load is by using on-off switching devices called *thyristors*. The name is derived from the thyratron gas-tube predecessor with the same characteristics. A **thyristor** turns on and remains on after it is activated by a momentary input signal. This characteristic gives a thyristor the ability to control large amounts of power to the load while consuming very little activation power.

Thyristors are used for loads that are powered by both DC and AC voltages.

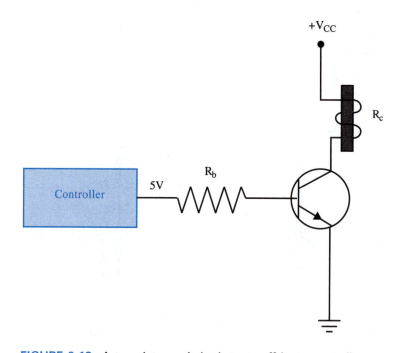

FIGURE 2-18 A transistor switched on or off by a controller

DC Thyristors

Silicon-Controlled Rectifier (SCR)

One type of load that uses DC power is the DC motor. DC motors are often used in applications that require precise speeds and constant torque as the physical demand on the motor varies. Regulating or varying the speed and torque of a DC motor is performed by changing the level of voltage applied to the motor. Raising the voltage enables the motor to increase its speed or torque by drawing more current. Lowering the supply voltage will slow the motor speed, lower the torque, or both.

The most common method of varying the motor voltage is to change the supply voltage into clipped pulses of DC. The rectification and clipping is performed by a thyristor device called a silicon-controlled rectifier (**SCR**). The SCR is a four-layer device with three terminals: the anode, the cathode, and the gate, as shown in Figure 2-19. It operates similarly to a normal diode, where current flows only in the forward-biased condition and is blocked when reverse biased. Therefore, the SCR is able to perform rectification. However, conduction of the SCR does not occur until it is both forward biased and triggered by the gate terminal. Once the gate has triggered the SCR into conduction, it operates similarly to a latched switch. The gate no longer has control of the SCR, and it remains on as long as current does not fall below a predetermined value called the holding current.

Figure 2-20 shows how an SCR controls the power delivered to a universal motor. A technique called phase control is used to vary the SCR's conduction angle, thereby causing an SCR to switch in such a way that the ratio of ON (conducting) time to OFF (non-conducting) time may be varied, allowing average power to the load to be changed. The term *phase control* refers to the time relationship between two events. In this case, it is the time relationship between the occurrence of the trigger pulse and the point at which the conduction alternation begins.

When the negative alternation occurs, the SCR is reverse biased and will not conduct. Diode D_2 is forward biased, allowing current to bypass R_1 and charge the capacitor. Due to the opposition of the charging capacitor, the current through the motor is small enough to be ineffective.

When the positive alternation occurs, the alternation begins to rise, the SCR becomes forward biased, and the capacitor begins to discharge the polarities it received during the negative alternation. Once discharged, it recharges at the opposite polarity. The discharging and charging rates are determined by the component values of the RC network. When the positive charge at the top plate of the capacitor is high enough, D_1 forward biases and allows current to flow through to the gate and fire the SCR. The result is that the supply current flows through the motor during the remaining portion of the alternation. During the next negative alternation, the SCR is reverse biased and is off, and the voltage drop across the motor is zero. The SCR turns on again at the point when the next positive alternation occurs. If the variable resistor is in position A, there is zero resistance and the capacitor discharges and charges almost immediately. The result is that current will flow through the motor for nearly 100 percent of the positive alternation. As the variable resistor arm is moved downward, resistance increases, which causes the capacitor to discharge and charge more slowly. This results in a firing time delay, which decreases the amount of current delivered to the motor. Diagrams A through D in Figure 2-20 show the waveform patterns displayed by

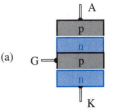

(a)

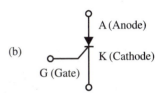

(b)

FIGURE 2-19   Silicon-controlled rectifier

the oscilloscope as the RC time constants are changed by varying the wiper arm position of R_1.

Unijunction Transistor There are two disadvantages of using the RC network to perform the phase control function for the SCR. First, the capacitor must discharge before charging in the opposite direction. This prevents firing during the entire 180 degree positive alternation. Second, it requires the same AC supply that provides power to the load.

A semiconductor component that was developed for turning on an SCR is the *unijunction transistor* (**UJT**). In practical applications, control devices such as computers, programmable controllers, and sensors produce a low voltage DC signal to turn on the SCR at a precise moment. UJTs are used to interface the controller to the high powered SCRs.

The structure of a UJT is shown in Figure 2-21(a). It is made of an N-type silicon bar called a base, to which a small amount of P-type material is diffused. The base terminals connected to each end are labeled B_1 (base 1) and B_2 (base 2). The P material is the emitter, and is labeled E. A power supply is connected across the base with the negative lead at B_1 and the positive lead at B_2. The internal resistance of the base causes a voltage gradient to form, as shown in Figure 2-21(c). Since the bar is a semiconductor, it only allows a small current to flow. Together, the emitter and B_1 form the single PN junction of the UJT. An equivalent circuit in Figure 2-21(d) presents the

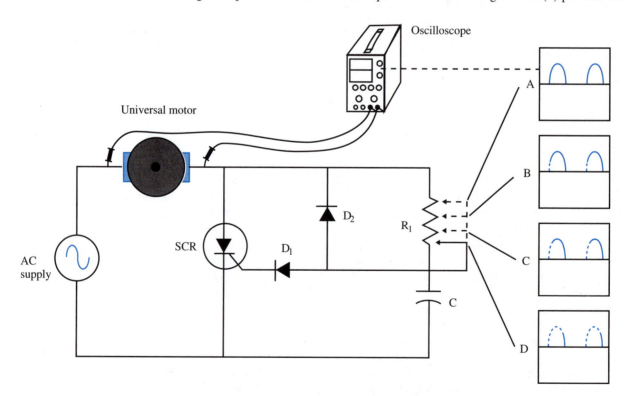

FIGURE 2-20 The phase control of an SCR varies the speed of a motor

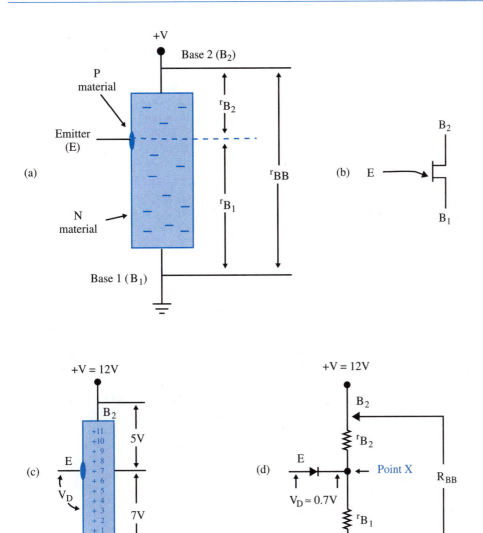

FIGURE 2-21 Unijunction transistor

internal characteristic of the UJT. The resistance between B_1 and point X is r_{B1}, and the resistance between B_2 and point X is r_{B2}. The sum of these two resistances is R_{BB}. The diode represents the PN junction between the emitter and the N material.

A bias voltage is connected across the junction with the emitter more positive than B_1. In its normal state, there is no current flow from B_1 to the emitter. When the bias voltage reaches 7.7V (V_{rB_1} + diode breakover), the UJT turns on. Increased current flows from B_1 to B_2, and from B_1 to the emitter. Current continues to flow until the bias voltage across the junction drops to about 3 volts. The UJT turns off until the bias voltage again becomes 7.7 volts.

The circuit in Figure 2-22 illustrates the operation of the UJT. The variable resistor controls the rate at which the capacitor charges. When the capacitor voltage reaches 7.7 volts, the depletion region across the junction collapses. A discharge path for the capacitor forms through the junction and resistor R_1. The current surge causes a pulse to develop across R_1 that is large enough to trigger the SCR on. When the capacitor is discharged to about 3 volts, the UJT turns off, permitting the capacitor to begin charging again. This circuit, called a relaxation oscillator, produces a continuous train of pulses. Its frequency is determined by the rheostat setting and is calculated by the formula located to the left of the diagram.

Figure 2-23(a) shows a controller that is used for triggering a UJT to fire an SCR. The control signal is synchronized with the beginning of each positive alternation applied to the SCR. As the AC waveform crosses the zero volt level in the positive direction, a timer in the microprocessor-based controller begins. The controller is programmed to produce the pulse at a specific moment in the alternation to generate a desired average voltage. Figure 2-23(b) provides waveform samples showing when in an alternation the pulse occurs, the resulting half-wave signal, and the average DC voltage the pulse produces.

AC Thyristors

Triac   The fact that the SCR will conduct current in only one direction makes it a unidirectional device. Some types of loads use AC power, which requires a supply current that flows in both directions.

In applications where full control of an AC signal is required, a bidirectional thyristor device called a *triac* is used. The **triac** is basically a two-way SCR with one gate.

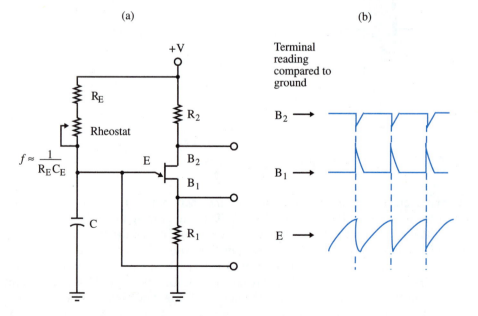

(a)   (b)

FIGURE 2-22   Relaxation oscillator

(a)

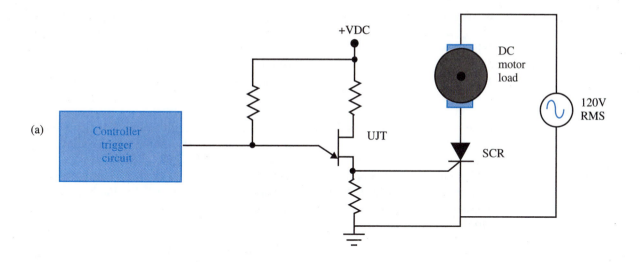

(b)

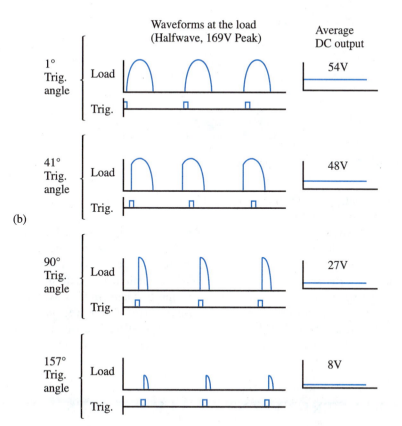

FIGURE 2-23 UJT controller

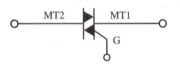

FIGURE 2-24 Schematic symbol of a triac

Its switching characteristics are the same in both directions. A triac symbol is shown in Figure 2-24. Its terminals are labeled MT1 (main terminal 1), MT2, and G (gate). When a voltage is developed across MT1 and MT2 at either polarity, it will begin conducting when a positive or negative voltage is applied to the gate. The RMS voltage applied to the load can be varied by changing the moment in the alternation when the triac gate is pulsed.

Diac A thyristor that was developed to fire a triac is the **diac**. The diac is a bidirectional device that does not have a gate. Its symbol is shown in Figure 2-25(a). The terminals are labeled anode 1 and anode 2. When a sufficient voltage across its terminals in either direction is reached (about 15 volts), the diac turns on. During conduction, it has very little resistance to current flow. Upon firing, the diac's resistance decreases and it conducts at a lower voltage (about 5 volts), as shown by the waveform in Figure 2-25(b).

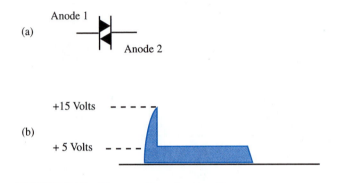

FIGURE 2-25 The semiconductor diac

The basic phase control circuit in Figure 2-26(a) shows how a diac and triac control current through an AC motor. An RC network performs the same type of phase control function that was used in the SCR power circuit in Figure 2-20. By varying the setting on the rheostat, the voltage charging time across the capacitor changes. When the charge on the capacitor reaches the breakover voltage of the diac, it fires. When the diac turns on, it triggers the triac into conduction. This action occurs during both the negative and positive alternations. The approximate circuit waveforms at strategic points in the circuit are shown in Figure 2-26(b). The power handling capabilities of the triac are much smaller than those of the SCR. The frequency at which it operates is limited to approximately 400Hz. These characteristics restrict the applications for which triacs are used.

2-5   OPTOELECTRONIC INTERFACE DEVICES

The voltage used by one element of a closed-loop system is not always equal to the voltage used by another element. Therefore they cannot be directly connected to one another. Optoelectronic devices are used to make the output of one section compatible with the input of another section. Optoelectronic devices pass electrical signals from

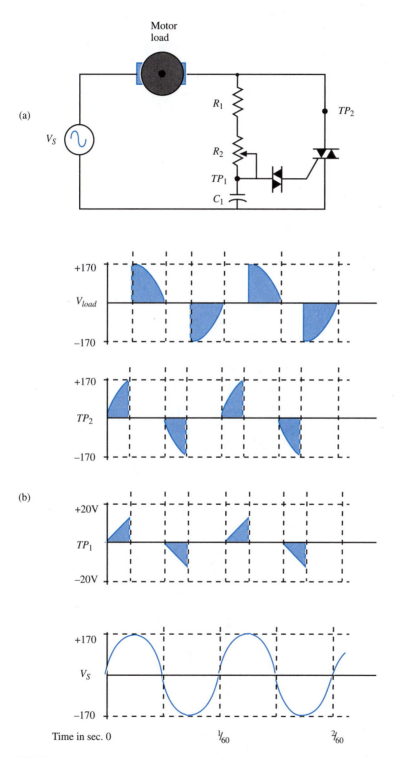

FIGURE 2-26 Diac-triac motor control by phase control

one element to another by means of light energy and semiconductors. An optoelectronic device consists of a light source and a photo detector. The light source converts electrical energy to light. The detector converts light energy to electrical energy.

The light source is usually a semiconductor light emitting diode (LED). In the forward biased state, light emission occurs when electrons combine with holes around the PN junction, as shown in Figure 2-27(a). During this process the electrons fall to a lower energy level and energy in the form of photons is released. Photons are light particles that travel in a waveform pattern.

Light detection of photons is accomplished by semiconductor devices. As the photons strike the semiconductor material at the PN junction, valance electrons are released, as shown in Figure 2-27(b). The valance electrons available in the semiconductor then enable current to pass through the PN junction.

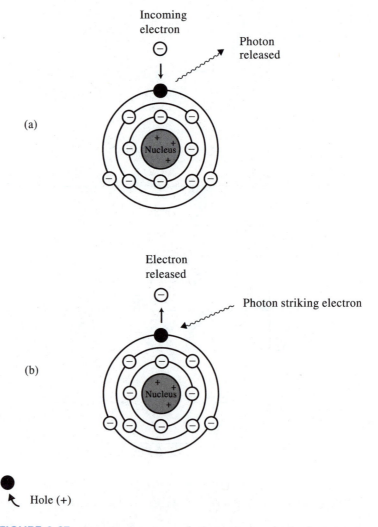

FIGURE 2-27 Atomic structure of photoelectronic devices

The detection of light and conversion into current is performed by light-activated devices such as photodiodes, phototransistors, photo SCRs, and phototriacs.

PHOTODIODES

The photodiode, shown in Figure 2-28, is a PN junction device that operates in the reverse bias mode. When a PN junction is reverse biased, heat causes the freeing of minority carriers in the depletion layer, which contributes to a small leakage current. High-energy photons strike the PN junction of the diode when it is exposed to light from the LED. The impact of photons causes electrons to be dislodged from their orbit and leave holes behind. This action of electron-hole pairs due to light exposure causes minority carriers to increase. The resulting current flow through the diode increases proportionately with light intensity. Photodiodes are used in applications that require quick response time for fast switching detection. Their primary limitation is that they allow current only in the microamp range between 50 uA and 500 uA to flow.

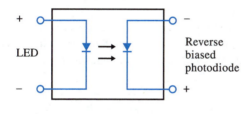

FIGURE 2-28   Photodiode

PHOTOTRANSISTORS

The phototransistor depends on a light source for its operation. Typically, the phototransistor has no external base lead, as shown in Figure 2-29. Therefore there is no bias source for external control. Instead, a light source operates the transistor in the same manner as a bias source. When photons from the LED strike the transistor's collector-base junction, the flow of minority carriers increases. This action causes the emitter-collector current to rise. If light intensity increases, more emitter-to-collector current will flow. Because the transistor amplifies, the amount of the output current it produces is much higher than the photodiode under the same illuminating conditions. However, its response time is slower than that of the photodiode.

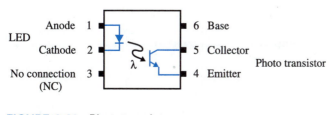

FIGURE 2-29   Phototransistor

PHOTO SCR

The photo SCR is also referred to as a light-activated SCR, or LASCR. The operation of the LASCR is similar to the conventional SCR except that it is usually activated by light instead of by a gate voltage. The LASCR symbol is shown in Figure 2-30.

The SCR is normally in the off condition. Its three leads enable the SCR to be triggered in one of three ways:

1. By light shining on the PN junction;

2. By a positive voltage applied to the gate; and

3. By a combination of the gate voltage and light intensity.

The output power an SCR controls is much higher than the amount required to trigger it. The level of light intensity used to turn on the LASCR can be controlled by adjusting the gate-cathode bias resistance. For example, a larger value resistance prevents the LASCR from turning on until a large amount of light intensity is reached. The LASCR remains on even after the light or the gate voltage is removed. When the current flowing through it is reduced below its holding current value, the SCR turns off and effectively blocks any current.

Because its power handling capacity is far beyond that of other optoelectronic devices, the LASCR is a superior high power switch. Photo SCRs are capable of switching current of 2 amperes and withstanding voltages as high as 200 volts.

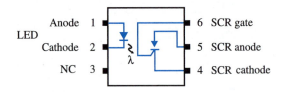

FIGURE 2-30 Photo SCR (LASCR)

PHOTO TRIAC

The photo triac is a bidirectional device designed to switch AC signals and pass current in both directions. Its symbol is shown in Figure 2-31. The photo triac is normally off if its PN junction is not exposed to light radiation of a certain density. During each alternation, it turns on when triggered by a specified light intensity, and turns off when the conducting current falls below a certain level. The current capacity of the photo triac is not as high as the LASCR.

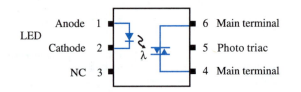

FIGURE 2-31 Photo triac

OPTOELECTRONIC PACKAGING

Optoelectronic devices are often constructed so that the light emitter and detector are fabricated into a six-pin integrated circuit package, as shown in Figure 2-32. This package, often referred to as an optocoupler, does not allow any external light to enter. The input, usually a +5 voltage, is applied to two pins of the IC package. These two pins are connected to the terminals of the internal light emitting diode. A different voltage source, e.g., +12 volts, +100 volts, or 120 VAC, is connected to two detector output leads of the IC. If the LED is turned on, its light illuminates the photodetector, which initiates an output current. The insulation resistance between the emitter and the detector is great enough to withstand an output voltage 5000 times greater than the input voltage. Some devices are capable of operating as high as 100KHZ. Since there is no electrical connection between the emitter and detector, the package is often called an *optoisolator*. IC packages are often used as an interface between a low voltage microprocessor and a high voltage AC motor that the microprocessor controls. They also protect against unwanted signals being induced into control circuitry due to power line noise that can improperly turn on a machine.

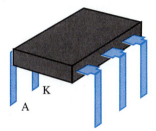

FIGURE 2-32
Photoelectronic
(optoisolator) package

2-6   DIGITAL-TO-ANALOG CONVERTERS

Digital-to-Analog converters (**DAC**s or D/A converters) are used to convert digital signals representing binary numbers into proportional analog voltages. Although these devices are now available in IC packages, they are analyzed here in a discrete form to better describe how they function.

A 4-bit input DAC is shown in Figure 2-33. It consists of a summing amplifier with its feedback resistor (R_F), four summing resistors, and four switches that are used to provide a 4-bit binary input. A switch in the open position represents the 0 state. In the

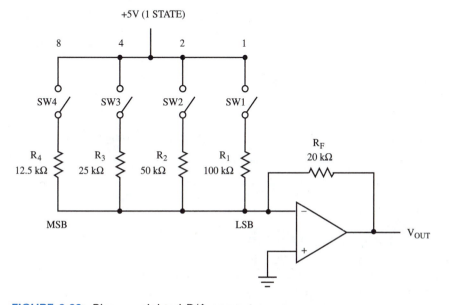

FIGURE 2-33   Binary-weighted D/A converter

closed position, it represents the 1 state. The placement of each switch corresponds to the same 8-4-2-1 weighted values of a 4-bit binary number. Resistors R_1 through R_4 are also selected with a weight proportional to the next. The 12.5 kilohm resistor R_4 is connected at the MSB (Most Significant Bit) input line. The values of the remaining resistors are selected by making each progressive resistor twice the size of the preceding one. The analog voltage is always at the op-amp output. The circuit is designed to operate so that a 4-bit binary number represented by the four switches is converted into voltages. Because 16 different combinations of switch positions are possible, (0-15), 16 different analog voltage levels proportional to the digital number applied are produced. The circuit in Figure 2-33 is designed so that it develops an analog output voltage equivalent to the binary number applied. For example, when all switches are in the open position to represent a binary input of 0000, the output is 0 volts. If SW1 is moved to the closed position, (binary 0001), the op amp output will be −1 volts. If SW1 and SW3 are in the closed position, (binary 0101), the op-amp output will be −5 volts. If all four switches are in the closed position (binary 1111), the analog output voltage will be −15 volts. The analog output voltage for each combination of switch setting can be determined by the same formula used for the summing operational amplifier.

EXAMPLE

What is the analog output voltage of the DAC in Figure 2-33 when a binary 1001 is applied?

SOLUTION

$$I_{R_1} = \frac{V_{R_1}}{R_1} = \frac{5V}{100 \; kohm} = .05 \, mA$$

$$I_{R_4} = \frac{V_{R_4}}{R_4} = \frac{5V}{12.5 \; kohm} = .4 \, mA$$

$$I_{R_F} = .05 \, mA + .4 \, mA = .45 \, mA$$

$$V_{OUT} = I_{R_F} \times R_F$$

$$= .45 \, mA \times 20 \; kohms = 9V$$

The table in Figure 2-34(a) provides all possible digital inputs and the corresponding output voltages for the circuit in Figure 2-33. Figure 2-34(b) provides the same information in a graphic format. The 4-bit DAC divides the reference analog output into fifteen equal divisions.

DACs in IC form are available with 8, 12, and 16 binary inputs. As the number of inputs increases, the reference analog voltage is divided into smaller divisions. For example, 8-bit DACs divide the analog output voltage into 255 equal parts, 12-bit converters into 4095 equal parts, and 16-bit converters into 65,535 equal divisions.

(a)

SW4 8	SW3 4	SW2 2	SW1 1	V_{OUT} (−V)
0	0	0	0	0
0	0	0	1	1
0	0	1	0	2
0	0	1	1	3
0	1	0	0	4
0	1	0	1	5
0	1	1	0	6
0	1	1	1	7
1	0	0	0	8
1	0	0	1	9
1	0	1	0	10
1	0	1	1	11
1	1	0	0	12
1	1	0	1	13
1	1	1	0	14
1	1	1	1	15

(b)

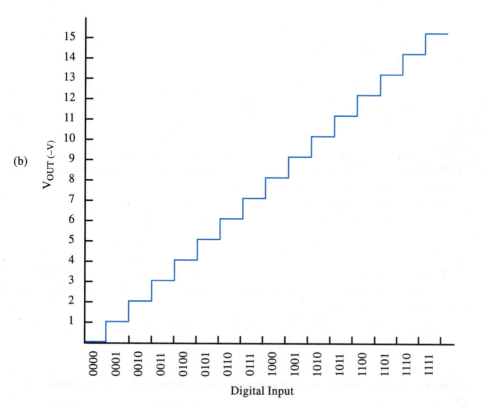

FIGURE 2-34 Analog output vs. digital input for the circuit of Figure 2-33

The number of equal divisions into which a DAC divides the reference voltage is called the *resolution*. The resolution of a DAC can be determined by the following formula:

$$\frac{V_{REF}}{2^n - 1}$$

- ◯ The 2 in the formula represents the binary number system.
- ◯ The n is the exponent that specifies to what power 2 is raised. It is determined by the number of binary inputs used at the input of the DAC. By taking 2 to the nth power, the maximum binary (equivalent decimal) number is determined.
- ◯ A 1 is subtracted from the maximum binary number to determine the number of equal steps (resolution) between the maximum binary number and minimum binary number.

PROBLEM

Find the resolution of a DAC with a reference voltage of 30 volts and 4 inputs.

SOLUTION

- ◯ Determine that the reference voltage is 30 volts.
- ◯ Because there are four digital inputs, $n = 4$.
- ◯ Raise $2^4 = 16$.
- ◯ Subtract 1 from $16 = 15$.
- ◯ Divide 30 volts/15 = 2-volt resolution.

Normally, resolution is expressed in terms of the number of binary input bits that are converted. A DAC with high resolution requires an accurate reference voltage because any variation can cause an error.

Resolution is obviously an important factor to consider when purchasing a DAC. Also important are its accuracy and operating speed.

INTEGRATED-CIRCUIT DIGITAL-TO-ANALOG CONVERTER

One popular DAC is the 8-bit DAC0808. The internal components supply proportional currents to its output lead.

Figure 2-35 shows the DAC0808 connected to an external 741 op amp. The current range is dictated by the 10-volt 5-kilohm combination connected to pin 14. The 2 mA flowing through resistor R_f is the maximum amount of current that can flow through output pin 4 (I_{OUT}). When the digital input is $0000\ 0000_2$, the minimum current of 0 mA flows through pin 4. When the digital input is $1111\ 1111_2$, the maximum current of 2 mA flows through pin 4. By using a 5-kilohm feedback resistor (R_f), the analog output voltage at the op amp output ranges from 0-10 volts. The 10 volts is produced when I_{OUT} is 2 mA. If a different analog output voltage range is desired, the gain of the op amp is adjusted by changing resistor R_f to a different value.

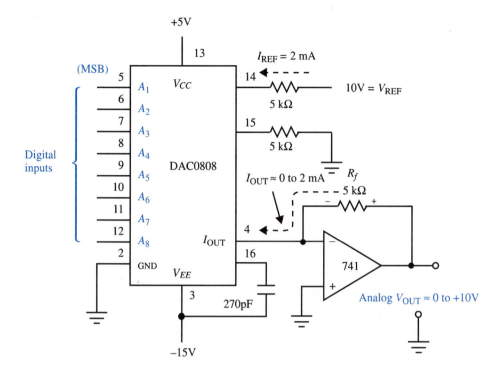

FIGURE 2-35 The DAC0808 and the 741 op amp connected to form a DAC

2-7 ANALOG-TO-DIGITAL CONVERTERS _____

The analog-to-digital converter (**ADC** or A/D converter) is capable of converting analog input voltages into proportional digital numbers. Analog-to-digital converters that operate at high speeds employ a circuit called a successive-approximation-register (SAR).

Figure 2-36(a) shows a simplified block diagram of an ADC that uses a SAR. Its eight output lines D_0–D_7 cause the D/A converter to produce different voltages, as shown in Figure 2-36(b). These voltages will result if 10 volts is supplied to the V_{REF} input line. Its operation is as follows:

1. When the START button is pressed, the SAR is reset on the negative edge of the pulse applied to the $\overline{WR}$ input.

2. The conversion is begun on the leading edge of the conversion pulse after the START button is released.

3. When the positive transition of the first clock pulse occurs, the SAR produces a high at its MSB output, D_7. This causes the D/A converter to produce an analog voltage that is one-half its maximum value.

4. If the D/A converter output is higher than the unknown analog voltage (analog V_{IN}), the SAR output returns low. If the D/A converter output is lower than the analog input voltage, the SAR leaves bit 7 high.

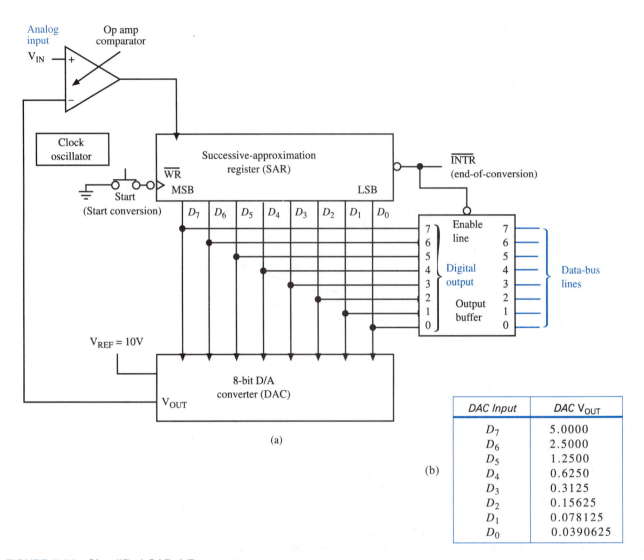

DAC Input	DAC V_{OUT}
D_7	5.0000
D_6	2.5000
D_5	1.2500
D_4	0.6250
D_3	0.3125
D_2	0.15625
D_1	0.078125
D_0	0.0390625

(b)

FIGURE 2-36 Simplified SAR A/D converter

5. The second clock pulse causes the next lower bit, D_6, to produce a high. If it causes the D/A converter output to be higher than the analog input, it returns to a low. If not, the SAR leaves D_6 high.

6. This process continues with the remaining six bits, D_5 to D_0.

7. At the end of the process, the SAR contains an 8-bit binary output that causes the D/A converter to produce an analog output equal to the unknown analog input. This occurs at the end of the eighth clock pulse. The 8-bit binary number contained by the SAR represents the analog input present at the eight output lines.

8. At the moment the eight-step conversion process is complete, the End-of-Conversion $\overline{INTR}$ line goes low. Because the ADC outputs are often shared with other devices on a common data-bus line, an 8-bit tri-state buffer is often connected to the digital outputs. When low, the $\overline{INTR}$ signal is used to enable the buffer to pass the digital count of the ADC to the bus lines. When the $\overline{INTR}$ output is high, the buffer outputs go into a high impedance state which allows another device to use the data-bus lines.

EXAMPLE

Show the waveforms that would occur if the successive-approximation-register A/D converter in Figure 2-36a were used to convert a 5.59-volt analog voltage to an equivalent 8-bit digital output.

SOLUTION

See Figure 2-37. The SAR is fast because an 8-bit SAR only requires eight clock pulses to perform the entire process.

INTEGRATED-CIRCUIT ANALOG-TO-DIGITAL CONVERTER

Figure 2-38 shows the block diagram of the ADC0804 analog-to-digital converter IC. The circuit shown is capable of converting the analog voltage into a proportional 8-bit digital output. The analog voltage range to be converted is determined by applying the desired maximum voltage to V_{DC}. For fine tuning, half of the V_{DC} voltage is applied to input $V_{REF/2}$. If necessary, a slight voltage change at $V_{REF/2}$ will then bring the ADC into calibration. By applying 5.12 volts to V_{DC} and 2.56 volts to $V_{REF/2}$, the circuit is capable of converting an analog voltage connected across $V_{IN}(+)$ and $V_{IN}(-)$ ranging from 0 to 5.12 volts. With eight output leads, there are 256 different analog voltage levels that are converted into digital outputs. Therefore, the resolution of this device is 0.39 percent ($1/255 = .0039 = 0.39\%$). With 5.12 volts as the maximum input voltage, each 0.02 volt ($5.12 \times .0039$) increase causes the binary count to increase by 1.

The ADC0804 IC contains an internal clock. To operate, a resistor and capacitor are connected to the CLK R and CLK IN inputs. The ADC0804 IC also contains an 8-bit successive approximation register for its conversion process. The SAR is reset on a negative edge of a pulse to the $\overline{WR}$ input lead by the closure of the START push button. When the push button is released, the pulse applied to the $\overline{WR}$ input returns high and the conversion process begins. At the end of this process, which takes eight clock pulses, output $\overline{INTR}$ goes low. The eight outputs that represent the analog input voltage will be present at the active-high output lines DB_0 to DB_7. To continue updating the applied analog input voltage, the $\overline{INTR}$ pin is connected to the $\overline{WR}$ input line. By doing so, 5000 to 10,000 conversions can be made per second.

The ADC0804 IC is a CMOS device that is designed to interface directly with some types of microprocessors. Therefore, some of its pins such as $\overline{RD}$, $\overline{WR}$, $\overline{CS}$, and $\overline{INTR}$ correspond to leads of the similarity-labeled microprocessors.

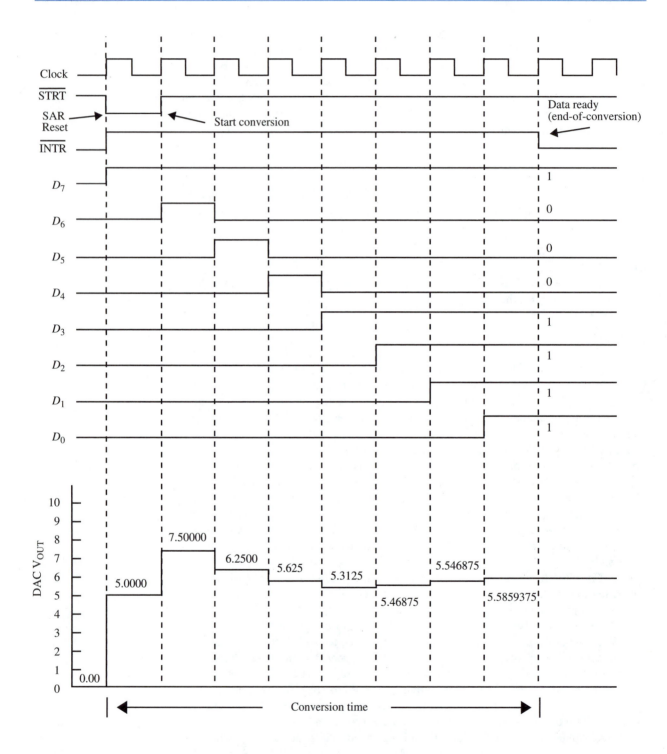

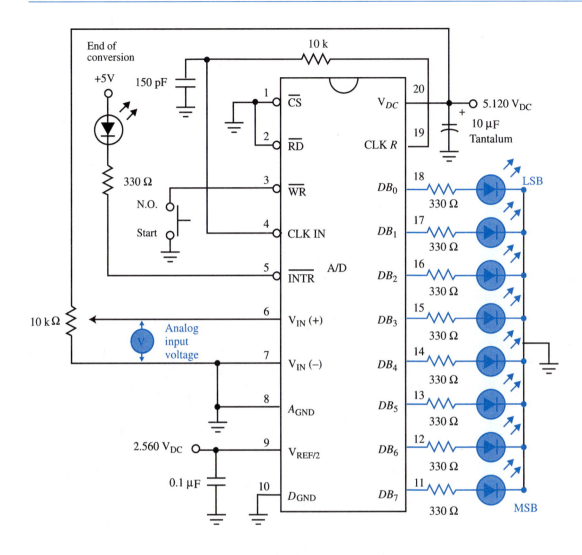

FIGURE 2-38 The block diagram of the ADC0804 analog-to-digital converter

2-8 TIMING DEVICES

Timing devices are used to produce rectangular signals referred to as square wave signals. Timing devices may generate either a single pulse or a continuous string of pulses. Single pulses are used to preset data into memory devices or to clear data. These signals are produced by **monostable multivibrators**. Continuous pulses are used as clock signals that are the heartbeat in computer devices. As they are fed through computer-based equipment, all events throughout the computing systems are properly timed and synchronized. These signals are produced by **astable multivibrators**.

A linear integrated circuit specifically designed for timing applications is the 555 monolithic IC chip. A pin diagram of this chip is shown in Figure 2-39.

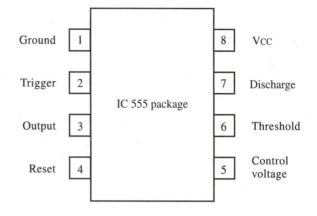

Ground	1		8	Vcc
Trigger	2		7	Discharge
		IC 555 package		
Output	3		6	Threshold
Reset	4		5	Control voltage

FIGURE 2-39   555 IC package

When a minimal number of external resistors and capacitors are connected to various pins of the 555 IC, it operates as an astable or monostable multivibrator. Figure 2-40 shows a schematic diagram of the 555 IC. It consists of the following sections:

Voltage Divider Network.   Resistors R_1, R_2, and R_3 are all 5K ohms. They form a voltage divider which biases the inverting (−) input of comparator A at 2/3 the power supply voltage (3.3v), and the non-inverting (+) input of comparator B at 1/3 the power supply voltage (1.65V).

Voltage Comparators.   Each comparator has one of its inputs connected to an external pin. The non-inverting input of comparator A is connected to external pin 6, called the *Threshold terminal.* The inverting input of comparator B is connected to external pin 2, called the *Trigger terminal.* The output of comparator A is Low if the voltage at the threshold terminal is lower than 3.3 volts. The output of comparator B is Low if the voltage at the trigger terminal is greater than 1.65 volts. The logic levels at the comparator outputs control the flip-flop.

R-S Flip-Flop.   The output of comparator A is connected to the R input of the flip-flop, and the output of comparator B is connected to the S input. The outputs of the two comparators are never on simultaneously. Only output $\overline{Q}$ of the R-S flip-flop is used. The $\overline{Q}$ lead is connected to the base of the transistor and the input of the output buffer. When the output of comparator A goes High, it causes the flip-flop to reset, generating a High at the $\overline{Q}$ output. When the output of comparator B goes High, it causes the flip-flop to set, generating a Low at output $\overline{Q}$.

Transistor.   The NPN transistor operates like a switch. When the $\overline{Q}$ output of the flip-flop is High, the transistor turns on and operates like a closed switch. When the $\overline{Q}$ output is Low, the transistor turns off.

Output Buffer.   The function of the output buffer is to produce a high current voltage to provide a sufficient signal for external circuitry. The buffer goes Low when $\overline{Q}$ is High, and goes High when $\overline{Q}$ is Low.

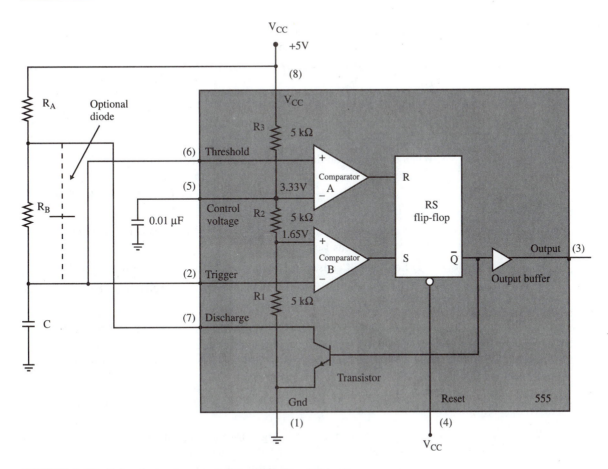

FIGURE 2-40 Schematic diagram of the 555 IC astable timer

555 ASTABLE MULTIVIBRATOR

The astable multivibrator diagrammed in Figure 2-40 has no stable output state. It is triggered by its own internal circuitry; therefore, it has no input lines. When power is applied, it switches back and forth at a desired rate between two states, producing a square wave at its output. The operation of the astable multivibrator is as follows:

Assume:

○ The capacitor is discharged.

○ Comparator A output is Low.

○ Comparator B output is High.

○ Flip-flop $\overline{Q}$ output is Low.

○ Transistor is off.

Therefore:

○ When power is applied to the circuit, current flows through the RC network of R_A, R_B, and C. When the capacitor charges to 1.66 volts, this potential is felt at the Trigger input (2) and causes the comparator B output to go Low.

○ When the capacitor charges to 3.34 volts, it is felt at the Threshold input (6) and comparator A goes High.

○ With a Low at flip-flop input S, and a High at input R, the $\overline{Q}$ output goes High.

○ A High at $\overline{Q}$ causes the output line of the output buffer to go Low.

○ A High at $\overline{Q}$ turns the transistor on, which allows the capacitor to discharge through the transistor and R_B.

○ When the charge on the capacitor goes less than 3.33 volts, the Threshold potential causes the comparator A output to go Low.

○ When the discharging capacitor goes less than the 1.65 volts, the Trigger input causes Comparator B to go High.

○ When comparator A output is Low and comparator B output is High, the flip-flop $\overline{Q}$ output goes Low.

○ A $\overline{Q}$ Low output causes the output line of the output buffer to go High.

○ A Low turns the transistor off, which opens the discharge path of the capacitor and starts the charging phase of the next cycle.

The rate at which the IC's internal components turn on and off is determined by the values of the external components connected to the IC.

The frequency of the output can be determined by the following formula:

$$f = \frac{1.44}{(R_A + 2R_B)\ C}$$

Initially, the external capacitor charges through R_A and R_B and then discharges through R_B. These charging and discharging times affect what is called a *duty cycle*. The duty cycle is the ratio of time the output terminal is High to the total time of one cycle. The duty cycle is set precisely by the ratio of these two resistors. The charging time (output buffer is High) is T_1. The discharging time (output buffer is Low) is T_2. The total period of time for one cycle is T. These values are calculated as follows:

$$T_1 = 0.693\,(R_A + R_B)\ (C)$$

$$T_2 = 0.693\,(R_B)\ (C)$$

$$T = T_1 + T_2 = 0.693\,(R_A + 2R_B)\ (C)$$

The duty cycle is:

$$DC = \frac{T_2}{T} \quad \text{or} \quad DC = \frac{R_B}{R_A + 2R_B}$$

Because the capacitor charges up through R_A and R_B and then discharges only through R_B, the duty cycle is always greater than 50 percent, as shown in the top waveform of Figure 2-41(a). However, it may be desirable to have a symmetrical square wave, which means that the time duration of the positive alternation equals that of the negative alternation, as shown in Figure 2-41(b). This would result if the duty cycle is 50 percent. This situation is possible only if the charging and discharging time

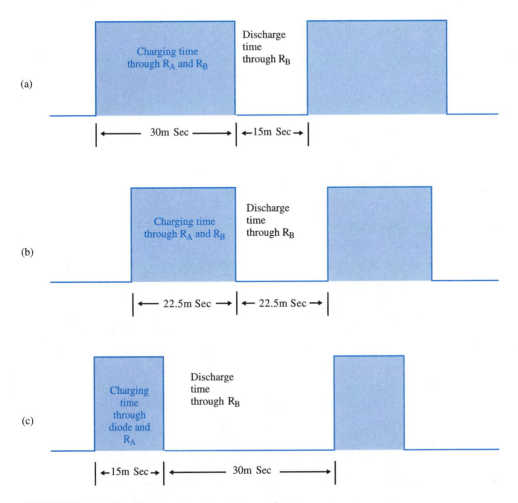

FIGURE 2-41 Waveforms showing duty cycles

durations of the capacitor are the same. By making R_A and R_B the same, and placing a diode across R_B with the anode connected to pin 7, and the cathode to pin 6, a symmetrical square wave is possible. The placement of the diode bypasses R_B and allows the capacitor to charge only through R_A. When the capacitor discharges, its current path is blocked by the reverse-biased diode, and only flows through R_B. Therefore, the charge and discharge paths are through resistances of the same value. Depending on the resistance ratios of R_A and R_B, this configuration allows the duty cycle to vary over a range of 5 percent to 95 percent, as shown in Figure 2-41(c).

555 MONOSTABLE MULTIVIBRATOR

The *monostable multivibrator,* also known as a one-shot, is characterized as having only one stable state. Its output is normally 0. When a triggering signal is applied to its input, the output changes from its normal stable state to a logic 1 (unstable state) for a specified length of time before automatically returning to its stable state. The triggering signal comes from either a mechanical switch or from another circuit. The period of time the monostable multivibrator remains in its unstable state is determined by an external RC timing circuit. The output pulse generated can be either longer or shorter than the input pulse.

Figure 2-42 shows the required connections for a 555 IC to operate as a one-shot. Its operation is as follows:

Assume:

    ○ The capacitor is discharged.

    ○ Comparator A output is Low.

    ○ Comparator B output is Low.

    ○ Flip-Flop $\overline{Q}$ output is High.

    ○ The transistor is on.

    ○ The output buffer is Low.

    ○ A +5 volt High is applied to the Trigger input.

Therefore:

    ○ While the Trigger signal is brought from a High to a temporary 0 volt potential by a push button closure, the comparator B output goes High.

    ○ A Low applied to the flip-flop's R input from comparator A and a High applied to the flip-flop's S input from comparator B cause the flip-flop's $\overline{Q}$ output to go Low.

    ○ A Low at the $\overline{Q}$ output of the flip-flop causes the Buffer output to go High.

    ○ A Low at the $\overline{Q}$ output turns off the discharge transistor which enables the capacitor to begin charging up toward $+V_{CC}$.

    ○ When the capacitor charges to 3.34 volts, the comparator A output goes High.

    ○ A High at the output of comparator A and a Low at the output of comparator B causes the RS flip-flop to reset and develop a High at its $\overline{Q}$ lead.

    ○ A High at the $\overline{Q}$ output causes the output buffer to go back to a normal Low state, and the one-shot pulse time duration is complete.

    ○ The High $\overline{Q}$ output turns on the discharge transistor which provides a discharge path for the capacitor.

    ○ When the capacitor is discharged, the one-shot awaits another negative-going pulse at the trigger input.

The capacitor reaches a 3.34 volt charge after 1.1 time constants. This time period determines the width of the output pulse of the one-shot. The time duration of the pulse is expressed in the following formula:

$$T = 1.1\ RC$$

where,   T is in seconds,
             R is in ohms,
             C is in farads.

The one-shot pulse duration can range from microseconds to several minutes.

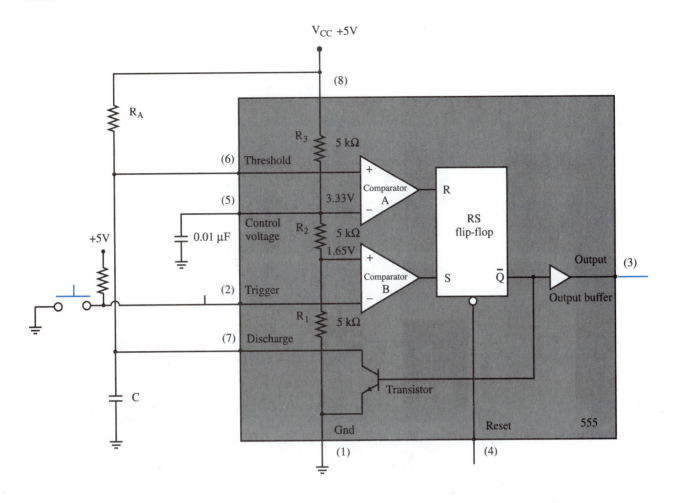

FIGURE 2-42 Schematic diagram of a 555 timer with the external timing components to form a monostable multivibrator

---

CHAPTER PROBLEMS

1. The NPN transistor will turn on harder by _____ (increasing, decreasing) the positive voltage applied to the base lead.

2. A _____ (negative, positive) potential applied to the base of an

   NPN transistor will drive it into the cutoff mode; a high _____ (negative, positive) potential applied to the base of a PNP transistor will drive it into the saturation mode.

3. An inverting op amp circuit has R_F = 5 kilohms and R_{IN} = 1 kilohm. What is the gain of this circuit?

4. What is the output voltage of the circuit in Figure 2-43?

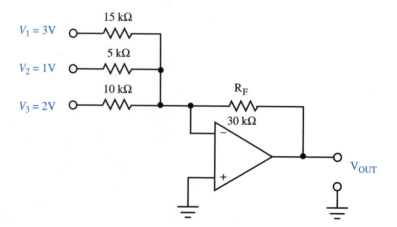

FIGURE 2-43 Circuit for Problem 4

5. A non-inverting op amp with an input resistor of 10k ohms and a feedback resistor of 50k ohms has 0.4 volts applied to its input. What is the output voltage?

6. At the moment an input signal applied to an integrator changes from 0 to +3 volts, it has a _____ (minimum, maximum) gain.

7. When a sawtooth-shaped signal is applied to the input of a differentiator, a _____-shaped signal is produced at the output.

8. The process where the switching threshold on a positive-going input signal applied to a Schmitt trigger is higher than the negative-going signal is referred to as _____.

9. Fill in the parenthesis of each of the following equations with <, >, or = symbols to describe how an op amp comparator operates:
   a. Inverting input voltage (___) noninverting input voltage = positive output voltage.
   b. Inverting input voltage (___) noninverting input voltage = zero output voltage.
   c. Inverting input voltage (___) noninverting input voltage = negative output voltage.

10. Assuming the values of all resistors connected to a difference op amp are the same, what is the output voltage when +3 volts are applied to the inverting input and +1 volt is applied to the noninverting input?

11. What three decisions does a magnitude comparator perform when comparing two different binary numbers?

12. __T __F   Triacs are used to control the speed of an AC motor.

13. The _____ (SCR, triac) thyristor is capable of handling the highest amount of current.

14. _ T _ F   The UJT can be connected directly to an AC or DC power source.

15. _ T _ F   A diac is a bidirectional device.

16. Describe the meaning of the term optoisolator.

17. When light strikes the base of a _____ (forward, reverse) biased photoelectric detector, it turns on.

18. DAC resolution is determined by the number of digital _____ lines available.

19. What is the resolution of a DAC with a maximum voltage of 15 volts and five input lines?

20. What is the resolution of a 5-bit ADC that has a maximum analog input voltage of +10 volts?

21. An 8-bit "successive-approximation-register" ADC requires how many clock pulses for each conversion?

22. In Figure 2-38, if the output reads 10000000, the analog voltage applied to the input is _____ volts.

23. The duty cycle of a square wave is the ratio of time the output terminal is _____ (low, high) to the total time period of one cycle.

24. How can a 555 astable multivibrator be constructed to produce a square wave that ranges from a 5 percent to 95 percent duty cycle?

25. Referring to Figure 2-40, what is the output frequency if R_A = 100k, R_B = 10k, and C = 10 ufd?

26. Referring to Figure 2-42, what is the amount of time the one-shot is in the unstable state if R_A = 2k, and C = 1 ufd?

CHAPTER 3
The Controller Operation

CHAPTER OBJECTIVES

After completing this chapter, you should be able to:

◯ List the four control modes used by the controller section.

◯ Define the following terms associated with control modes of a closed-loop system:

Cycle Time	Proportional Band
Full Range	Stable/Unstable
Differential Gap	Steady-State Error
Tuning	Predictive/Anticipatory

◯ Describe the operation of each type of mode control function.

◯ Explain the operation of the operational amplifier circuitry that performs each of the three PID mode functions.

◯ Define the following terms associated with Fuzzy Logic systems:

Fuzzy Set	Antecedent Block
Membership Function	Consequent Block

◯ List and describe the three stages of Fuzzy Logic Control.

◯ Describe each instruction of a simplified Fuzzy Logic program.

◯ List a practical application of a PID and Fuzzy Logic control system.

INTRODUCTION

The controller is an element of the closed-loop system that processes information to perform the decision-making function. The controller can be considered the brain that enables automated systems to operate without human intervention.

The input applied to the controller is the error signal, which is the difference between the desired set point and the feedback signal. The controller calculates changes needed in the controlled variable to compensate for disturbances that upset the process, or changes in the set point. The controller responds to these changes by producing an output signal that drives the actuator to alter the controlled variable until the error signal is reduced to zero.

The controller may be as simple as a spring-balanced mechanical lever or as complicated as a computer. It may control one process or several simultaneously. It may be analog, digital, or a combination of both.

3-1 CONTROL MODES

Figure 3-1 illustrates the operation of the controller. The input applied to the controller is called the error signal. It represents the difference between the setpoint signal and the feedback signal. The input error signal is expressed by the following formula:

$$e(t) = \text{Set point} - \text{Feedback Signal}$$

The error signal is not constant; it changes through time. Therefore (t) is used with symbol e in the equation.

The controller output signal is expressed as $x(t)$. Because the controller output also changes with time, (t) is used with the output symbol x. The time required for the controller to respond to the error signal depends on the control mode used.

There are four control modes of operation that are commonly performed by the controller section of a closed-loop system.

1. On-Off
2. Proportional
3. Proportional-Integral
4. Proportional-Integral-Derivative

All four modes of control respond to error signals. They differ in the speed and accuracy with which they eliminate the error between the set point and the controlled variable.

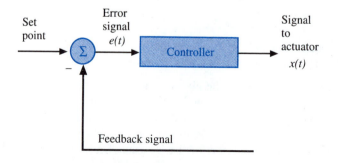

FIGURE 3-1 Controller representation

3-2 ON-OFF CONTROL

The On-Off control mode is the most basic type of control system. Its output has only two states, usually fully on and fully off. One state is used when the controlled variable (e.g., temperature, fluid level, voltage) is above the desired value (set point). The other state is used when the controlled variable is below the set point. The On-Off controller is also referred to as the two position control, or bang-bang control.

The home heating system shown in Figure 3-2 illustrates this mode of control. The thermostat is the measurement device. When the room temperature (controlled variable) falls below the setting (set point), the thermostat closes a switch that is connected to a fuel valve in the furnace, as shown in Figure 3-2(a). With the switch closed, the valve is fully opened. The furnace turns on and begins to generate heat. When the room temperature rises above the setting, the thermostat opens the switch connected to the furnace fuel valve. An open switch closes the fuel valve to extinguish the flame. With the furnace off, the temperature in the room begins to fall. When the temperature has gone low enough, the furnace turns back on.

Since the controlled variable must deviate from the set point to cause control action, the process response will continually cycle. The cycling occurs because of two factors:

1. Process disturbances cause the output to deviate from the set point.

2. The corrective action of the On-Off controller cannot adjust the output to exactly match the process demand. Instead, by being either fully on or fully

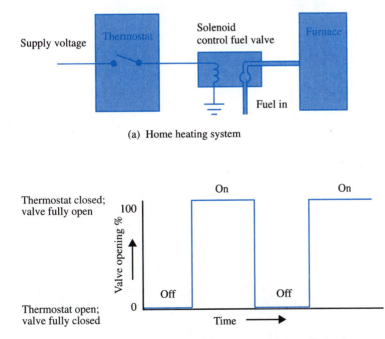

(a) Home heating system

(b) Graph illustrating operation of thermostat and furnace fuel valve

FIGURE 3-2 On-Off controller

off, the actuator's response is too large to return the process to the set point. The temperature is said to oscillate as it continually rises above and below the set point, as graphically shown in Figure 3-3.

The inherent cycle condition is detrimental to most final correcting devices, such as the fuel valve, pumps, relays, etc. By turning the output on and off so frequently, the rapid oscillation wears equipment and shortens its life.

To prevent rapid cycling, the time between the oscillations can be lengthened by adding an On-Off **differential gap** to the controller function. Also referred to as the *deadband,* the differential gap forces the controlled variable to move above or below the set point by a certain amount before the controlled action will change again. Figure 3-4 illustrates the differential gap function added to the thermostat device. The temperature must rise 2 degrees above the set point before the furnace turns off. To turn the furnace back on, the temperature must fall 2 degrees below the set point. Differential gap is defined as the smallest change in the controlled variable that causes the value to shift from on to off, or off to on. Therefore the differential gap for the thermostat is 4 degrees.

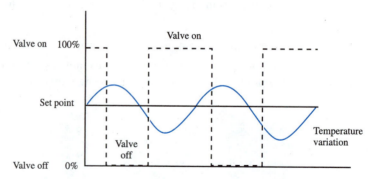

FIGURE 3-3 Graph illustrating temperature oscillation above and below set point as fuel valve is opened and closed

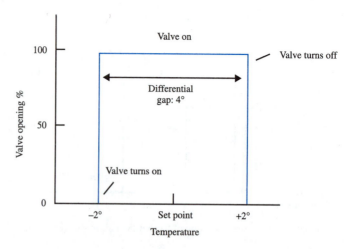

FIGURE 3-4 Differential gap of a thermostat

The differential gap is also expressed as a percentage of the full range of the controlled device. For example, the temperature range on a typical home thermostat is between 40 and 90 degrees. Therefore, the full range is 50 degrees (90 − 40 = 50). A temperature variance of 4 degrees represents 8 percent of full control range, because

$$\begin{array}{c}\% \text{ Differential Gap} \\ \text{(Deadband)}\end{array} = \frac{\text{Differential Gap}}{\text{Total Control Range}}$$

$$= 4/50$$

$$= 0.08 \text{ or } 8\%$$

The graph in Figure 3-5(a) illustrates the operation of the thermostat, the fuel valve in the furnace, and the resulting room temperature. It shows that the room temperature does not respond instantly after the fuel valve is turned on or off. For instance, after

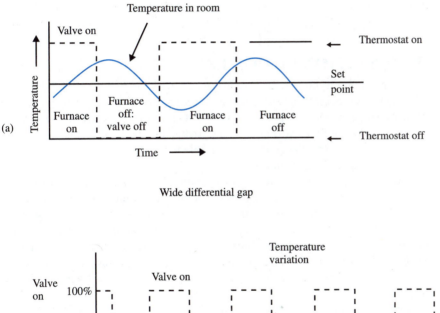

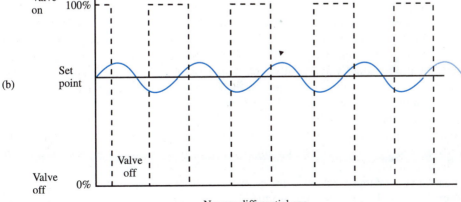

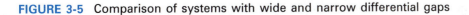

FIGURE 3-5 Comparison of systems with wide and narrow differential gaps

the thermostat turns the fuel valve off, the furnace and ducts contain enough heat so that the temperature in the room will not immediately begin to fall. Likewise, the temperature in the room does not rise as soon as the furnace turns on. A certain amount of time passes before the heat generated inside the furnace travels to the location of the thermostat. This lagging effect of the temperature behind the thermostat switching action is called *hysteresis*. Figure 3-5(b) shows the effects of narrowing the differential gap. The narrow gap causes rapid cycling with a small deviation from set point. The wider differential shown in Figure 3-5(a) causes less frequent cycling, but at the expense of greater deviation from the set point. A compromise is made between frequency of cycling and amplitude.

Because the On-Off control mode is simple, inexpensive, and inherently reliable, it is the most common type of feedback system. On-Off controllers are widely used in applications that can tolerate the cycling and deviation from set point. For example, they control thermostatic furnaces, refrigerators, and open closed valves for liquid tanks. They would never be used to control precision devices, such as a robot.

3-3 PROPORTIONAL CONTROL

Some situations require tighter control of the process variable than On-Off control can provide. Proportional control provides better control because its output operates linearly anywhere between fully on and fully off. As its name implies, its output changes proportionally to the input error signal. The greater the error, the more the output responds. This action returns the controlled variable to the desired setpoint value without the rapid cycling of On-Off control.

To illustrate the operation of proportional control, the operation of a furnace is again used. To obtain this type of control, two modifications must be made to an On-Off system. First, the On-Off switch in the thermostat is replaced by a thermistor in a bridge network. The output of the bridge produces a variable voltage in response to temperature changes. Second, the solenoid-type fuel valve in the furnace must be replaced by a proportional valve. The proportional valve opens proportionally to the input voltage from the bridge. The larger the voltage, the more fuel it supplies so that a higher temperature is produced.

Figure 3-6(a) shows the proportional control furnace system. The graph in (b) plots the percentage of the proportional valve opening versus room temperature. The temperature of 70 degrees is the set point for the system. At a 70 degree room temperature, the proportional valve is 50 percent open. At this point, the bridge is balanced and a zero volt feedback signal is produced. The positive voltage error signal produced by the summing op amp causes the proportional valve to be half open when the set point of −5 volts is not offset by the feedback signal. If the temperature drops below 70 degrees, the resistance of the thermistor increases. As a result, the voltage at the inverting input of the difference op amp becomes greater than the voltage at the non-inverting input. The feedback signal produced by the difference op amp becomes a negative voltage that is added to the negative setpoint voltage at the summing op amp. Therefore, as the error signal produced by the summing op amp goes more positive, the proportional valve opens more than 50 percent. For example, if the room temperature suddenly drops to 60 degrees, the error signal goes higher and causes the

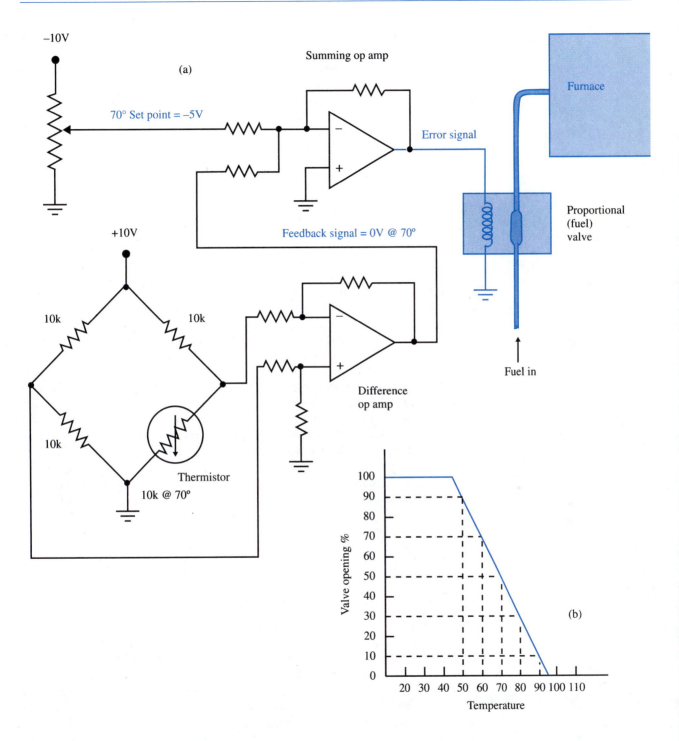

FIGURE 3-6 Proportional control heating system

valve to open from 50 percent to 70 percent. The room temperature rises back to its set point. If the temperature rises above 70 degrees, the thermistor resistance decreases, causing the voltage at the inverting input to become less than the non-inverting input at the difference op amp. Therefore, a positive voltage feedback signal is produced that cancels the negative setpoint voltage at the summing op amp. The canceling effect produces a less positive error signal, causing the proportional valve to be less than 50 percent open. As the valve closes to less than 50 percent, the room temperature decreases back towards the set point.

PROPORTIONAL BAND

If a disturbance causes the room temperature to drop to a specified point (e.g., 45 degrees), the proportional valve becomes fully open. If the room temperature rises to, e.g., 95 degrees, the valve becomes fully closed. Therefore, in this example the full operating temperature range of the proportional valve is 50 degrees (95 − 45 = 50). This range is called **proportional band**. Within the band the valve response is proportional to the temperature change. Most proportional controllers have an adjustable proportional band. In a system that has a narrow proportional band, the response to a disturbance is rapid. The temperature is adjusted to the set point quickly. The response to a system with a wider proportional band will take longer.

It would appear that the system with the narrow proportional band is better because the setpoint temperature would be restored more quickly. However, the characteristic of a narrow proportional band is that a system has a tendency to oscillate. When the system responds quickly, it tends to overshoot the set point. The system tries to correct itself by shifting the valve in the opposite direction. However, the system overshoots again in the opposite direction. The oscillations normally die out at which time the system becomes *stable*. If the proportional band is too small, oscillations will not stop. A proportional band of zero percent will cause the system to operate almost the same as the On-Off control. When the system continues to oscillate, it is *unstable*.

The size of the proportional band is simply the inverse of the proportional gain.

$$\text{Proportional Band } (\%) = \frac{100\%}{\text{Proportional Gain}}$$

To boost the error signal for driving the proportional valve, an amplifier may be used in the system. By increasing its gain, the proportional band narrows. For example, a gain of 20 produces a proportional band of 5 percent; a gain of 1 produces a band of 100 percent; a gain of infinity produces a proportional band of 0 percent.

The proportional band can be expressed as a percentage of the total adjustment possible for the set point. In the proportional controlled furnace, the proportional band is 50 degrees. The full control range shown on the graph in Figure 3-6(b) is 100 degrees minus 0 degrees, or 100 degrees. To find the percentage value of the proportional band, divide the proportional band by the full control range, then multiply by 100:

$$\% \text{ Proportional Band} = \frac{\text{Proportional Band}}{\text{Full Control Range}} \times 100$$

$$= \frac{50}{100} \times 100$$

$$= .50 \times 100$$

$$= 50\%$$

This formula is the same one used to find the differential gap. The difference between the proportional band and differential gap is the manner in which the controller output responds within the band or gap. In the On-Off system, the output of the controller does not change as the measured value passes through the gap. In the proportional system, the controller output changes proportionally with the error as the measured value passes through the proportional band.

STEADY-STATE ERROR

The proportional controller is tuned so that the set point causes the proportional valve to open 50 percent with a given load. The 50 percent figure is desirable because the controller has equal amounts of corrective action from the set point to the maximum and minimum temperature settings. When the temperature produced by the furnace is at the 70 degree set point, the voltage supplied to the proportional valve is 50 percent.

If the load changes, the 50 percent valve position can no longer maintain the same temperature. Figure 3-7 illustrates what action then occurs. Suppose a disturbance causes the temperature to drop. A more positive error signal voltage is produced. The condition causes the proportional valve to open more than 50 percent. As the temperature rises, the thermistor resistance decreases and the proportional valve starts to close. If the disturbance continues for a long time, the proportional valve cannot return to the 50 percent open position. Instead, it must remain open more than 50 percent to offset the disturbance. For the valve to remain open above 50 percent, the error signal

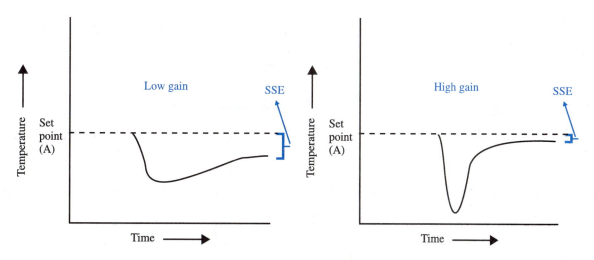

FIGURE 3-7   Relationship between gain and steady-state error

must be slightly more positive. Thus, the actual measured temperature could never climb to the 70 degree set point. Instead, it may stop at approximately 69.9 degrees in order to maintain the error voltage necessary to keep the valve open slightly more than 50 percent. The difference between the set point and the measured value is called **steady-state error**, or *offset*.

The example of steady-state error in the home heating system is similar to that found in many process control applications. Some process control systems allow some degree of steady-state errors. In other systems—especially motion control applications—proportional control does not provide the necessary level of control. In a position type of motion control application, steady-state error cannot exist because precision is required.

Figure 3-8 shows the schematic diagram for a proportional position control robotic system. The output of the system is connected to the arm of a robot. The arm is attached mechanically to the wiper of a potentiometer. As the robot's arm moves, the output voltage of the pot at the wiper varies. The potentiometer is the feedback device that supplies the negative feedback signal in the system. The voltage produced indicates the position of the arm and is applied to the inverting input of an op amp. The command setpoint signal is supplied by a computer. Since a computer's output signal

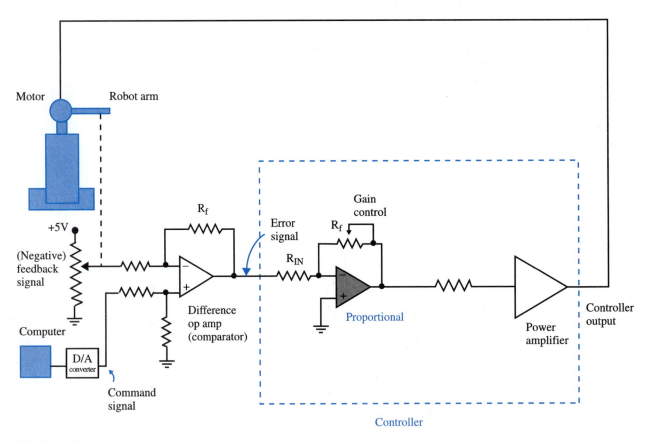

FIGURE 3-8 Proportional mode control system

is digital, a D/A converter is needed to change the value to an equivalent analog voltage. The output of the D/A converter is connected to the non-inverting input of the op amp. The op amp used is a difference type that operates as the comparator. Its function is to compare the command (setpoint) signal with the feedback signal, and produce an appropriate error signal.

The output of the difference op amp is connected to an inverting op amp which amplifies the error signal. This is called a proportional op amp because its gain is proportional to the ratio of the resistor values for R_{IN} and R_f. Since the output power of a standard op amp is seldom high enough to drive a motor, the error signal is further amplified by a power amplifier. The output of the power amp is connected to the motor which drives the robot's arm.

When the computer digital output is zero, the potentiometer will be zero volts, and the robot arm will be in the lowest position. Suppose the computer supplies a new position command to move the arm upward. The computer data and the resulting voltage of the D/A converter are shown in Figure 3-9. The computer outputs a series of numbers that increment until a value is reached that represents the desired position. The analog output voltage of the D/A converter ramps upward in small steps in a positive direction. The voltage change stops when the computer stops incrementing. This signal is compared to the feedback signal from the potentiometer by the difference op amp. Since the feedback signal lags behind the setpoint signal (because it does not respond immediately), a positive error signal voltage is produced by the difference op amp. The error signal is amplified by the proportional op amp and is also inverted to a negative voltage. The output of the proportional op amp is further amplified by the power amp and is also inverted to a positive voltage. With a positive voltage

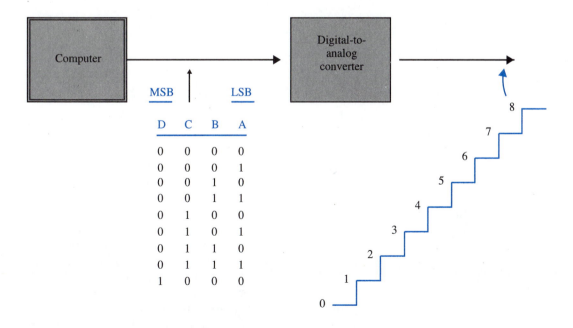

FIGURE 3-9 Computer command output signal converted to a proportional analog voltage

applied to the motor, the arm moves upward. The arm stops when it reaches the command position. At that position, the voltage of the potentiometer will equal the voltage of the D/A converter. This condition causes the comparator output to go to zero volts and stop the motor from turning the arm.

A closed-loop proportional motion control system is unlikely to have precise accuracy. The arm might never reach the desired position because the closer it approaches the location, the closer the voltage on the wiper comes to the command signal voltage. Therefore the output voltage applied to the difference op amp becomes very small and causes the motor to slow down. Eventually the mechanical friction of the robot arm cannot be overcome by the small amount of current that flows through the motor from the power amp. It falls short of its desired position, and a residual condition of steady-state error exists.

To reduce offset, the proportional band can be made smaller by increasing the gain of the proportional operational amplifier. This adjustment will also speed response to the command signal. However, the proportional band can be narrowed only so far before *instability* occurs. Instability is when the device being positioned oscillates because of overshooting. The system will try to correct the overshoot error by reversing the direction of the arm. The arm oscillates above and below the position before it dampens out and stops, as illustrated by the graph in Figure 3-10.

The friction of the load is not the only cause of steady-state error. Offset depends on three factors:

1. Load or demand on the process.
2. The low gain or wide band of the controller.
3. The set point at which the controller is set.

Changes in any of these three factors can result in some offset.

To overcome offset, the control mode known as **integral** control is used.

3-4   PROPORTIONAL-INTEGRAL CONTROL

The integral (or *reset*) mode of control is designed to eliminate the offset inherent in proportional-mode control. It develops a control signal that depends on the absolute value of the offset. The integral mode does not function by itself. It is used along with the proportional control mode in the controller section of a closed-loop system.

When an error signal first appears, the controller is tuned so that the proportional-control signal returns the process to the desired control point. This proportional-control signal is immediate and fast acting. If a deviation between the set point and controlled variable is present after the operation of the proportional control mode is completed, an additional corrective signal is required, which is supplied by the integral control mode function. A small corrective action is developed slowly to reduce the deviation to zero only after it is certain that there is a definite steady-state error.

The operation of the integral mode of control is illustrated in the robotic closed-loop system shown in Figure 3-11. It shows the same circuitry as the proportional-only controller, with an additional amplifier that performs the integral action. This second op amp is called an *integrator*.

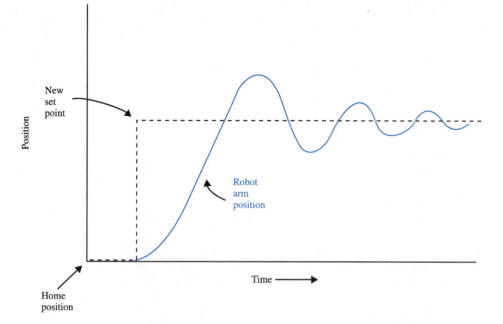

FIGURE 3-10   Instability of a proportional mode control system when gain is too high

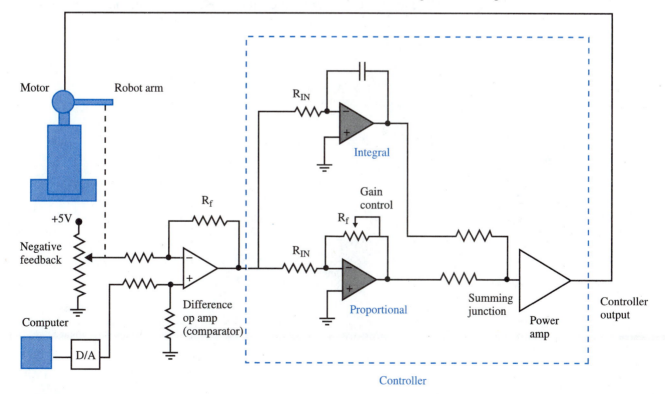

FIGURE 3-11   Proportional-integral mode control system

The integrator resembles an inverting op amp. The difference is that the feedback resistor is replaced by a capacitor. At the first instant a DC voltage is applied to its input, the capacitor operates like a short circuit. Recall that the gain of an inverting op amp is dependent on the ratio of the feedback resistance (R_f) and the input resistance (R_{IN}): Gain = R_f/R_{IN}. Therefore, since the capacitor initially provides low impedance in the feedback loop, the gain of the integrator is very low. The output voltage is also low. However, as the capacitor begins to charge, the current charging the capacitor reduces. Its impedance increases until it is fully charged, at which time it acts like an open switch. The result is that the R_F/R_{IN} ratio increases, the gain of the op amp increases, and the output voltage reaches saturation. The magnitude of the integrator output is proportional to the input voltage and the length of time the voltage is applied.

The operation of an integrator is further illustrated in Figure 3-12. This shows a graph that compares the input voltage with the output voltage. At T_1, a positive DC voltage is applied to the input of the integrator. Its inverted output voltage increases in a negative direction until saturation is reached at T_2.

Suppose the computer sends out a command signal for the robot to move. The proportional portion of the controller immediately responds to the setpoint change and drives the motor. The robot arm moves in the direction commanded by the computer, but stops just short of the desired position. The proportional mode has completed its response to the command setpoint change. Since the arm is out of position and does not achieve the desired location, the setpoint voltage and the feedback voltage are not the same. The result is that a steady-state error is present at the difference op amp output.

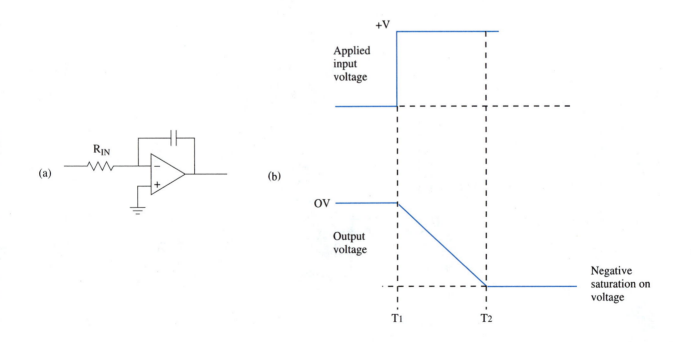

FIGURE 3-12 Operation of an integrator operational amplifier: (a) Schematic diagram; (b) Waveform diagram

This voltage is not sufficient to drive the robot arm the remaining distance. At this point the integral control mode takes over.

The steady-state error is also fed to the integrator op amp. The longer the error exists, the greater the output voltage of the integrator op amp becomes. The increasing integrator error signal is summed with the small amplified offset voltage of the proportional amplifier output. In time, the power amp receives enough input energy to turn the motor shaft until the robot arm attains the desired position. The steady-state error is then eliminated. Since the integral action continues to reset the amplifier gain until the process variable equals the setpoint value, it is also referred to as the *reset* mode of control. Although the two names are synonymous, reset is the older term.

The proportional-integral control mode is used in applications where load disturbances occur frequently and setpoint changes are infrequent. It is also used when load changes are slow, to allow enough time to elapse before it is necessary for the integral function to aid the proportional operation.

3-5 PROPORTIONAL-INTEGRAL-DERIVATIVE CONTROL

It is usually desirable to move the robotic arm quickly from one position to another. However, rapid movements are not possible using the proportional mode without excessive gain. If the gain setting of a proportional amplifier is too high, an instability condition develops where overshoot and subsequent oscillations occur. To reduce the overshoot and bring the controlled variable to the set point rapidly, a control mode called **derivative** or *rate* control is used.

The term *derivative* refers to the rate of change. A derivative controller produces an output that is proportional to the rate that the error signal changes. How much the error signal changes depends on the difference between the set point and the measured variable. If the error signal is changing very rapidly, the derivative output is large. When the error signal is changing slowly, the derivative output is small. If the error signal is stable, the derivative output is zero.

The function of the derivative controller is to provide a proportional correction to the error signal. For example, if the error signal gap increases, the derivative mode control gives a boost to the system to stop the error from increasing any further. The faster the error signal increases, the larger the boost. When the error signal gap decreases, the derivative mode control provides braking action. The faster the error gap closes, the stronger the braking action. The braking action reduces overshoot and dampens out any oscillations of the controlled variable.

In Figure 3-13(a), the derivative function is performed by a differentiator op amp circuit. Like the integrator op amp, the differentiator op amp resembles an inverting op amp. The difference is that the input resistor is replaced by a capacitor. Figure 3-13(b) provides a graph that compares the input voltage with the output voltage. If the input voltage applied to the differentiator is a constant DC voltage, the output is zero volts, as shown during time period W. If the input changes slowly at a constant rate, the output will be a small steady DC voltage, as shown during time period X. Time period Y shows that if the input voltage rises at a rate that is constant, but more rapidly, the differentiator output amplitude will be higher. Time period Z shows that if the input

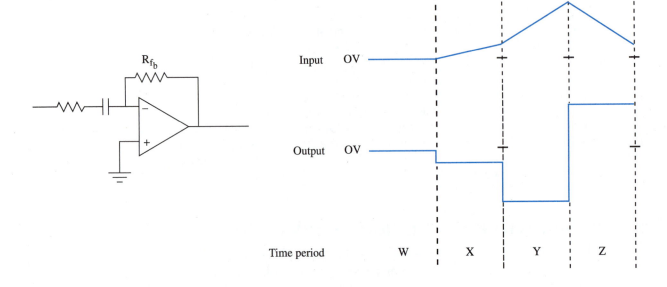

FIGURE 3-13   Operation of a differentiator operational amplifier

voltage decreases at the same rate, the differentiator will produce an equal output voltage of opposite polarity.

Figure 3-14 shows a robotic control system that contains proportional, integral, and derivative (PID) circuits. Figure 3-15 graphically shows the operation of the PID mode control system. Suppose an application requires that the robot arm move quickly from one position to another. The computer outputs a rapidly-incrementing series of numbers. The analog output waveform of the D/A converter shown between points A and D rises quickly at a steady rate.

The analog signal is fed to the non-inverting input of the difference op amp. Since the arm of the robot initially does not move, the output of the difference amplifier starts to rise and develop an error signal. As the error is fed to the input of the proportional amp, it is further amplified by the summing power amp. This action causes the motor to drive the robot arm toward the desired position. As it does, a voltage from the potentiometer, which is the feedback signal, begins to rise, as shown soon after time period A begins. However, the amplitude of the error signal continues to grow in the positive direction, as shown between points A and B. This happens because the stationary inertia of the robot arm has to be overcome, causing it to move slowly at the start. Therefore the measured variable from the feedback pot does not change as fast as the command signal from the computer.

The output of the difference op amp is also feeding the derivative amplifier. As the error signal voltage increases its amplitude, as shown between points A and B, a negative voltage is created by the derivative network. The derivative voltage is added to the proportional voltage by the summing power amp. The combined voltages cause the power amp output to increase, which makes the robot arm move faster. Eventually it moves fast enough that the measured variable is changing as fast as the command setpoint signal, as shown at time period C. This boost by the derivative function prevents the error signal from increasing any further.

Between points C and D, the error signal does not change. The output of the derivative amp goes to zero and the proportional function operates alone.

When the command signal from the computer reaches the value that represents the desired position, it stops changing. The output voltage of the D/A converter also stops increasing, as shown at time period D. Since the arm has not yet reached the desired position, the set point and measured variable are unequal. Therefore the difference amp continues to produce a voltage, causing the arm to continue moving. Because the

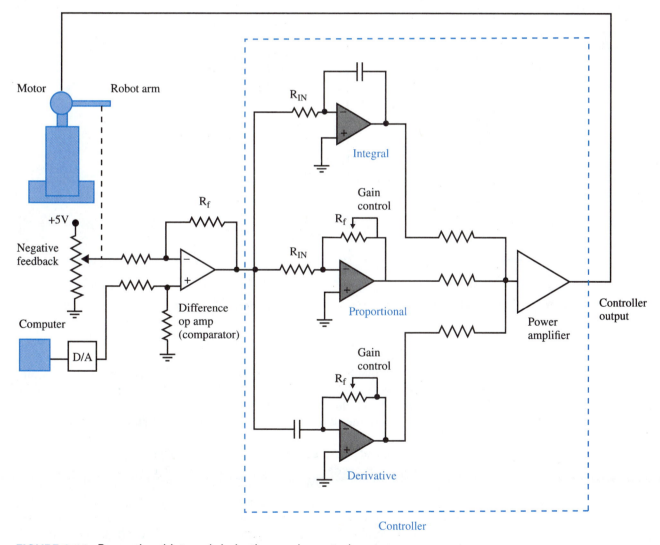

FIGURE 3-14 Proportional-integral-derivative mode control system

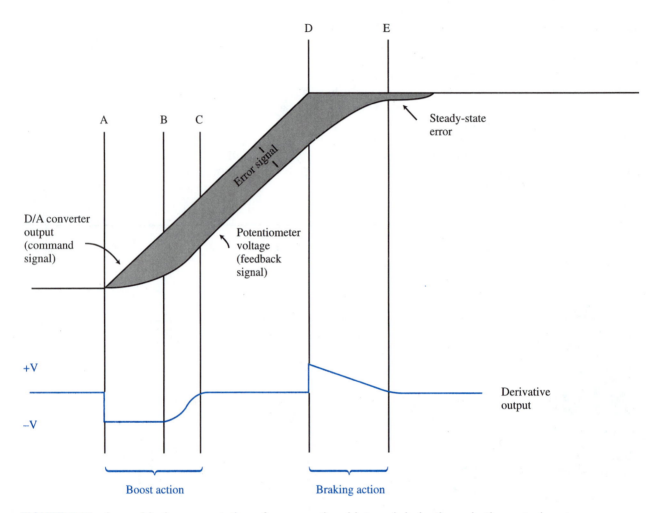

FIGURE 3-15   A graphical representation of a proportional-integral-derivative robotic control system

error signal decreases in amplitude, as shown between time periods D and E, a positive voltage is produced by the differentiator op amp. This voltage is subtracted from the proportional output by the summing power amp. Since the combined voltages cancel, the power amp output decreases. The result is that the motor causes the arm to slow down enough so it does not overshoot.

As the arm nears the desired position, the error signal stops changing and approaches zero. The result is that the proportional and derivative outputs go to zero. Since the set point and measured variable are not exactly equal, as shown in part E, the integral op amp takes over and causes the arm to move the remaining distance.

Whenever there is a large setpoint change, the controlled variable will usually lag behind and cause a rapid change of the error signal. Because the derivative controller detects this trend, it responds by compensating for large system changes before they fully develop. Therefore, derivative control is sometimes referred to as *anticipatory* or *predictive* control. Because derivative control tends to reduce system oscillation, the proportional gain can be set at higher values to further increase the speed of response of the controller to system disturbances.

Derivative-mode control is never used alone. It is usually combined with the proportional and integral modes for systems that cannot tolerate offset error and require a high degree of stability. This type of system—which combines the advantages of proportional, integral, and derivative action—is known as a three-mode **PID** controller. PID controllers are best suited for systems that need to react quickly to large disturbances. They are also recommended in systems where load changes frequently occur.

Derivative control is rarely combined with proportional control only. When it is, proportional-derivative action is used in applications where lag times vary and where offset error is tolerated.

To obtain the best possible PID control for a particular application, the gain settings for each mode must initially be made. These settings are different for each system. While the system is actually running, tuning adjustments are often made to the gain settings to attain optimal performance. Gain adjustments can be performed by trial and error, or automatically by autotune controllers.

In a conventional PID system, the process being controlled is seldom performed by op amp circuits. Instead, these three modes are performed by computer software packages that calculate a set of differential equations. When a setpoint change or disturbance occurs, feedback devices send signals that cause the numerical values of the mathematical equations to change. Calculations are made to produce a solution which tells the PID controller how to adjust the system parameters to meet the new requirements.

PID control is an industrial standard well understood by many control engineers. It is a popular control technique that has been proven through many years of use.

3-6   INTRODUCTION TO FUZZY LOGIC

PID control is the most common type of control technique presently used in automated systems. It utilizes mathematical equations or Boolean expressions to perform the process. Some types of applications that use mathematical models are often difficult, or impossible, to calculate because of the complexity involved. The math function required is often too difficult to write, or takes too long to solve because it requires massive computations. An example is a system where one or more controlled variables change completely or irregularly. These situations typically require human intervention to make the necessary adjustments to correct the operation.

A revolutionary control technique called **fuzzy logic** is capable of performing some of the operations that are too complex for PID systems. Fuzzy logic is a form of artificial intelligence that enables a computer to simulate human reasoning.

When people make decisions, the present conditions are observed by our biological sensory inputs. The human response is based on rules that have been formulated through personal knowledge and experience. However, instead of using hard and fast rules, each rule is weighed based on its importance. The human thinking process differentiates significant conditions from insignificant conditions to decide on the appropriate action to take. The operation of fuzzy logic is designed to make decisions in a similar manner.

There are several attractive features associated with fuzzy logic technology. Because of its programming structure that allows the use of estimates, control programs can be constructed with as little as one-tenth the rules of conventional systems. This results in shorter program development time and much faster program execution. Also, instead of developing complex mathematical formulas, the engineer converts personal knowledge and experience into words as the computer is programmed. The operational and control laws of the system are also expressed linguistically, which make it intuitively easier to understand and justify.

3-7   FUZZY SETS

A key element of fuzzy logic is the *fuzzy set*. In conventional mathematics, which is based on Aristotelian logic, related items are grouped together in *crisp sets*. An item is either a part of a set or not. There is no middle ground; there is no partial membership. This strictness is a characteristic of binary logic and is often considered a virtue. However, the all-or-nothing characteristic of crisp sets is very restrictive and often impractical. For example, a set does not allow inexact, subjective concepts such as *partial, slightly, average,* or *faster.*

Consider using the concept of classifying people as being middle-aged. Let us assume that a 45-year-old person can reasonably be thought of as in the center of middle age. Also assume that the person begins being middle-aged at 35 and ends being middle-aged at 55 years old. Figure 3-16 shows a graphical representation of the boundary of middle-aged persons. It has square sharp edges, similar to a conventional square wave which has two grades. Those individuals 34 and younger, or 56 and older are not considered middle-aged. The grade of the graph indicates they are defined by a value of 0. Everyone within the age group of 35 to 55 is a member of a set. According to the graph, the grade of 1 indicates that all people in this range are considered equally middle-aged. The classification method provided by crisp sets is not capable of illustrating the degree to which a person is a member of a set. For example, it does not show that a person who is 45 is more middle-aged than a person who is 35.

The structured membership classification of a crisp set is overcome by the variable boundary of fuzzy sets. The graph in Figure 3-17 shows a bell-shaped curve that represents membership in a fuzzy set. The height at any point along the curve, known as degree of membership, corresponds to the amount at which a person is middle-aged. A person at age 45 is located at the peak and is given a degree-of-membership value of 1. People who are 35 and 55 are located at the bottom portion of the curve and are given a degree-of-membership value slightly higher than 0. The degree-of-membership of a middle-aged person gradually becomes stronger as the individual gets older, and then becomes weaker past 45 until age 55 is reached.

3-8   MEMBERSHIP FUNCTIONS

The square wave and the bell-shaped curves are called membership functions. The binary crisp sets have degrees of membership that are either 0 or 1. The variable fuzzy set has degrees of membership that range from 0 to 1. The fuzzy type of membership

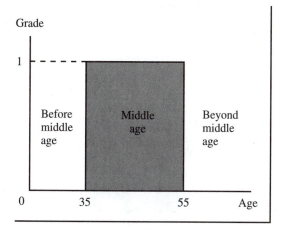

FIGURE 3-16 Crisp-set representation of middle age

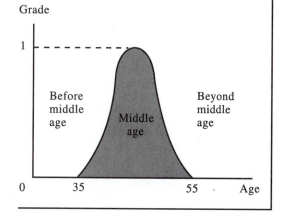

FIGURE 3-17 Fuzzy-set representation of middle age

function provides a method of dealing with gradual changes, that is, situations that fall between absolute values.

THE FUZZY RULE-BASED SYSTEM

Fuzzy logic control can be broken down into three stages:

1. A crisp-to-fuzzy transformation of inputs, called the *fuzzifier.*
2. Applying the fuzzy inputs as conditions to a rule base, called *inferencing.*
3. Combining the resulting action and transforming it from a set of fuzzy outputs back to executable crisp outputs, called the *defuzzifier.*

The basic structure of a fuzzy rule-based system is shown in Figure 3-18.

3-9 A FUZZY LOGIC APPLICATION

An experiment that was demonstrated at the Second Congress of the International Fuzzy System Association held in Tokyo, Japan in 1987 will be used to help understand the operation of fuzzy logic. It involves an inverted pendulum—the equivalent of balancing a broomstick on the palm of a hand (Figure 3-19). A pole is attached to a vehicle by a hinge. The pole can only fall to the right or left from its upright position. By monitoring the error (the angular position of the pole) and the error rate (the speed it is falling), the movement of the vehicle is controlled so as to keep the pole upright. To measure the position of the rod, an encoder is used. A tachometer measures the speed with which the rod is moving. Both types of feedback measuring devices provide input values to the fuzzy controller. The data it represents is calculated by the fuzzy controller. Its solution is converted into an electrical signal that causes the DC servo motor to move the vehicle so that it keeps the stick balanced.

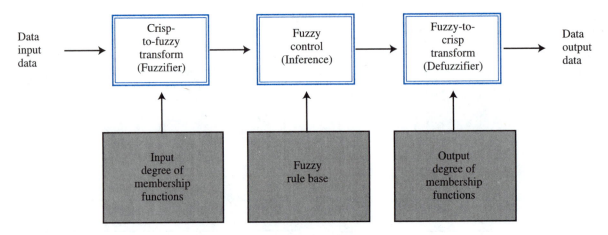

FIGURE 3-18   Three major components of a fuzzy rule-based system

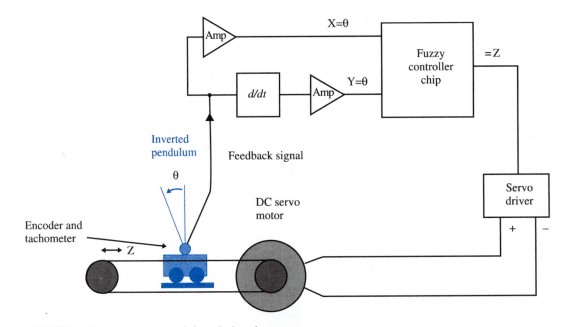

FIGURE 3-19   Inverted pendulum balancing system

FUZZY LOGIC PROGRAMMING

The operation of the controller is based on a programming scheme that imitates the way people process information. Instead of using complicated equations to program the controller, such as the one shown in Figure 3-20, If-Then statements called *production rules* are used.

The production rules are used to determine what set of conditions is present in the inputs. They also describe how the output should respond to the input information. In

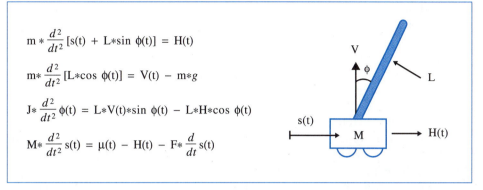

$$m * \frac{d^2}{dt^2} [s(t) + L*\sin \phi(t)] = H(t)$$

$$m* \frac{d^2}{dt^2} [L*\cos \phi(t)] = V(t) - m*g$$

$$J* \frac{d^2}{dt^2} \phi(t) = L*V(t)*\sin \phi(t) - L*H*\cos \phi(t)$$

$$M* \frac{d^2}{dt^2} s(t) = \mu(t) - H(t) - F* \frac{d}{dt} s(t)$$

FIGURE 3-20   Mathematical model of balancing stick example (Courtesy of Omron)

the case of the balancing rod, *If* phrases called *antecedent* blocks describe the states that the stick can be in, and *Then* phrases called *consequent* blocks describe how the vehicle should move in response to those states. Each production rule statement contains one or more key words that relate inputs and outputs to fuzzy set membership functions. They are ambiguous terms that provide a method of expressing the degree to which an input or output condition matches a fuzzy set. The rules of operation the programmer writes are based on the programmer's knowledge.

Table 3-1 contains the production rules that are needed to keep the rod balanced. To write the production rules, the input and output data are used. The inclination of the rod compared to the vertical position is represented by the label θ. The angular velocity of the rod is $d\theta$. The change in the speed of the vehicle is ΔV. The ambiguous words

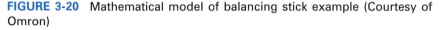

TABLE 3-1   PRODUCTION RULES FOR THE BALANCING STICK EXAMPLE

Rule	Antecedent Block	Consequent Block
Rule 1	If the stick is inclined moderately to the left and is almost still	then move the hand to the left quickly
Rule 2	If the stick is inclined a little to the left and is falling slowly	then move the hand to the left slowly
Rule 3	If the stick is inclined a little to the left and is rising slowly	then keep the hand as it is
Rule 4	If the stick is inclined moderately to the right and is almost still	then move the hand to the right quickly
Rule 5	If the stick is inclined a little to the right and is falling slowly	then move the hand moderately to the right slowly
Rule 6	If the stick is inclined a little to the right is is rising slowly	then keep the hand as it is
Rule 7	If the stick is almost vertical and is almost still	then keep the hand as it is

contained in the rules are *moderate, almost, a little, slowly,* and *quickly.* When more than one (input) variable is used in the antecedent fuzzy block, it is linked by the word *and.*

The Fuzzifier Stage

Once the rules are programmed into the fuzzy controller, fuzzy logic is put into operation. Fuzzy set theory is then applied as the validity of each input is determined by being compared to the sets established by the rules.

The first step is to transform crisp logic input data into fuzzy logic information. An example of this operation is to convert speed into fuzzy sets (Table 3-2). A speed of 70mph may have degrees of membership in each of several different fuzzy sets that represent different ranges of speeds.

To determine if the 70mph input condition falls within the range of a fuzzy set, the input signal is compared with algorithms that output a degree-of-membership value.

In the balancing stick application, the rod may incline in multiple positions. These positions are described by the key words in the production rules. They also can be graphically shown by the multiple fuzzy set membership functions shown in Figure 3-21. Seven membership functions with corresponding labels are used, representing three ranges in the positive direction, three in the negative direction, and a zero. The labels represent the positions described by the key words in the production rules. Table 3-3 shows the relation between the key words in the production rule (Table 3-1) and the labels with abbreviations in the membership function diagram (Figure 3-22). Because calculations required for bell-shaped curves are so complex, triangular shaped fuzzy sets are used instead.

Figure 3-22 graphically shows how the fuzzy controller evaluates whether the position of the rod falls within a membership function, and also its degree of membership. Suppose that the vertical line represents the angle of inclination at a given instant. This input value is superimposed over the antecedent membership function on the top row of the graph. The seven rows relate the graph in Figure 3-21 and the production rules in Table 3-1.

In rule 1, the key word in the phrase *moderately to the left* describes the angle of inclination θ. In the membership set, this equates to *positive medium,* or label PM. The grade where the current input value crosses the PM membership function is 0.7. This value corresponds to the degree of membership (fuzzy variable between 0 and 1) the rod position has within the PM fuzzy set.

TABLE 3-2   **SPEED CONVERSION INTO FUZZY SETS**

Fuzzy Set Ranges	Input Value		Degree of Membership
Very Slow	(70mph)	=	0
Slow	(70mph)	=	0
Medium Fast	(70mph)	=	0.3
Fast	(70mph)	=	0.8
Very Fast	(70mph)	=	0.5

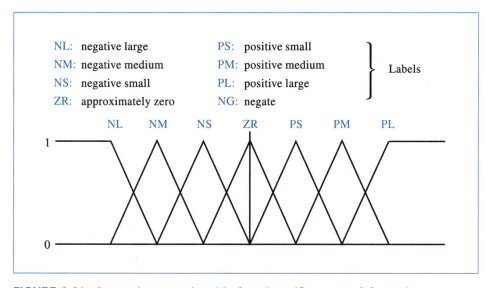

NL: negative large   PS: positive small
NM: negative medium   PM: positive medium   } Labels
NS: negative small   PL: positive large
ZR: approximately zero   NG: negate

FIGURE 3-21 Antecedent membership functions (Courtesy of Omron)

TABLE 3-3 PENDULUM ANGLE INPUT DESCRIPTORS

Linguistic Description	Abbreviation	Label
FAR LEFT	NL	negative large
MEDIUM LEFT	NM	negative medium
SMALL LEFT	NS	negative small
VERTICAL	ZR	zero
SMALL RIGHT	PS	positive small
MEDIUM RIGHT	PM	positive medium
FAR RIGHT	PL	positive large

Next, the rod position is compared to the condition described by Rule 2. The input signal indicates the rod position falls within the range of *a little to the left,* which equates to *positive small,* or PS membership at a value of 0.3. In Rule 3, the current input is compared to the *positive small* or PS membership set to arrive at a value of 0.3. The same input data, that is, the rod position, is processed by each production rule. Since each rule processes the information using different parameters, the output of each rule will be different.

Rules 4 through 7 are evaluated in a similar manner. Since the input does not compare to any of the remaining rod position membership functions, the remainder of the degrees of membership values are zero.

The same multiple fuzzy sets shown in Figure 3-21 can be used to evaluate the membership function of the rod velocity. Figure 3-23 shows a vertical line that represents the rod speed at the same time the rod position was evaluated in Figure 3-22. In Rule 1, the speed *almost still* describes the angular velocity (dθ), which equates to the *approximate zero* or ZR membership function. The grade where the current input crosses the ZR membership function is 0.8. In Rule 2, the dθ input intersects with the PS membership set for a value of 0.2. In Rule 3, the phrase *very slowly* describes the *negative small* (NS) membership set. Since the dθ input does not intersect with the NS membership set, the degree of membership is given a value of 0. The procedure is repeated for Rules 4 through 7.

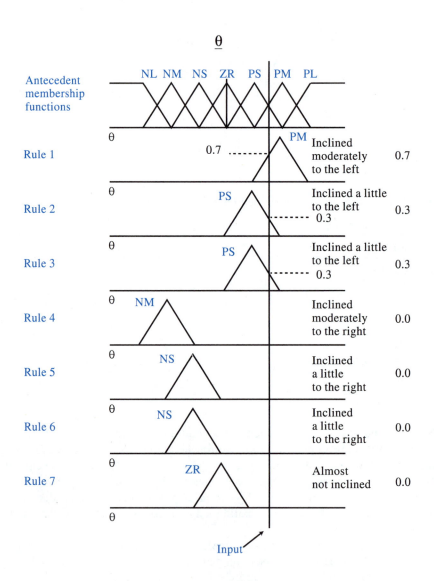

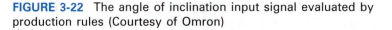

FIGURE 3-22 The angle of inclination input signal evaluated by production rules (Courtesy of Omron)

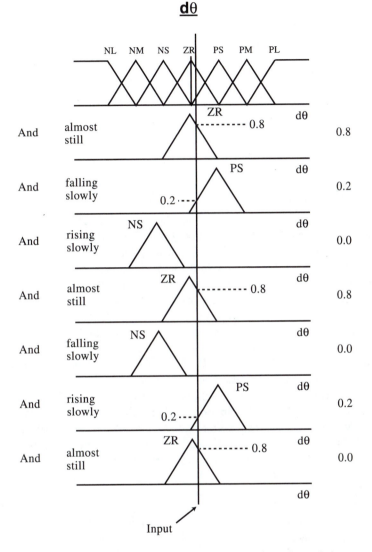

FIGURE 3-23 Evaluating the membership function of the rod velocity (Courtesy of Omron)

Referring back to Figure 3-21, observe that there are seven possible states for each input in the system. Since there are two inputs in the balancing stick application, there are 49 possible rule combinations if all seven labels are used. The value of 49 is determined by squaring 7 (7^2). The 7 represents the largest possible numbers of input ranges. The superscript 2 indicates that there are two different inputs that provided signals to the balancing stick controller.

Instead of using the maximum number of production rule combinations, the programmer determines that the seven listed in Table 3-1 are sufficient to control the process. The remaining combinations can be discarded. By eliminating all but seven of the

49 possible rules, the fuzzy logic steps are simplified. This process of selecting only the relevant rules is similar to the thinking process used by humans.

Note that the membership functions for each of the variables are arranged so that they overlap each other by approximately 50 percent. In this way, a reliable reading can be taken even when the input level is not clear or when it changes continuously.

The Inference Stage

The next step in the operation of fuzzy logic technology is called *Inferencing*. This procedure determines which antecedent degree-of-membership values are most relevant at a particular moment. The ones that are used then provide consequent output commands, which cause the vehicle to move properly.

To determine which degree-of-membership values are most valid, it is necessary to take the logical product of the two grades in each rule. The logical product of a group of grades is simply the lowest degree of membership value. Thus in rule 1, the logical product of 0.7 for θ and 0.8 for $d\theta$ is simply 0.7. This procedure, called *Determining MIN* (minimum), is repeated for each of the rules. The logical products for each rule are listed in Figure 3-24. Since variable θ and $d\theta$ are linked by an "and," one variable is dependent upon the other. Therefore, by selecting a minimum value, a value (product) is chosen that satisfies both conditions.

The logical product of the antecedent block is used as input data evaluated by the membership functions of the consequent block. For the balancing stick application, the consequent membership function in Figure 3-24 is the same as the antecedent membership function in Figure 3-21. However, since the consequent membership is designed on the basis of the programmer's experience, it does not necessarily have to be identical to the antecedent membership functions.

Each consequent membership function receives input data from the antecedent logical product value of the corresponding rule. As a result, the shape of the consequent membership function for each rule is usually altered. This adjustment is accomplished by chopping off the tops of the consequent membership functions to match the minimum grades of their antecedent parts. Figure 3-24 shows both the antecedent blocks and the corresponding consequent blocks of the balancing stick.

In Rule 1, the consequent membership function is cut off above 0.7 of the positive medium (PM) membership function. The result is the shape of the PM membership function changes from a triangle to a trapezoid. In Rule 2, the PS membership function is cut off above 0.2. In the remaining rules, the minimum grade is zero, and the consequent membership function is cut off completely at the base. Once the consequent blocks are derived for each rule, the operation is ready for the final step of the Inference stage.

This operation begins by combining the consequent portions of each rule by the logical sum method. All of the area outside the shaded portion is cut. The shaded area represents the combined consequent block values that remain after all the rules have been processed.

The technical term for this process is called *Determining MAX* (maximum). The maximum value is found because the maximum accepts all conclusions. Because the rules are related, but separate, selecting maximum will allow all of the rules to influence the final outcome. Even if it is only a small influence, it should effect the

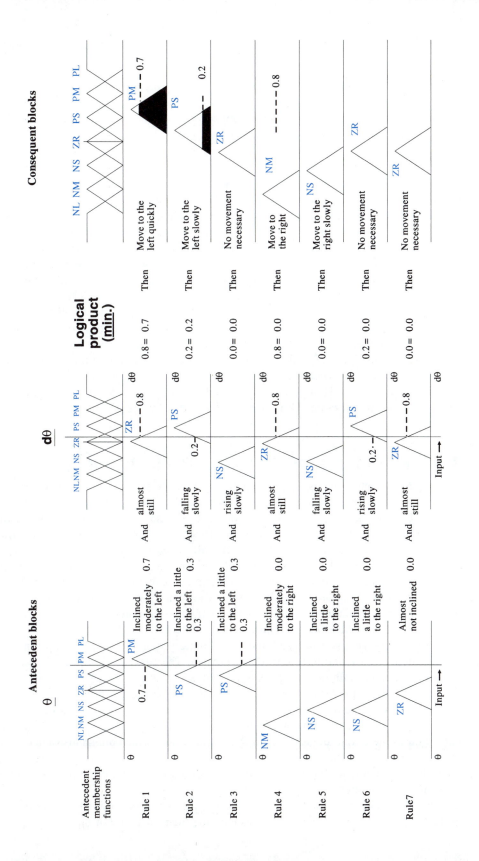

FIGURE 3-24 Determining the logical sum of the consequent blocks in balancing stick example (Courtesy of Omron)

movement of the vehicle. This concept is based on *OR* relationships between the rules, called *taking the logical sum* in set theory. Only the consequent fuzzy sets that have a degree of membership above zero are summed. The logical sum is graphically shown in the lower right corner of Figure 3-25(a).

Defuzzifier Operation

The final phase of the fuzzy processing sequence is to perform defuzzification. The defuzzier step calculates the center of gravity of the rod by mathematically determining the weighted average numerical value of the summed trapezoids. The consequent block with the greatest influence on the calculation is the one with the highest antecedent input. The mathematical answer becomes the controller's output value that is converted into a voltage represented by signals ΔV as shown in Figure 3-25(b). This voltage is applied to the servo motor that moves the vehicle. A positive voltage causes the vehicle to move in one direction, and a negative voltage causes it to move in the opposite direction. The higher the voltage amplitude, the higher the velocity.

Figure 3-25 illustrates the entire fuzzy processing sequence for a given moment. The process by which the conclusion was formed took into account several variable factors, and considered only those that were valid to cause the output to provide a proper response, thus operating in a manner similar to the human thinking process. The fuzzy control steps are performed continuously. As input values constantly change, the fuzzy processing occurs hundreds of times per second to update its operation.

3-10   FUZZY LOGIC MICROCONTROLLER

There are many methods used by computing devices to perform fuzzy logic control. One is the technique used by a fuzzy microcontroller chip shown in the block diagram of Figure 3-26. There are 16 Fuzzifiers. Each one stores a fuzzy membership function of a production rule. Since there are 16 separate fuzzifiers, 16 different terms can be stored for a rule. A one of eight selector allows each fuzzifier to input data from any one of eight variable input signals, such as the rod inclination or velocity. Each fuzzifier evaluates the applied input and determines a degree-of-membership. Next, the outputs of the fuzzifier are applied to the Minimum Comparator for minimal comparisons. Once the minimum value for a rule is found, it is stored in a temporary maximum register in the Maximum Comparator.

After the first rule is processed, the Rule Memory transfers membership function data of the next production rule to the fuzzifiers. The fuzzification procedure is repeated and another minimum degree-of-membership value is stored in the Minimum Comparator. After processing up to 64 rules that can be stored in the Rule Memory section, the Maximum Comparator determines which rules have valid degrees of memberships. By using sum-of-products calculations, an optimal action value is passed on to the output devices. All of the valid action value commands (antecedent blocks) used in the calculations were transferred along with their corresponding consequent block rules from the fuzzifier to the Minimum Comparator, and then to the Maximum Comparator.

Once all of the rules have been processed and the command information transferred to the Output Register, the device begins entering the next group of inputs.

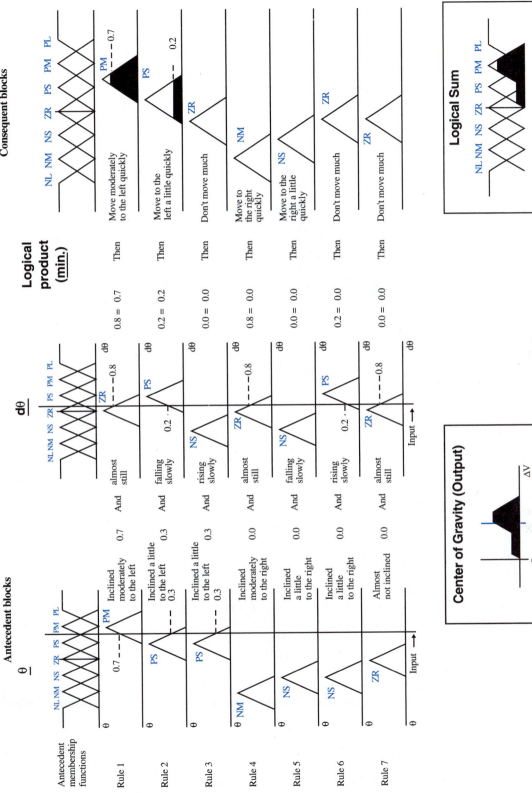

FIGURE 3-25 Defuzzification by calculating center of gravity in balancing stick example (Courtesy of Omron)

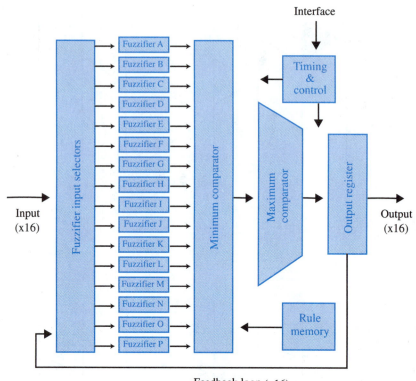

FIGURE 3-26 Fuzzy microcontroller module

3-11   CONCLUSION

Fuzzy logic gets its name from the way it operates. It accepts vague input data and transforms it into precise mathematical form which is then used by a machine to perform solving actions.

Fuzzy logic theory was created in 1965 by UC–Berkeley professor Lofti Zadeh. His ideas originated by studying work on multivalued logic by a number of mathematicians earlier in the century. The purpose of their research was to develop mathematical problem-solving models that copied the methods used by humans. People can gather imprecise and ambiguous information from their environment and use the technique of approximate reasoning to make accurate decisions quickly and efficiently.

When fuzzy logic was first introduced to the engineering community, it was met with indifference and hostility. However, in Japan, Europe, and China, it was accepted more openly, and immediately began to produce positive results. In electronics control, for example, it was found that these techniques resulted in faster, smoother, more efficient, and easier to design solutions.

Fuzzy logic has been used to remove the unwanted vibration of a video camera while it is recording a scene from a helicopter. It is used in automotive applications such as anti-lock braking systems and automatic transmissions. The Panasonic

Wet/Dry electric shaver uses a fuzzy logic microchip to determine whether it is shaving a slightly whiskered area or a heavy patch, after which it responds by using one of 576 shaving patterns. This technology is now being applied to industrial applications where its use will likely grow well into the twenty-first century.

Fuzzy technology can be applied through computer or industrial programmable controller software, dedicated controllers, or through fuzzy microprocessors.

CHAPTER PROBLEMS

1. Which section of the closed-loop system performs the four control modes and fuzzy logic operations?

2. In an On-Off heating system an error signal is produced when the measured temperature is _____ (above, below) the set point.

3. List two factors that cause the controlled variable to deviate from the set point in an On-Off system.

4. By _____ (increasing, decreasing) the On-Off differential gap, the cycle time is lengthened.

5. Calculate the differential gap percentage if a full temperature control range is 80 degrees and the differential gap is 8 degrees.

6. The lagging effect of the controlled variable behind the error signal is called ____ .
   a. time delay   b. hysteresis   c. lag time   d. all of the above

7. __ T __ F   By narrowing the proportional band, a closed-loop system may oscillate.

8. Steady-state error is also referred to as _____ .

9. Offset is reduced by _____ (increasing, decreasing) the system gain of the closed-loop system.

10. __ T __ F   A setpoint change can also produce offset.

11. The ____ mode is designed to eliminate offset.
    a. proportional   b. integral   c. derivative

12. Proportional-integral control is used in which of the following applications: ____
    a. Where load disturbance occurs frequently and setpoint changes are infrequent.
    b. Where load disturbances occur frequently and setpoint changes are frequent.
    c. Where load changes are slow.
    d. Where load changes are fast.

13. The term *derivative* means _____ of change.

14. __ T __ F   A derivative controller produces an output that is proportional to the amplitude of the error signal.

15. The output of the derivative function _____ (adds to, subtracts from) the output of the proportional output when the error rate is increasing.

16. The output of the derivative function _____ (adds to, subtracts from) the proportional output when the error rate is decreasing.

17. __ T __ F   In a PID system, after the derivative function is complete and the proportional signal is ineffective, the integral function is performed.

18. List which type of operational amplifier performs the following PID mode functions:

Proportional:

Integral:

Derivative:

19. __ T __ F   Fuzzy sets allow partial truths.

20. Match the three fuzzy logic terms with their definitions.
   ____ A crisp-to-fuzzy transformation of inputs.
   ____ Applying the fuzzy inputs as conditions to a rule base.
   ____ Combining resulting actions and transforming fuzzy sets to crisp outputs.
   a. Inferencing   b. Fuzzifier   c. Defuzzifier

21. _____ (If, Then) phrases are called antecedent blocks.

22. List the order in which the following three stages occur in a fuzzy logic operation by placing 1st, 2nd, 3rd to the left of each term:
   ____ Defuzzifier
   ____ Fuzzifier
   ____ Inferencing

SECTION

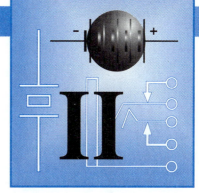

ELECTRIC MOTORS

The electric motor is the most common device used to perform the actuator function in an industrial control loop. It converts electrical energy into mechanical power. The electric motor is the workhorse in both commercial and industrial applications. In the home, the furnace, refrigerator, washer, and dryer are all powered by electric motors. They also drive manufacturing industry. It is estimated that 58 percent of all electrical power generated is used to supply industrial electric motors. Because they are used so extensively, electric motors are an important area of study in the field of industrial electronics.

CHAPTER 4
DC Motors

CHAPTER OBJECTIVES

After completing this chapter, you should be able to:

○ Describe the operating principle of a DC motor.

○ List the major components of a DC motor.

○ Define the following terms:

Motor Action	Armature Reaction
Main Field	Neutral Plane
Commutation	CEMF
Rotary Motion	Full Load
Torque Force	Overload
Holding Torque	Partial Load
Speed Regulation	No Load

○ Make the following calculations for a DC motor:

Speed Regulation	Power
Torque	Horsepower
Work	Efficiency

○ Describe the operation of the following DC motors and identify their characteristics:

Shunt	Compound
Series	

○ Reverse the direction of a DC motor.

○ Choose the types of DC motors needed for specific applications.

INTRODUCTION

A direct current (DC) motor converts DC electrical energy into mechanical energy. As direct current is used by the motor, it produces a mechanical rotary action at the motor shaft. The shaft is physically coupled to a machine or other mechanical device to perform some type of work.

DC motors are highly versatile mechanisms. They are well suited for many industrial applications. For example, they are used where accurate control of speed or position of the load is required. They can be accelerated or decelerated quickly and smoothly, and their direction easily reversed. This makes them very useful in machine tool operations and in robotics. They provide higher starting torque than other motor types. Because the DC battery is the best portable power supply, DC motors are used for electric tools, carts, tow motors, and other forms of mobile equipment. In this chapter, the operation and characteristics of the DC motor are described.

4-1   PRINCIPLES OF OPERATION

The conversion of electrical energy to mechanical energy is accomplished by a principle called **motor action**. There are two requirements for motor action to exist. The first requirement is that there is a current flow through a conductor. As it does, a circular magnetic field develops around the wire. Those lines of force go in a direction described by the *left-hand rule* shown in Figure 4-1. The thumb points in the direction of electron current flow. The fingers point in the direction of the circular flux lines around the wire.

The second requirement is that a force on the conductor develops. The force is produced when the conducting wire is placed inside the magnetic field formed between two magnet poles, as shown in Figure 4-2. This magnetic field is referred to as the **main field**. The direction of the force depends upon the direction of current through the wire and the direction of the flux lines between the poles.

Figure 4-2 illustrates this concept. The magnetic field develops between two poles of either a permanent magnet or an electromagnet. Normally, these lines are straight and go in a north-to-south direction. However, when the conductor is placed between the poles, the lines become distorted. On one side of the wire the two fields combine and become very concentrated. On the other side they go in the opposite direction. The effect is that they cancel each other, making a weak force. The distorted lines on

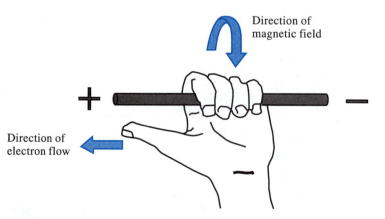

FIGURE 4-1 The Left-hand Rule showing the direction of electron flow and the magnetic field around a conductor

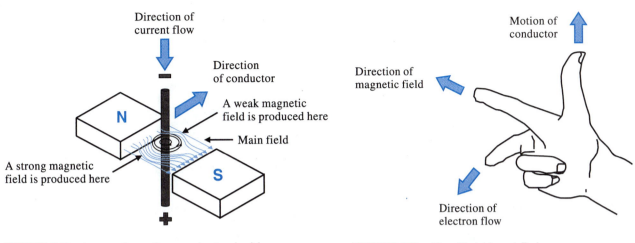

FIGURE 4-2 Interaction of a conductor inside a magnetic field causing movement of the wire

FIGURE 4-3 The Right-hand Rule

the stronger side are elastic like rubber bands. Since they are stretched, they tend to straighten out. Straightening exerts a force on the conductor and pushes in the direction of the weak side until it moves out of the field. Figure 4-3 illustrates the *right-hand rule* for motors. It shows the direction that a conductor carrying current will be moved in a magnetic field. The index finger points in the direction of the magnetic field lines (north-to-south). The middle finger points in the direction of the current in the wire. The thumb points in the direction of the wire movement. This is the fundamental principle of motor action.

4-2 ROTARY MOTION

A current-carrying conductor in a magnetic field tends to move at right angles to the field. Once it moves out of the field, the force is reduced to zero and no further action takes place. Since a motor produces a continuous rotary motion, it is necessary to make the current-carrying conductor into a single loop of wire. Figure 4-4 shows the loop placed between two magnetic poles. When the loop is connected to a DC supply, current flows from point A to point B on one side of the loop, and from point C to point D on the other side of the loop; that is, the current flows in opposite directions through the sides of the loop across from each other. Therefore, one side is pushed upward and the other side is pushed downward. Because the loop is designed to pivot on its axis, the combined force results in a twisting action called **torque**.

This action is illustrated by the cross section view in Figure 4-5. The x indicates the point at which current flows into the page. The dot indicates the point at which current flows out of the page. The large arrows show the direction of each wire segment. The loop rotation is counterclockwise (CCW). When it reaches a position perpendicular to the field, there is no interaction of the magnetic fields. This is called the **neutral plane**. Due to inertia, the loop continues to spin CCW. However, since the direction of current flow through the loop doesn't change, the interaction between the conductor segments and the flux lines develops a force in the opposite direction. Instead of

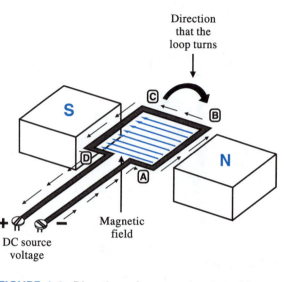

FIGURE 4-4 Direction of torque developed by a loop of wire

continuing in the CCW direction, the loop stops and then changes direction. An oscillating motion is produced until the armature settles at the neutral plane.

A continuous rotation is achieved in Figure 4-6 by reversing the direction of current through the loop the instant it passes through the neutral plane. The current change is accomplished by a switching device called a **commutator**. Sometimes referred to as a *mechanical rectifier,* the commutator is in the shape of a ring that is split into two segments. Each segment is connected to an end of the loop. The commutator and loop

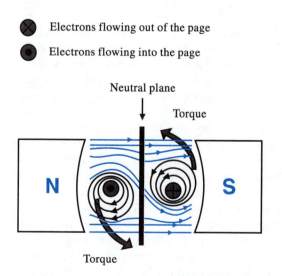

FIGURE 4-5 Cross-section view of the loop of wire inside the main field

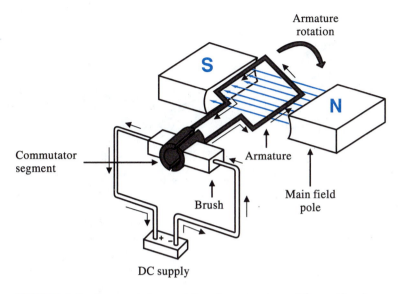

FIGURE 4-6 Commutator and brush arrangement for a simple DC motor

rotate together and are referred to as the **armature**. A pair of carbon brushes supply current to the armature windings. The brushes are sliding connectors that make contact with the commutator segments as the armature rotates. Each brush is connected to a terminal of the DC supply.

The drawings in Figure 4-7 illustrate how the switching action of the brushes and commutator cause the armature to rotate one revolution. The direction of the flux lines in the main field is from the north pole to the south pole. The switching action is called **commutation**.

1. In position 1 the current enters the loop through the negative brush and exits through the positive brush. The torque developed causes the armature to rotate in a counterclockwise direction.

2. When the armature is in position 2, the brushes make contact with both commutator segments. The armature loop shorts out and current flows from one brush to the other through the commutator segments. The result is that no torque is produced. However, inertia causes the armature to continue rotating past this position.

3. When the armature rotates past the neutral position in position 3, the sides of the loop are in the opposite position than they were in position 1. The switching action of the commutator reverses the direction of current flow through the armature loop. This causes current to flow into the armature segment closest to the south pole, as it did in position 1. The torque developed causes the armature to continue rotating in the CCW direction.

4. In position 4, the armature is again in the neutral position. Since inertia carries the armature toward the position in position 1, the cycle is repeated.

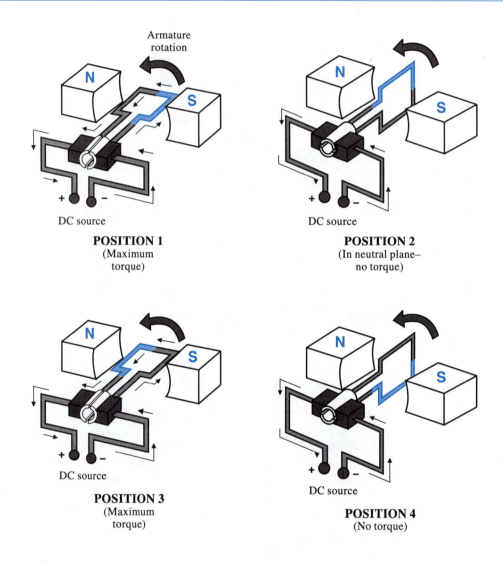

FIGURE 4-7   The operation of a simple DC motor as it rotates 360 degrees

The rotation of the armature continues in one direction because the commutation keeps reversing the current direction through the loop. This way the armature always interacts the same way with the main field to maintain a continuous torque in one direction.

There are two disadvantages of using a motor with one armature. One problem is starting the motor when the armature is in the neutral position. Since the armature loop is shorted, no torque is developed to cause movement. To start, the armature must be physically moved out of the neutral position. The other disadvantage is that when the motor runs, its speed is erratic because its torque is irregular. Maximum torque is produced when the armature loop is at a right angle to the main field, and minimum when it is located in the neutral plane.

Both problems are corrected by using a two-loop armature with four commutator segments, as shown in Figure 4-8. The ends of the loop are connected to opposite segments of the commutator, and the loops are electrically connected in parallel. When one loop is in the neutral position, the other is in the position of maximum torque. As the armature turns, the commutator switches current to the loop that approaches the neutral plane. The disadvantage of this configuration is that during brief moments during the revolution only one loop is connected, while the other rotates as dead weight. This situation occurs at the moment when commutation takes place. The loop that is at a right angle to the main field creates the torque. The loop in the neutral plane is dead weight because there is no interaction of magnetic fields.

By connecting loops of the armature to adjacent commutator segments, this problem is corrected. One commutator segment per loop is used instead of two segments per loop. Electrically, the armature is two series circuits connected in parallel, as shown in Figure 4-9. When current flows through the brushes, all four loops carry the current and contribute to the torque.

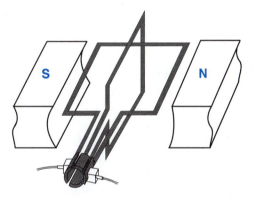

FIGURE 4-8   A two-loop armature provides self-starting and steadier and stronger torque

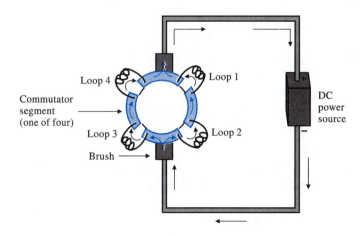

FIGURE 4-9   A four-loop armature that uses four commutator segments

4-3 PRACTICAL DC MOTORS

In a practical motor, more than four armature loops and commutator segments are used. Since each conducting wire develops torque, a larger number of loops and commutator segments produce more turning force. Additional torque is also developed by adding more turns to each armature loop. A further improvement is the use of more than one set of field poles. Adding more poles makes it possible for an armature conductor to develop maximum torque several times during a revolution.

4-4 CONTROL OF FIELD FLUX

(a)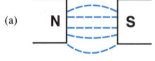

(b)

FIGURE 4-10 Flux lines between poles

Magnetic flux lines have a tendency to repel each other, even if they run parallel. Figure 10(a) shows how flux lines between two poles of a magnet bow away from each other. To eliminate bowing, magnets curved at the end of the poles are used, as shown in Figure 4-10(b). The outer magnetic lines have a greater intensity. Therefore, the stronger flux lines force the other lines inward so that they run straight between the poles.

4-5 COUNTERELECTROMOTIVE FORCE

As the armature conductors rotate, they cut through the main field. These conditions cause an electromotive force, or EMF, to be induced into the armature coils the same way that voltage is produced by a generator. The faster the armature turns, the more EMF it generates. The induced EMF opposes the EMF applied to the armature by the DC power source. For this reason, it is called a **counterelectromotive force** (CEMF), or *back EMF*. To the power source, the CEMF appears as another power source connected series-opposing. The CEMF does not supply an opposing current; however, it reduces the current that flows through the armature. The CEMF cancels out a portion of the applied voltage, and the difference between the two forms a net voltage that produces the resultant armature current. The voltage applied by the DC source is always greater than the CEMF.

The amount of CEMF produced is proportional to three factors:

1. *The physical properties of the armature.* These include the number of turns in the coil, its diameter, and its length. The induced EMF increases as the size gets larger.

2. *The strength of the magnetic field supplied by the field poles.* The induced EMF increases as the flux becomes stronger.

3. *The rotational speed of the armature.* A rapidly moving conductor will induce more CEMF than a slowly moving wire.

4-6 ARMATURE REACTION

At the moment the switching action of the commutator takes place, the armature loop is at a right angle to the field flux lines and midway between the pole pieces. This axis is called the *geometric neutral plane*. Because the loop is not cutting flux lines, it will not generate a CEMF.

In practice, the actual neutral plane of the motor shifts from the geometric neutral plane, as shown in Figure 4-11. The shift takes place because there are two magnetic fields between the poles. One is the main field, and the other is the flux lines built up around the armature conductors. Their interaction distorts the main field. The perpendicular neutral plane becomes shifted in the direction opposite the armature rotation. This shifting of the neutral plane is known as **armature reaction**. Armature reaction varies depending on the armature current and speed of the motor. As more current is applied the faster the motor runs and the larger the armature reaction becomes.

With the neutral plane shifted, commutation is disrupted because it no longer takes place when the corresponding armature loop is perpendicular to the main field. Instead, the armature cuts through the tilted flux lines the moment the brushes make and break contact with the commutator segments. As a result, EMF is induced into the loop which causes arcing to occur at the commutator segments that move under the brush. Sparking causes the brushes and the commutator to pit, increasing the wear on both.

The arcing due to armature reaction adversely affects the motor in three ways:

1. It reduces torque.

2. It makes the motor less efficient.

3. The continuous sparking shortens the life of the brushes and damages the commutator.

INTERPOLES

The effect of armature reaction is corrected by using special windings called **interpoles**, sometimes called *commutating poles*. Shown in Figure 4-12(a), they are smaller poles placed between the main poles. Interpole windings are connected in series with

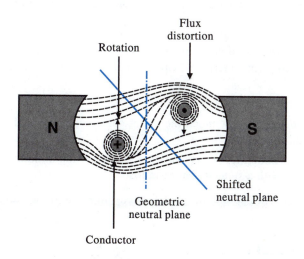

FIGURE 4-11 Shifted neutral plane due to armature reaction

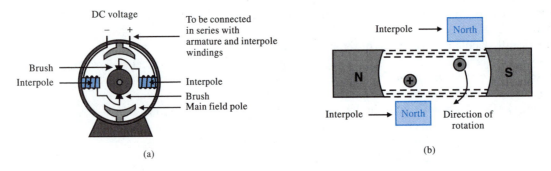

FIGURE 4-12 Interpoles

the armature windings. The magnetic fields formed around the interpoles oppose the magnetic field around the armature coils, and push back the distorted flux lines so that they are in a straight line between the poles. Therefore, the neutral plane is shifted back to the original position, as shown in Figure 12(b). Interpole windings are self-regulating, because they are in series with the armature. If the armature current increases, so does the canceling effect of the interpoles.

Another method sometimes used to correct armature reaction is by using **compensating windings**. These windings are embedded into the metal core of the main field poles and are electrically connected in series with the armature conductors. Like the interpoles, their function is to cancel out the distorted effects of the armature field.

4-7   MOTOR SELECTION

DC motors are available in different sizes. The larger they are, the more power they have. Motor manufacturers also produce models of DC motors that are designed to operate at different speeds. When selecting a motor for a particular application, the engineer should find one that can supply the speed and mechanical power required by the load being driven.

The engineer must also make a selection based on the motor type that best fits an application. There are several types of DC motors to choose from. Each type of motor has definite operating characteristics. Operational requirements determine what type of motor to use.

Two characteristics used in this selection process are:

Speed Regulation.   How much the motor speed will vary with a change in the mechanical load.

Torque.   How much torque is available when starting a motor, or how much it will vary with a sudden change in load.

Each type of motor will operate differently when subjected to the various load conditions it encounters.

SPEED REGULATION

A motor is designed to operate at **full load**. Full load is the maximum power it can provide to run its rated mechanical load all of the time. It is possible for the motor to run above full load, but not for a sustained period of time: It will overheat and likely become damaged. This situation is called an **overload** condition. The overload condition becomes excessive if the motor stalls because it is unable to move the load. If it stops, the current draw from the power source is maximum and a circuit protection device will deactivate the motor. When the physical load is reduced from the full-load condition, this situation is called a **partial-load** condition. The motor operates at **no load** when the physical load is disconnected from the motor shaft.

When the mechanical load connected to the motor is reduced, the motor speed will increase. The amount it increases depends on the type of motor employed. The ability of a motor to maintain its speed when the load is changed is called **speed regulation**. The speed regulation of a motor is calculated by comparing its no-load speed to its full-load speed. It is usually expressed as a percentage of its full-load speed by using the following formula:

$$\text{Speed Regulation} = \frac{\text{No-Load Speed} - \text{Full-Load Speed}}{\text{Full-Load Speed}} \times 100$$

PROBLEM

The no-load speed of a motor is 1800 RPM. When the rated load is connected to the shaft, the speed drops to 1720 RPM. What is the speed regulation in percent?

$$\frac{1800 - 1720}{1720} \times 100 = 4.65\%$$

An example of speed regulation is the operation of a hand drill that uses permanent magnets to develop the main field. When the drill is turned on, it runs at no load. As the drill bit cuts through the material, it slows down to the full-load condition. The amount it slows down at full load compared to the no-load speed is its speed regulation value.

If the speed of the motor is relatively constant over its normal operating range, the motor has good speed regulation. It will perform well as a constant speed motor. A motor whose speed varies greatly from no load to full load has poor speed regulation.

TORQUE

Force is a form of energy that can cause motion. Torque is a particular type of force. It is a twisting action that causes an object to rotate. This type of force causes a motor shaft to turn.

The load a motor is driving may rotate like a fan or a pump. It may also be a mechanism that moves in a straight line, like a conveyer belt. Even though these mechanical loads move differently, they are all powered by the turning action of a motor.

When a load is pushed or pulled in a straight line, the force that moves it is measured in pounds. Torque, however, is measured differently. The amount of torque a motor produces is measured by multiplying the force it will exert times the distance between the center of the shaft and the point where the force is being applied, as determined by the following formula:

$$T = F \times r, \text{ where}$$

- F is the magnetic force acting on the conducting armature, measured in pounds.
- r is the radius in feet, measured from the axis of rotation to the point where the force is applied.
- T is the rotary action exerted by the motor shaft, measured in pound-feet (*lb-ft*).

When a load is connected to the shaft of the motor, it exerts a resistance, or opposing torque, in the opposite direction from the one in which the motor turns. If the torque produced by the motor is greater than the counter-torque of the load, the motor shaft will turn. If the counter-torque is greater than the torque produced by the motor, the shaft will not turn. Even though the load does not move, torque is still produced.

The magnitude of the force is determined by the following factors:

- ○ Strength of the main field, θ.
- ○ The strength of the armature field. This value is expressed by the value of the armature current, I_a.
- ○ The physical construction of the motor, K. These include:
  1. The active length of the conductors.
  2. The number of active conductors.
  3. The radius of the armature.

The physical properties of the motor are a fixed constant because they are left unchanged. The torque of the motor can therefore be controlled by changing the magnetic strengths of the main field and the armature field.

Figure 4-13 illustrates the concept of torque. It shows a motor turning a winch to lift a weight of 100 pounds. To determine how much torque is required to lift the weight, the radius of the winch drum is multiplied times the force exerted by the weight being lifted. If the diameter of the winch drum is three feet, 150 pound-feet of torque is needed.

$$\begin{aligned}
\text{Torque} &= \text{Force} \times \text{Radius} \\
&= 100 \text{ lbs} \times 1.5 \text{ ft} \\
&= 150 \text{ lb–ft}
\end{aligned}$$

PROBLEM

Find the torque of a motor that has a tangential force of 240 pounds at the surface of a pulley six inches in diameter.

SOLUTION

$$\begin{aligned}
T &= F \times r \\
&= 240 \times 3/12 \\
&= 60 \text{ lb–ft}
\end{aligned}$$

Suppose the load in Figure 4-13 is doubled. The motor will respond to the change by producing just enough torque to satisfy the demands of the new load. In this situation, the motor will have to exert 200 foot-pounds of torque. As long as the torque requirements of the load are within the capabilities of the motor, it will always move the load.

The torque developed by the motor when driving its rated mechanical load is called the *rated load torque*. This is a constant torque that drives the load at a steady speed.

When starting the motor from a dead stop, it takes more effort to get it started than to keep it running. The same concept applies to starting a car. First gear is used when starting to provide the extra torque needed to overcome the inertia of starting. Less torque is required to keep the motor or car moving.

Electric motors are designed to supply the extra torque needed to start the load. The starting torque of DC motors ranges from 200 to 450 percent of the rated load torque. Speed torque curves for different types of DC motors will be provided throughout the remainder of the chapter.

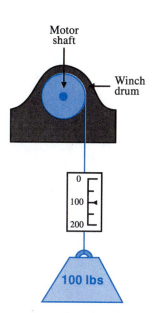

FIGURE 4-13 Winch lifting a 100-pound weight

WORK

The primary function of a motor is to perform work. The motor does mechanical work when it supplies a force to move a physical object through a distance. The force acting on the object must overcome some resisting force. For example, work is done when the weight in Figure 4-14 is pulled. Work is calculated by multiplying distance times force, as shown by the following formula:

$W = D \times F$

D = Distance in feet

F = Force in pounds produced by the rotary torque of the motor shaft

W = Work in foot-pounds

Work is not done unless the load is moved a distance.

The concept of work is illustrated in Figure 4-14. A motor exerts a torque to turn a winch drum so that it pulls a 250-pound weight a distance of ten feet. Therefore, the motor performs 2500 foot-pounds of work.

$$W = D \times F$$
$$= 10 \text{ ft} \times 250 \text{ lbs}$$
$$= 2500 \text{ ft–lbs}$$

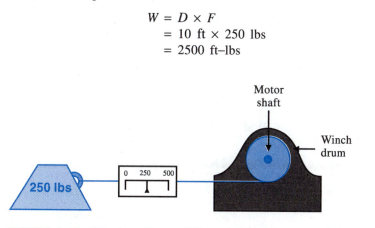

FIGURE 4-14 Winch pulling a 250-pound weight

If a small motor operates long enough, it will perform a lot of work. A powerful motor will do the work quickly.

HORSEPOWER

Placed on the housing of a motor is a nameplate that provides relevant information for the engineer or technician. It does not contain information about the torque the motor exerts, or the amount of work it will perform. Instead, it lists a power rating in units of horsepower that the motor delivers to the load.

Power is defined as the rate of doing work. It describes how fast a particular amount of work is accomplished, and is calculated by the following formula:

$$Power = \frac{Work}{Time}$$

Work is in foot-pounds
Time is in minutes
Power is in foot-pounds per minute

Suppose 5000 pounds of cargo is lifted by a winch to a height of fifty feet. The work required is 5000×50, or 250,000 foot-pounds. If the time it takes to raise the cargo is two minutes, the power required is:

$$\frac{250,000 \text{ ft–lb}}{2 \text{ min}} = 125,000 \text{ ft–lb per min}$$

When 33,000 pounds are moved one foot in one minute (or an equivalent combination), one horsepower (HP) of work is performed. In the example of the cargo winch, the horsepower required to raise the load is:

$$\frac{125,000}{33,000} = 3.79 \text{ hp}$$

The combination of speed the motor runs and the output torque it exerts determines the horsepower it is capable of producing. In the example of the cargo winch, twice the horsepower is required to move a load of twice the weight in two minutes, or the same weight at twice the speed.

Suppose a pulley with a radius of one foot is connected to the end of a motor shaft that produces an output torque of ten pounds at a rate of 1000 RPM. The formula for determining the horsepower of a rotary output is:

$$HP = \frac{\text{Speed (in RPM)} \times \text{Distance (in feet)} \times \text{Torque (in pounds)}}{33,000}$$

In one revolution, the circumference of the pulley moves a distance of $2 \times \pi \times 1 \text{ft}$, or 6.28 ft. Therefore,

$$HP = \frac{1000 \text{ RPM} \times 6.28 \text{ ft} \times 10 \text{ lb–ft}}{33,000} = \frac{62,800}{33,000} = 1.9 \text{ hp}$$

POWER CONSUMPTION OF A MOTOR

Another rating of a motor is *wattage*. This value identifies the amount of power consumed by the motor without overheating at its rated voltage and speed as it performs work. Power is measured in units of watts; one watt equals 0.737 foot-pounds per second. If the horsepower rating of the motor is known, its value can be converted to watts by multiplying it by 747, since there are 747 watts to a horsepower. Until recently, motor power ratings were given exclusively in HP. Now, more manufacturers are rating motors by watts, or kilowatts (KW).

MOTOR EFFICIENCY

The mechanical output power of a motor used to drive a load is always less than the power supplied to its input. A part of the energy supplied to the motor is dissipated into heat and is therefore wasted.

The heat losses of motors consist of copper losses and mechanical losses. Examples of both types of losses are as follows:

1. Copper losses
   A. Armature I^2R losses
   B. Field losses
      (1) Shunt field I^2R losses
      (2) Series Field I^2R losses
      (3) Interpole field I^2R losses

2. Mechanical losses
   A. Iron losses
      (1) Eddy-current
      (2) Hysteresis
   B. Friction losses
      (1) Bearing friction
      (2) Brush friction
      (3) Windage (air friction)

These unavoidable losses are expressed as **efficiency**. The efficiency rating of a motor is simply the ratio of the power produced by the output shaft to the power supplied by the source. It is expressed in percentage by the formula:

$$\text{Percent Efficiency} = \frac{\text{Power out}}{\text{Power in}} \times 100$$

PROBLEM

Suppose the wattage consumed by the motor is 3.75KW and it produces an output of 4.75HP.

SOLUTION

1. Calculate the wattage at the output by multiplying

$$4.75\text{HP} \times 746 = 3543.5\text{W}$$

2.   Use the efficiency formula:

$$\frac{3543.5}{3730} \times 100 = 95\%$$

4-8   INTERRELATIONSHIPS

The ultimate function of the motor is to drive a mechanical load. The energy required to run the motor is drawn from the power source at the same rate mechanical power is being used. Therefore, the rate of electrical power consumption is directly proportional to the mechanical requirements of the load plus heat losses.

A change in the mechanical load has an effect on armature current, torque, speed, and CEMF, all of which are related to each other. The armature current produces a magnetic field around the armature. The interaction with the main field causes the armature to turn. The rotating armature produces a CEMF. The CEMF regulates the armature current. At any normal operating speed, the exact amount of CEMF produced will limit the armature current to a value just sufficient to produce the torque required to drive the load.

The motor is also a self-regulating device. If the load varies, the speed changes, which affects the CEMF. The new CEMF adjusts the armature current until the torque matches the load's new requirements. With all of the factors balanced, the motor is in the state of equilibrium.

4-9   BASIC MOTOR CONSTRUCTION

Mechanically, all motors have two main parts or assemblies: the armature and the field poles. The horsepower developed by a motor results from the reaction between the magnetic fields created by these two parts.

The main parts of a DC motor are shown in Figure 4-15. The field poles are core pieces mounted inside a non-moving, hollow drum-shaped housing. These field pole pieces are either permanent magnets or electromagnets. If interpoles are used, they are placed between the main pole pieces. End covers (also called bells) that support bearings are placed at each end of the housing. Together, all of these parts make up the field pole assembly. The housing is made of steel, which conducts magnetic flux better than air and allows stronger magnetic fields to be established. Its strength also physically supports the stresses that develop inside the motor as it drives the load. The field pole assembly is also referred to as the *stator*.

The moving portion of the motor is the armature, which rotates inside the housing. It consists of a cylindrical core made of sheet-steel laminations that are attached to the shaft. The outer surface of the core has slots where the armature loops are placed. The armature windings are soldered to the commutator, which is also mounted on the shaft. A fan attached to the end of the shaft keeps the internal parts of the motor cool as the armature rotates. The bearings mounted on the stator's end plates support the shaft at both ends. The brushes are pressed against the commutator by specially-designed

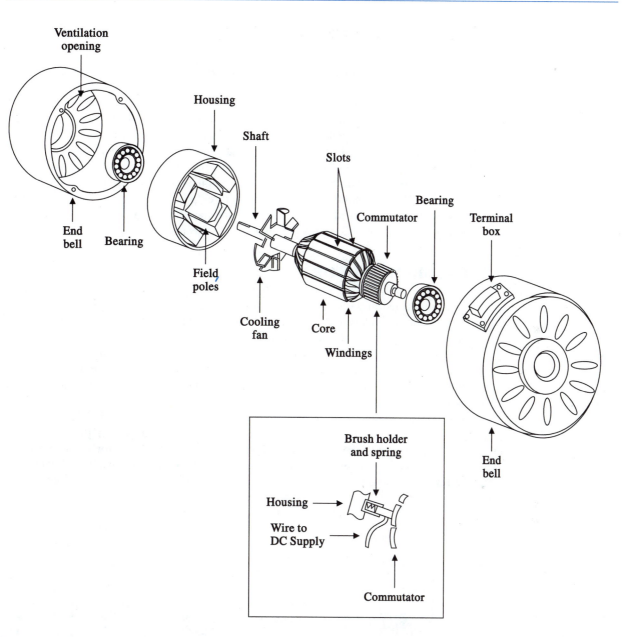

FIGURE 4-15 Parts of a DC motor

tension springs. The brushes and springs are placed inside holders that mount to the stator housing. The armature assembly is also called the *rotor*.

The flux produced by the field windings passes through the motor housing, field poles, armature core, and any air gaps, all of which is known as the *magnetic circuit* of a motor. Electric circuits of a DC motor are made up of the armature winding, commutator, brushes, and field winding (if it is an electromagnet).

4-10   MOTOR CLASSIFICATIONS

The most common way to classify DC motors is by describing how the flux lines of the main field are supplied. For example, the motor described in Figure 4-7 utilized a permanent magnet.

In the other types of DC motors, the field is supplied by an electromagnet. The field assembly consists of coils wrapped around laminated pole pieces that are mounted on the inside of the drum housing. Therefore, these types of motors are often referred to as *wound-field* motors. The power source that supplies the electromagnetic current for the armature is also used for the field coils. The flux lines supplied by electromagnets are much stronger than those of permanent magnets. Also, the field strength can be varied to achieve desired results. There are three principle types of wound-field DC motors: shunt, series, and compound. They are classified by how their field windings are connected to the DC supply in relation to the armature.

THE SHUNT MOTOR

The shunt motor, shown in Figure 4-16, gets its name from the fact that the field winding is connected in parallel—or shunt—with the armature windings. This configuration provides an independent path for current flow through each coil. Because the two windings are in parallel, the applied voltage connected to each of them is the same. The shunt field coil is wound with many turns of fine wire. Therefore the shunt field has a higher resistance than the armature circuit and draws less current. Because the current is low, the field coil requires a large number of turns to produce a magnetic field of sufficient strength.

The interaction between the magnetic fields of the shunt field coil and the armature produces the torque that causes the motor shaft to rotate. The strength of the shunt field with respect to the armature field will determine both the motor's torque and the speed at which it rotates. The stronger the magnetic fields, the greater the torque and the faster the rotation.

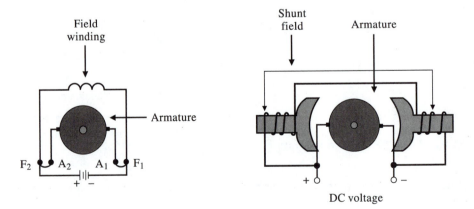

SCHEMATIC DIAGRAM   WIRING DIAGRAM

FIGURE 4-16 DC shunt motor with the field connected in parallel with the armature

Since the shunt coil is connected across the fixed-line voltage terminals, its magnetic field strength is constant. Even though the CEMF in the armature varies as the speed changes, it has no effect on the field strength of the shunt coil. Therefore the flux field does not change significantly as the physical loading conditions vary. The speed of the motor is primarily proportional to the current that flows through the armature.

Suppose that the motor is operating in the under-load condition when the physical load is increased:

1. The motor will begin to slow down.

2. The reduction in speed proportionately reduces the CEMF.

3. The lower CEMF allows the current through the armature to increase.

4. The larger armature current creates a stronger magnetic flux that interacts with the fixed flux of the shunt field, resulting in a stronger torque to accommodate the added load requirements.

5. The speed of the motor stops decreasing.

These reactions will result in only a slight reduction in speed.

If the physical load is removed from the shunt motor, it goes into the unloaded condition:

1. The motor speed will begin to increase.

2. As the speed rises, the CEMF increases proportionately until it nearly equals the line voltage.

3. When enough armature current flows to overcome bearing, brush, and wind friction, the motor speed stops changing.

The no-load speed is slightly higher than the rated speed.

Since the speed regulation from no load to full load of the shunt motor does not exceed 12 percent, it is considered a constant speed motor. Because of these constant speed characteristics, DC shunt motors are used for applications requiring exact control, such as numerical control machines.

Torque Characteristics of Shunt Motors

The amount of starting torque the motor produces determines how fast it accelerates. It speeds up as long as the developed torque is more than the load's resistance.

When the motor is turned on, the shaft is not rotating. Since the RPM is zero, there is no CEMF. The net voltage equals the applied voltage, so the current flow through the armature is as high as possible. At starting, all types of DC motors produce their maximum torque because the interaction between the rotor and stator magnetic fields is at the highest level.

Consider a shunt motor with an armature resistance of 5 ohms and 115 volts applied. The armature current is equal to the applied voltage divided by armature resistance.

$$I_a = \frac{V_a}{R_a} = \frac{115V}{5 \text{ohms}} = 23A$$

The magnitude of torque for a shunt motor is illustrated by the torque formula:

$$T = k_t \cdot I_a \cdot \theta$$

Motor Constant K_t = .5
Armature Current I_a = 23A
Field Flux θ = 1

$$T = .5 \; 23A \; 1$$
$$= 11.5 \; \text{lb–ft}$$

The starting armature current is the highest of all DC motors because the opposition to the supply current is primarily the armature resistance. Yet, the shunt type has the lowest torque of all wound-field DC motors.

The reason for the low torque lies in the construction of the field coil. Its resistance is very high because it has many turns of fine wire. Therefore, the field current and field strength are very low. Since torque is proportional to armature current I_a and field strength θ, the resultant torque produced is relatively small. The starting torque of a shunt motor is approximately 250 percent of its full load torque rating.

Motor Speed Control in Shunt Motors

In many applications, the speed of the motor must be varied. The intentional control of shunt motor speed is accomplished by three methods: field flux control, terminal voltage control, and armature voltage control.

Field Flux Control   The RPM of a DC shunt motor can be controlled beyond its rated base speed by changing the strength of the main field flux. The field is varied by placing a rheostat in series with the shunt field, as shown in Figure 4-17. When resistance is increased, the speed goes up. Conversely, as resistance is reduced, the speed goes down. Therefore the shunt field acts as a magnetic brake on the armature.

Although it may seem more logical that a reduction in field flux will also reduce speed, the opposite occurs. The speed of the motor actually increases because the reduced field flux causes the CEMF in the armature circuit to decrease. Therefore armature current and torque increase. The armature speed continues to rise until the torque is balanced by the opposing torque of the mechanical load.

If the field circuit opens, the speed of the motor can become excessively high, causing it to break apart. The strong armature field interacts with the magnetic field created by residual magnetism of the field coil core to create the torque.

This method of speed control is only used for applications that require a constant horsepower in a partial-load condition. Other limitations of field flux control include relatively low starting torque and poor speed regulation.

Terminal Voltage Control   The RPM of a DC shunt motor can be controlled below normal speed by varying the terminal voltage. This method is seldom used because a reduction in speed is accompanied by a substantial loss of torque.

Armature Voltage Control   When the field is connected to the same power supply as the armature, it is called a *self-excited* DC shunt motor. It is also possible to connect

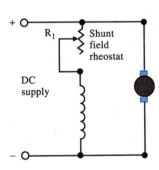

FIGURE 4-17 Shunt field rheostat added for speed control

separate power supplies to the field coil and armature circuit, called a *separately-excited* DC shunt motor.

The preferred method of controlling the speed of a separately-excited DC shunt motor is by adjusting the armature voltage while maintaining constant field voltage. This technique is used to decrease the motor speed below its rated base speed. Speed regulation and starting torque are generally not affected, except at the very lowest speeds.

THE SERIES MOTOR

The series motor, shown in Figure 4-18, gets its name from the fact that the field winding is connected in series with its armature. The field coil develops little resistance because it is wound with few turns. The small resistance allows a high current to flow through the windings. The field coil is wound with a large gauge (size) wire to handle the current that passes through the armature. Even though the coil has a small number of turns, the magnetic field that forms around the windings develops an adequate torque because its flux lines are concentrated by the pole pieces, and because the current is high.

Since the field coil and armature are connected in series, the same current flows through both coils. As the physical loading conditions change the speed, the CEMF causes the armature current to vary, affecting the magnetic field around each coil. Therefore, the torque and speed change by the square of the current. These conditions prevent the motor from maintaining a constant speed under changing load conditions. Therefore, a DC series motor is classified as a poor speed regulation machine.

If the load coupled to a series motor is disconnected, it goes into a no-load condition called *run-away*. In this situation, the motor will accelerate until it physically breaks apart. For example, suppose a normally loaded motor is running. The current flow through the armature and series coil develops a flux that produces just enough torque to turn the load.

1. At the moment the load is removed, the current flow is larger than that required by the load. Therefore the motor speed increases.

2. As the motor speed increases, the CEMF gets larger.

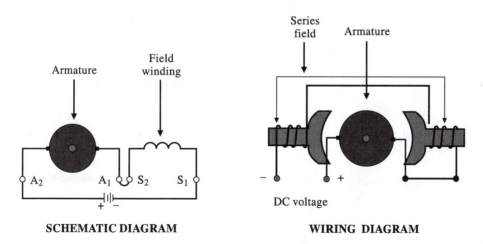

SCHEMATIC DIAGRAM   WIRING DIAGRAM

FIGURE 4-18 DC series motor with field connected in series with armature

3. A greater CEMF causes the current through the armature and field to reduce. If the resultant field strength reduction were directly proportional to the armature current, it would decrease at the same rate that the speed increased. Therefore, the CEMF would stop increasing, the current would become constant, and the speed would stabilize.

4. However, because the series field coil has few turns of heavy wire, its flux strength decreases at a faster rate than the armature current decreases. This condition keeps the CEMF from building at a rate as fast as the speed increases.

5. The CEMF is unable to reduce armature current fast enough to stop the motor from increasing its speed.

Even though the armature current continues to decrease, the torque it produces is enough to accelerate the unloaded motor until it breaks apart.

Due to their runaway characteristics, series motors are not recommended for belt or chain driven systems. A broken chain or belt could result in a no-load condition.

Torque Characteristics of Series Motors

Series motors have the highest starting torque of DC motors. The reason is that when the motor starts, the high current flows through *both* the field coil and armature winding connected in series. Therefore the torque increases by the square of the current. For example, if the starting current is 1.73 times the rated load current, the flux strength around the rotor and stator increases 1.73 times. Therefore, the starting torque is 1.73^2, or three times (300 percent) the rated torque. The starting torque of a series motor is typically 350 to 500 percent of its full load torque rating.

The starting armature current is lower in the series motor than the shunt motor because it is opposed by two series coils. The opposition to the supply current is the armature coil and the field coil. Consider a series motor with an armature resistance of 5 ohms, a field coil resistance of 10 ohms, and 115 volts applied. The armature current is equal to the applied voltage divided by the total resistance of the armature coil and field coil:

$$I_a = \frac{V_a}{R_a + R_f} = \frac{115V}{5 + 10} = 10A$$

The magnitude of torque for a series motor is illustrated by the torque formula:

$$T = K_t \cdot I_a \cdot \theta$$

Since the field strength also depends on armature current, the torque equation is rewritten:

$$T = K_t \cdot I_a^2$$

$$\text{Motor Constant } K_t = .5$$

$$\text{Armature Current} = 10A$$

$$T = .5 \cdot 10^2$$
$$= 50 \text{ lb–ft}$$

Compare this result to the torque calculation for shunt motors. The torque of the series motor is greater than that of a shunt motor even though its starting armature current is less. Therefore, one characteristic of a series DC motor is that it can provide a very high torque when starting, or when a sudden heavy load is encountered.

THE COMPOUND MOTOR

A compound motor, shown in Figure 4-19, has both a series field and a shunt field. The series field is connected in series with the armature circuit. The shunt field is connected in parallel with the armature circuit. This configuration combines the high starting torque of the series motor with the shunt motor's constant speed under changing load conditions. However, the torque and speed regulation performance characteristics of the compound motor are not as good as those of each individual motor. Yet the compound motor represents a viable alternative.

There are two types of compound motors. In a *cumulative* compound motor, the series field is connected so that the direction of electron flow through it causes the magnetic flux to add to the shunt field. In a *differential* compound motor, the series field is connected so that it opposes the shunt field. The overall flux is the difference (differential) of the two. The cumulative is the more common type of compound motor.

Suppose that the cumulative compound motor is operating in an underload condition when the physical load is increased. The motor slows down, the CEMF reduces, and the armature current increases. When enough current flows through the armature to create the necessary magnetic interaction (between the armature, series and shunt coils) to match the torque demand of the increased load, the motor speed stops changing. The compound motor speed decreases when the load increases, but less so than with the series motor.

If the cumulative compound motor encounters a no-load condition, the armature will speed up. However, it does not have the runaway characteristics of the series motor because a large enough CEMF is developed as the armature cuts through the series and shunt coil fields. When the armature current decreases to a certain level, the torque reduces so that it can no longer accelerate the motor and the speed stabilizes.

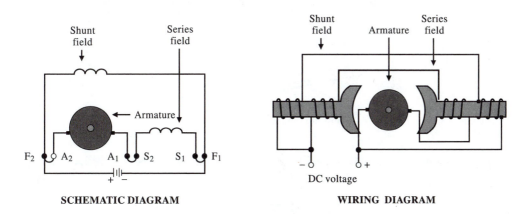

SCHEMATIC DIAGRAM   WIRING DIAGRAM

FIGURE 4-19 DC compound motor with field connected in both series and parallel with armature

Due to the reaction with the series field and the influence of the shunt field, the cumulative compound motor can achieve a speed regulation of about 25 percent. This percentage can be improved if the motor is connected in the differential compound configuration. The differential motor tends to have a more constant speed up to the full load rating of the motor. However, if overloaded, the series field may become strong enough to override the shunt field, which causes the motor to stop and sometimes reverse itself. Therefore, differential compound motors are not used unless it is unlikely an overload condition will be encountered.

Torque Characteristics of Compound Motors

The starting torque of the cumulative compound motor is approximately 300 to 400 percent of its full load rating, which is greater than that of the shunt motor but less than that of the series motor. It is not as strong as the series motor because of the influence of both the series field and the shunt field. The differential compound motor has less torque than either the series motor or the shunt motor. Because the series and shunt magnetic fluxes cancel, its overall field flux is weak.

If the physical load of a running cumulative compound motor is increased, its speed will reduce slightly more than with a shunt motor. The speed stabilizes when the increase in armature current causes an increase in torque to handle the added load. The speed of the differential compound motor, however, may rise as the load is increased before it stabilizes. Figure 4-20 shows the torque and speed characteristics of the three types of motors.

REVERSING DC MOTORS

The direction of rotation of wound-field DC motors is achieved by changing the direction of electron flow through the field (or fields) relative to the electron flow through the armature. Therefore the direction of rotation cannot be changed by simply reversing the negative and positive leads of the DC power source that feeds the motor. Instead, either the field windings or armature windings can be reversed, but not both. In a compound motor, both the shunt and series field coils must be changed, or else the motor will be switched from a cumulative to a differential configuration (or vice versa).

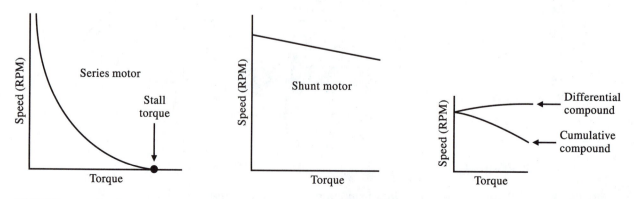

FIGURE 4-20 Graph of torque produced versus armature speed with line voltage held constant

In industrial practice, to reverse the direction of motor rotation, it is standard to change the armature connections. If the motor has commutating pole windings, they are considered a part of the armature circuit. Therefore current flow through them must also be reversed by changing their connections.

4-11   COIL TERMINAL IDENTIFICATION

The electrical parts of a DC motor consist of different types of windings that are marked for identification. The shunt field winding, which consists of many turns of fine wire with a resistance between 100 to 500 ohms, is marked as F_1 and F_2. The series field winding, which consists of a few turns of a larger gauge wire with a resistance of 1 to 5 ohms, is marked S_1 and S_2. The armature winding, which has a very low resistance, is marked A_1 and A_2. If the motor includes a commutating winding or interpole winding as part of the armature circuit, it is marked C_1 and C_2.

CHAPTER PROBLEMS

1. The twisting effect of a DC motor called _____ is produced primarily by the interaction of magnetic fields.

2. The main magnetic field in a motor comes from ____.
   a. the field coil   c. the commutator and brushes
   b. the armature coil

3. Maximum field interaction occurs when the main and armature fields are
   ____.
   a. parallel   b. at right angles

4. Which of the following statements are functions of the brushes and commutator: _____
   a. To provide a path for armature current flow.
   b. To connect and disconnect armature coils in sequence.
   c. To provide a path for field current flow.

5. __ T __ F   In a DC motor, there is a large inrush of current at first, which then drops off as the armature begins to rotate, generating CEMF.

6. __ T __ F   The interpoles are always connected in parallel with the armature.

7. When the motor speed increases, the CEMF in the armature _____ (increases, decreases).

8. Armature reaction is corrected by _____, which is/are used to shift the neutral plane back to the proper position.

9. If a DC series motor develops a starting torque of 450 percent of its full-load torque rating of 192 ounce-inches, the torque will be ____.
   a. 864 oz-in   c. 384 oz-in
   b. 2.5 lb-ft   d. 6.3 lb-ft

10. A motor rated at 3/4 HP can also be rated at ____ watts of output power.
    a. 384   c. 2.5
    b. 559.5   d. 1.253

11. To reverse the direction of rotation of a DC motor, it is best to reverse the ____.

    a. armature connections
    b. main field connections
    c. both the armature field and the main field

12. In an operating DC motor, the armature current depends on the applied voltage ____.
    a. minus the CEMF   c. both a and b
    b. minus the armature resistance

13. In a DC motor, an increasing mechanical load ____.
    a. increases armature current   c. has no effect on armature current
    b. decreases armature current

14. __ T __ F   A shunt motor's field winding has more resistance than the armature.

15. A DC shunt motor has ____.
    a. a high starting torque   c. zero speed at no load
    b. a constant speed rating   d. All of the above

16. To change direction of a DC shunt motor, you must interchange leads ____ and ____.
    a. A_1 and F_1 c. C_1 and C_2 e. F_1 and C_1
    b. S_1 and S_2 d. F_1 and F_2

17. When the load on a DC shunt motor in increased, its speed will ____ and the amount of torque developed will ____.
    a. increase, increase   c. increase, decrease
    b. decrease, decrease   d. decrease, increase

18. A DC shunt motor operating at 240V draws 4.5A. It has an output of 1 horsepower. Its efficiency is ____.
    a. 69   b. 73   c. 58   d. 43

19. __ T __ F   A series DC motor should never be connected to a load by a belt or chain drive.

20. The series winding of a DC series motor has _____ (low, high) reactance.

21. A DC series motor has ____.
    a. low starting torque   c. high starting torque
    b. low no-load speed   d. zero speed at no-load

22. In a DC series motor, if the armature current is reduced to one-half of its full-load rating, the torque is ____.
    a. constant   c. reduced
    b. doubled   d. increased

23. A compound motor has ____.
    a. higher starting torque than a DC motor
    b. a better constant speed rating than a shunt motor
    c. a higher starting torque than a shunt motor
    d. no interpoles

24. When connecting a compound motor for operation, which leads are wired in parallel with the power supply?
    a. A_1–A_2 b. C_1–C_2 c. S_1–S_2 d. F_1–F_2

25. A differential compound motor has ____ than a cumulative compound motor.
    a. higher starting torque   c. more constant speed
    b. higher RPM at no load

26. A cumulative compound motor with an open winding will ____.
    a. stop dead   c. race away
    b. stop and reverse   d. reduce its speed by half

27. __ T __ F   Neither a series nor a compound motor can be reversed simply by changing the input power leads.

CHAPTER 5

AC Motors

CHAPTER OBJECTIVES

At the conclusion of this chapter, you will be able to:

○ Describe the principles of the alternating and rotating magnetic fields.

○ List the different types of rotors and stators in AC motors.

○ List the factors that determine the speed of an AC motor.

○ Calculate the following for an AC motor:

    Synchronous speed   Slip

○ List the different types of AC motors and describe their operation.

○ Identify the characteristics of each type of AC motor.

○ Reverse the direction of an AC motor.

○ Choose the type of AC motor needed for specific applications.

INTRODUCTION

An AC motor converts AC electrical energy into mechanical energy, producing a mechanical rotary action that performs some type of work. Because alternating current is the standard power generated and distributed, AC motors are the most common type of motor used in commercial and industrial applications.

Generators at power plants develop **three-phase power** which is delivered to industrial plants. Huge motors use the three-phase electricity to provide the mechanical power for many types of production machinery, for example, pumps, cranes, and paper machines. Single-phase power is also delivered from the three-phase distribution to industry, residential, and small business customers. AC motors that use single-phase electricity typically produce less horsepower than three-phase motors. They drive such things as furnaces, air conditioners, washing machines, ovens, clocks, and fans.

There are many types of AC motors. Each one has different operating characteristics that provide the speed and torque capabilities for specific applications. Their durability enables them to operate 24 hours a day for many years without maintenance.

5-1 FUNDAMENTAL OPERATION

Figure 5-1 shows a simplified diagram of an AC motor. It has two pole pieces with a permanent magnet placed between them. The coil of wire that wraps around the pole pieces forms electromagnets. The electromagnets are stationary and are called the **field poles** or the **stator**. The permanent magnet is free to turn and is called the **rotor**.

ALTERNATING FIELD

The stator windings in Figure 5-1 are excited by AC power. The resultant field generated between the poles alternates with the applied alternating power. As the rotor magnet interacts with the poles of the stator, it pivots on its axis. The rotor will make one complete revolution for each complete AC cycle applied to the stator, as shown in Figure 5-2(a) through (e).

(a) At time T_0, no field is developed between the stator poles because there is no current.

(b) During time period T_1, the positive alternation of AC voltage occurs. As the field builds up around each stator piece, the polarity of the rotor ends closest to them are alike. The rotor begins to turn because the like poles are

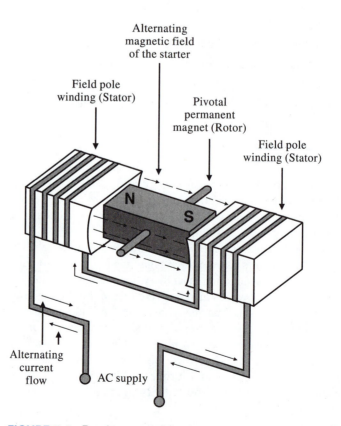

FIGURE 5-1 Fundamental AC motor

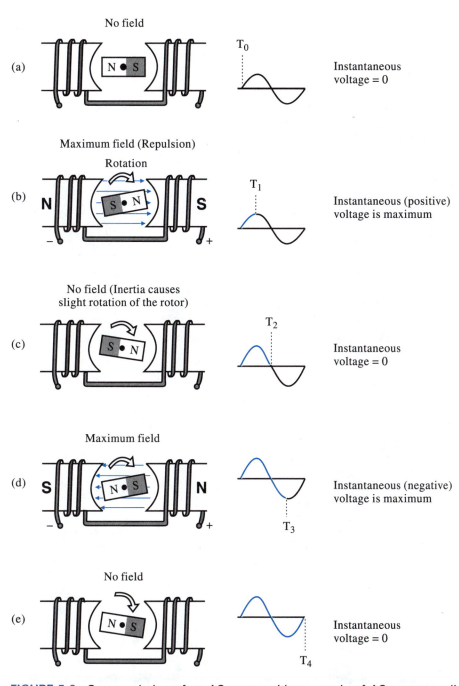

No field

(a)

T_0

Instantaneous
voltage = 0

Maximum field (Repulsion)

Rotation

(b)

T_1

Instantaneous (positive)
voltage is maximum

No field (Inertia causes
slight rotation of the rotor)

(c)

T_2

Instantaneous
voltage = 0

Maximum field

(d)

T_3

Instantaneous (negative)
voltage is maximum

No field

(e)

T_4

Instantaneous
voltage = 0

FIGURE 5-2   One revolution of an AC motor with one cycle of AC power applied

repelled. After the rotor goes past a quarter turn, it is attracted to the opposite poles of the stator. It continues to rotate until the N and S poles of the
rotor are aligned with the opposite poles of the stator.

(c) At time T_2, the applied current is zero, and there is no field between the poles. Due to inertia, the rotor continues to turn past 180 degrees.

(d) During the time period T_3, the AC current changes direction through the field coils. The polarity of the stator magnetic poles is reversed and the rotor is again repelled.

(e) After the rotor goes past three-quarters of a turn, the rotor ends are attracted to the unlike stator poles. Also, the AC current and resultant field strength drops until it reaches zero. The inertia carries the rotor past 360 degrees as it begins another rotation and the next AC cycle is repeated.

ROTARY FIELD

There are two disadvantages of the AC motor described in Figure 5-2. First, if the rotor was exactly parallel to the stator's flux lines, the magnetic repulsion would be equal and it probably would not rotate. It would start to turn only if the rotor was slightly offset.

Second, the rotor might not run in the desired direction. The direction it was offset from the stator's flux lines would determine the direction it turned.

Both of these disadvantages are corrected by making the stator's magnetic field rotate instead of alternate, as shown in Figure 5-3. As the field poles revolve in a clockwise direction, they attract the opposite poles of the rotor. The result is that the rotor turns by following the rotating field.

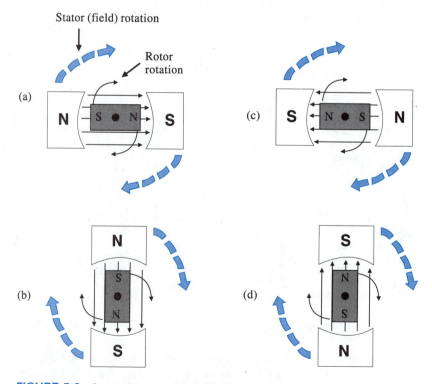

FIGURE 5-3 A rotating magnetic field of a stator

5-2   CONSTRUCTION AND OPERATION

STATOR CONSTRUCTION

Two-Phase

It is impractical to physically rotate the stator field poles, as illustrated in Figure 5-3. However, it is possible to rotate the fields electronically by applying two sinewaves to two stators that are 90 degrees apart. Figure 5-4 uses a series of drawings to illustrate how two AC sinewaves that are 90 degrees out of phase cause a rotor to make one revolution. Phase 1 is supplied to the vertical stator windings, and Phase 2 is supplied to the horizontal stator windings.

1. At time T_0, Phase 1 produces a maximum vertical magnetic field, while Phase 2 produces no horizontal field. The rotor aligns itself vertically with the two energized field poles.

2. At time T_1, equal amounts of current flow through both vertical and horizontal windings. A resultant flux develops between adjacent poles which causes the rotor to turn 45 degrees CCW.

3. At time T_2, no current flows through the vertical windings, while maximum current flows through the horizontal coils. The rotor turns another 45 degrees CCW and aligns itself between the horizontal poles.

4. At time T_3, current flow reduces through the horizontal windings. Meanwhile, the current flow through the vertical coils reverses direction. The resultant flux causes the rotor to turn another 45 degrees CCW.

Between time periods T_4 and T_8, the process continues and the rotor turns as it follows the rotating stator field. After the 360-degree rotation is completed, the next revolution will begin in the same direction. The rate at which the magnetic field in the stator rotates is called the **synchronous speed**.

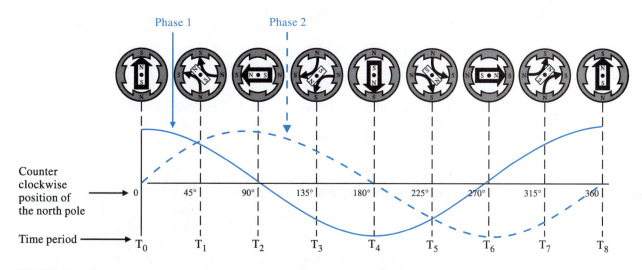

FIGURE 5-4 Two sinewaves used to create a rotating magnetic field

Three-Phase

Industrial factories commonly use three-phase power, in addition to single-phase voltages. Three-phase AC power consists of three alternating currents of equal frequency and amplitude, but each differing in phase from the others by one-third of a period. This characteristic makes three-phase AC power ideal for developing rotating stator fields for motors.

The creation of a rotating stator field using three-phase power is illustrated in Figure 5-5. The three phases of alternating current can be thought of as three different single-phase power supplies. These three-phase currents reach maximum values at different times. Each phase supplies one of three separate pairs of coils wound around stator poles. The phases are designated as A, B, and C. Phase A supplies poles A1 and A2, Phase B supplies poles B1 and B2, and Phase C serves poles C1 and C2. Each set of windings is equidistant from the others. Because the three-phase currents are displaced in time by 120 electrical degrees, and the three-phase windings are equally spaced 60 mechanical degrees apart, the resulting magnetic field will rotate in space as though the poles are rotating mechanically.

Figure 5-5(a) through (f) illustrates the sequence of events that occur during one 360 degree rotation of the stator field with three-phase power supplied.

T_1 Figure 5-5(a) shows the resultant magnetic field from all three currents during time period T_1. Because Phase A has the greatest amplitude, the greatest concentration of magnetic flux lines is between stator poles A1 and A2.

T_2 During time period T_2, Phase C has the greatest magnitude, causing the field to shift from poles A1 and A2 to poles C1 and C2.

T_3 During time period T_3, Phase B has the largest amplitude, and the field shifts another 60 degrees between poles B1 and B2.

T_4 During time period T_4, Phase A has the greatest amplitude, but current flow is in the opposite direction than it was during time period T_1. The field develops between poles A1 and A2, but at the opposite polarity.

T_5 During time period T_5, the field develops between poles C1 and C2, but in the opposite direction than it was during time period T_2.

T_6 During time period T_6, Phase B has the greatest amplitude and causes the stator field to rotate another 60 degrees between poles B1 and B2, but in the opposite direction than it was during time period T_3.

The changes in amplitude and direction of the current flow always occur in the same order, and at the same time interval, to create the rotating field. The direction of field rotation can be changed by reversing any two of the three-phase lines connected to the coils.

ROTOR CONSTRUCTION

If a permanent magnet was used as the rotor, it would turn as the magnetic field is rotated around the stator. AC motors do not use permanent magnets for their rotors. Instead, they use electromagnets. There are two methods of energizing the rotor so that it creates its own magnetic field. The first is to connect an electrical current to the rotor windings. This type of rotor is used for AC synchronous motors. In the second

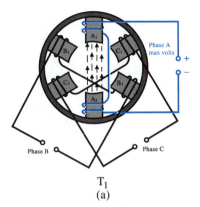

T₁
(a)

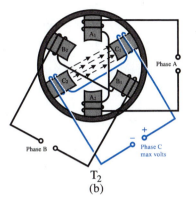

T₂
(b)

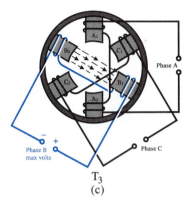

T₃
(c)

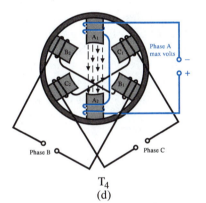

T₄
(d)

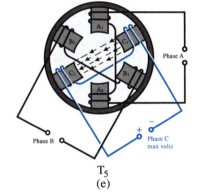

T₅
(e)

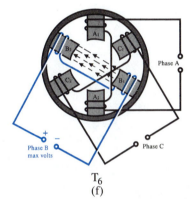

T₆
(f)

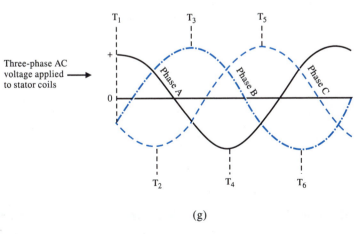

(g)

FIGURE 5-5   Three-phase AC power: magnetic fields and waveform

method, the rotor is not connected to any electrical source. Instead, it becomes an electromagnet through electromagnetic induction. This type of rotor is used for AC induction motors.

Induction Motor Rotors

Electromagnetic induction results from the rotating magnetic flux of a stator inducing a voltage into the rotor. If the rotor has a complete electrical path, current will circulate through the rotor and develop its own magnetic field around it. The stator and rotor magnetic fields interact at right angles and cause the rotor to turn. For comparison, the stator can be described as the primary of a transformer. The rotor can be compared to a secondary of a transformer.

Squirrel Cage Rotors

The induction motor with a squirrel cage rotor is the most common rotating electrical machine. The squirrel cage motor receives its name from the design of the rotor, which resembles a cage used for squirrels, hamsters, and similar pets. Figure 5-6(a) shows the portion of a squirrel cage rotor that carries current. Figure 5-6(b) shows the complete rotor with the iron core in place. The "cage" portion is made of aluminum or brass bars, embedded just below the surface of the core. They are joined to conducting *end rings* that are placed at each end of the core. The end rings short circuit the bars and provide a complete circuit path for current to flow through, regardless of the rotor's position. Note that a bar always forms a pair with another bar directly opposite it in the rotor. Along with the end rings, these pairs resemble the one-loop rotor that was described in the DC motor chapter.

Wound Rotors

Another type of rotor used by AC motors is the **wound rotor**. As its name implies, the rotor is constructed by using wound coils of wire in place of the conducting bars of the squirrel cage motor. Current flows through the wound coils and creates a surrounding magnetic field. The rotor turns as its magnetic field interacts with the stator field. The current that flows through the rotor is either induced by the rotating stator field or is provided by an external DC power source. The number of rotor coils must be the same as the number of stator coils. Each winding terminates at slip rings that are mounted on the shaft of the motor. The currents are carried by brushes that ride on the slip rings to an external connection. The brushes connect either to a DC power source or to an external resistor bank (if the currents are induced). This type of rotor is commonly used in three-phase motors.

PRINCIPLES OF OPERATION

When the stator winding is energized by a two- or three-phase supply, a rotating magnetic field develops at synchronous speed. As the field sweeps across the rotor, an EMF is induced in the conducting bars by transformer action. The resultant current flows through the complete circuit loops consisting of the bars and end rings. Because these loops are short circuits with very low resistance, the current flow is high, producing a strong magnetic field. As the rotor and stator field interact, motor action is created and the rotor turns, as shown in Figure 5-7. Unlike a DC motor, which has

Conductor bars

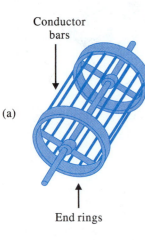

(a)

End rings

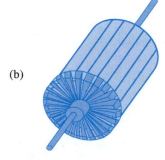

(b)

FIGURE 5-6 A squirrel cage induction motor rotor

a stationary main field, the AC motor has a rotating field. Instead of the rotor turning 45 degrees until it is out of the main field, as in the DC motor, it follows the main rotating field. However, though it will chase the main rotating field, it will never catch it.

Speed and Slip

The instant power is applied to the motor, current flows through the stator coils. The stator's magnetic field begins to revolve at synchronous speed. Two factors determine the speed at which the magnetic field rotates:

1. The frequency of the applied voltage
2. The number of stator poles per phase

The higher the frequency, the faster the motor runs. The more poles a motor has, the slower it runs. The smallest number of poles possible in an AC motor is two.

One cycle of the applied voltage is required for each pair of poles to cause the rotor to turn 360 degrees. For example, in a two-pole motor (one pair), the stator field makes one revolution per cycle of 60Hz power, or 3600 RPM. The formula for determining the synchronous speed of the stator field is:

$$N = \frac{f \times 60}{P}$$

where,   N = RPM
          P = Number of Pole Pairs
          f = Applied Frequency
          60 = Formula Constant based on Seconds per Minute

PROBLEM

Find the synchronous speed of a 4-pole motor (2 pole pairs) with 60Hz applied.

SOLUTION

$$N = \frac{60 \times 60}{2} = 1800 \text{ RPM}$$

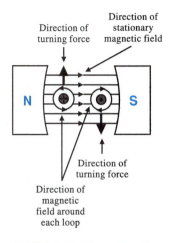

FIGURE 5-7   The interaction of the rotor and stator magnetic fields creates motor action, which causes the rotor to turn

AC motors are wound for synchronous speeds shown in Table 5-1.

The reason the synchronous speed of a four-pole motor is half that of a two-pole motor is described in Figure 5-8(a). The current from the AC source flows through all four coils simultaneously. Because the coils are wound around adjacent poles in the opposite direction, complementary north and south poles are formed 90 degrees apart.

Suppose that during a positive alternation, windings A and C develop north poles and windings B and D form south poles, as shown in Figure 5-8(a). As a reference, consider the north pole at winding A. When the negative alternation occurs, the current through the coils reverses and the polarity at each pole changes. Windings A and C develop south poles and windings B and D form north poles. The effect is that the north pole at coil A rotates 90 degrees clockwise to coil B. During the next alternation, currents reverse and the north pole rotates clockwise to winding C. Every 180 degree alternation causes the field to rotate 90 degrees. Therefore, it requires two cycles of

TABLE 5-1   AC MOTOR SYNCHRONOUS SPEEDS

Number of Poles	60 Hertz Synchronous Speed
2	3600
4	1800
6	1200
8	900
10	720
12	600

AC power to rotate the stator field one 360-degree revolution. The rotor follows the field in an attempt to lock in on it.

In addition to changing motor speed, the reason for having more than two stator poles in a motor is to make the field stronger, causing the motor to run more smoothly.

The rotor of an induction motor cannot run at synchronous speed. If it were possible for the rotor to attain the same speed as the rotating field, the flux lines of the stator could not be cut by the rotor. There would be no EMF induced into the rotor and no rotor current. Because its flux would be lost, there would be no torque developed to turn the rotor. However, this condition is not possible because there will be friction and windage losses. To induce an EMF, the rotor speed must be less than synchronous speed. This difference between rotor speed and synchronous speed is called **slip**.

If no weighted load is connected to the rotor shaft, the rotor and the stator rotating magnetic fields will spin at nearly the same rate. A minimal induced voltage will produce a very small amount of torque. In practical motors, the no-load slip is 2 to 10 percent. If a load is added to the motor shaft or the load is increased, the rotor will slow down and the slippage will increase. A larger amount of induced voltage will be developed and torque will increase. As the load increases, the percentage of slip increases. The amount of slip is also affected by the type of rotor bars used in the construction of the rotor.

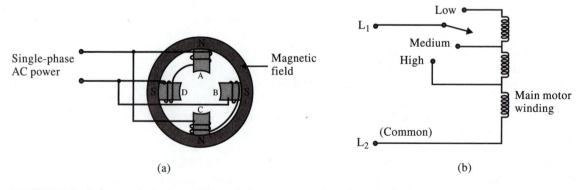

(a)   (b)

FIGURE 5-8 A four-pole stator with complementary poles placed 90 degrees apart, and a multi-speed control

The slip of an induction motor is expressed as the percentage of synchronous speed. The percent slip is determined by subtracting the synchronous speed from the speed of the rotor, and dividing the difference by the synchronous speed. Take, for example, a two-phase motor that has a synchronous speed of 3600 RPM and a rotor speed of 3450. The percent slip can be determined by using the following formula:

$$\text{Percent Slip} = \frac{\text{Synchronous Speed} - \text{Rotor Speed}}{\text{Synchronous Speed}}$$

$$= \frac{150}{3600} \times 100$$

$$= 4.16\%$$

The third method of varying the speed of a single phase AC motor is by changing the inductance of the motor windings. The schematic diagram in Figure 5-8(b) shows a three-speed motor. The permanent motor winding is located between the terminals marked *Common* and *High*. When the rotary switch is changed from the *High* position to the *Medium* position, additional coils are inserted. The increased inductive reactance reduces the current flow through the winding. The result is that the magnetic field is reduced, the motor produces less torque, the slip increases, and the speed decreases. The torque and speed decrease further by adding more turns when the rotary switch is changed to the low-speed position. This type of speed control is generally used only to operate low-torque loads such as fans and blowers.

5-3   SINGLE-PHASE INDUCTION MOTORS

Typically, single-phase commercial power is supplied to residential customers. Therefore, two- or three-phase AC voltages cannot be used by motor stators to produce a rotating field. To start a single-phase motor used in the home, some means must be provided for getting two phases from the standard single-phase AC source. One process of deriving two phases from one is *phase-splitting*. The two sets of independent out-of-phase magnetic fluxes develop the rotating magnetic field.

The process of phase-splitting is performed electrically inside the motor by the stator circuitry. There are several popular types of single-phase AC motors:

1. Resistance-Start Induction-Run Motor
2. Capacitor-Start Induction-Run Motor
3. Shaded-Pole Motor

Their names are derived from the types of components used to split the primary supply sinewave into a simulated secondary phase. These motors are discussed in the following text.

5-4   RESISTANCE-START INDUCTION-RUN MOTOR

One of the most widely used types of single-phase motors is the resistance-start induction-run motor. It has two separate windings connected in parallel to the power source, as shown in Figure 5-9(a). One coil, called the *main* or *run* winding, has a comparatively low resistance and a high inductance. The second coil, called the *auxiliary*

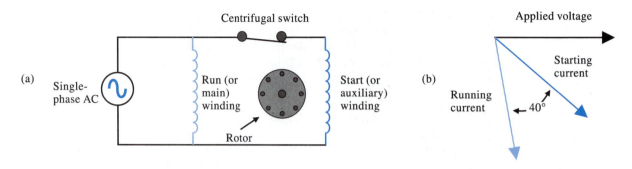

FIGURE 5-9   The resistance-start split-phase motor

or *start* winding, has a comparatively high resistance and lower inductance. To achieve the high value of resistance, the start winding is made of many turns of fine gauge wire. The motor receives its name from the fact that the start winding is more resistive than the run winding.

When power is first applied, both windings are energized. Because the start winding is more resistive, the current flow through it will slightly lag the line voltage. Since the run winding is more inductive, the current flow through it will appreciably lag the applied voltage, as in any inductive circuit. The resultant two out-of-phase currents resemble a two-phase power source. Ideally, the phase difference should be 90 degrees, because maximum starting torque is developed in this situation. In practical motors, however, the phase difference is much less. In the resistance-start motor, the phase difference is 35 to 40 degrees (Figure 5-9(b)). Nevertheless, the phase difference of the currents is enough to create two magnetic fields that are out of phase to form an overall rotating magnetic field in the stator. This condition applies torque to the rotor, thereby starting the motor.

The rotating field in the stator is necessary only to start the rotor turning. Once the rotor accelerates to about 80 percent of its normal speed, it is able to follow the alternating magnetic field created by the run winding. Since the field of the start winding is no longer required, it is removed from the circuit by a mechanical device called a *centrifugal switch,* which is connected in series with the start winding. The centrifugal switch contains a set of spring-loaded weights which push a fiber washer against a movable switch contact. At start-up, the contacts are closed, which electrically connects the start coil to the power source. As the shaft accelerates, the centrifugal force causes the weights to overcome the force of the springs. The washer retracts and the contacts open, which disconnects the start winding from the circuit.

OPERATING CHARACTERISTICS

The primary advantages of the resistance-start motor are that it is inexpensive, requires very little maintenance, and has constant speed characteristics. The no-load current is usually 60 to 80 percent of the current drawn by the motor at full load. Most of the no-load current consumed by the motor is used to produce the magnetic fields around the motor's coils. Only a small portion is used to overcome the mechanical friction and the copper and iron losses.

One disadvantage of a resistance-start induction-run motor is its low starting torque. Two conditions cause this characteristic. The first is that, when the motor starts, the inrush of current is four to six times the motor's rated value at full load. The high current causes a large line voltage drop, which reduces torque. The second condition is that the main winding current lags behind the auxiliary winding by a small amount, resulting in a weak rotating field. These conditions limit the starting torque to only 150 to 200 percent of the motor's rated running torque at full load.

Since the high starting current reduces almost instantly, this is not a major problem. However, resistance-start motors larger than $\frac{1}{3}$ horsepower are usually not approved by power companies for applications that require frequent starting and stopping.

Another disadvantage of this type of motor is its noise. Because of the varying magnitude of the magnetic fields that cut the rotor, the torque developed under load is pulsating and causes a 120-cycle vibration. This vibration can be reduced by using resilient rubber mounting supports.

Resistance-start motors are most commonly manufactured in sizes from $\frac{1}{30}$ hp to $\frac{1}{2}$ hp. They are widely used to drive loads that are fairly easy to start, do not require reversing, and do not need to be started and stopped frequently. For example, they are well suited for small machines such as drill presses, oil burners, sump pumps, some washing machines, and a number of other household appliances. These motors run on both 115 and 230 VAC.

5-5   CAPACITOR-START INDUCTION-RUN MOTOR

Another type of split-phase motor is the capacitor-start motor. Like the resistance-start motor, the capacitor-start motor has two windings: a start winding and a run winding. They are both connected across the line, and both are in parallel. However, the capacitor-start motor has a low reactance electrolytic capacitor in series with the start winding. The capacitor value ranges from 150 to 180 ufd. It is usually mounted in a metal casing located on top of the motor. A centrifugal switch is connected in series with the capacitor and start windings, as shown in Figure 5-10(a).

The purpose of the capacitor is to produce a starting torque that is substantially higher than that of the resistance-start induction-run motor. When the motor reaches 70 to 80 percent of full speed, the centrifugal switch opens. This disconnects both

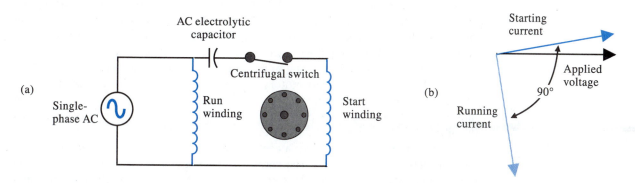

FIGURE 5-10   The capacitor-start induction-run motor

the start winding and the capacitor from the circuit. The capacitor-start motor differs from the resistance-start motor only during the starting period. After the machine reaches its normal operating speed and the auxiliary winding is removed, their performance becomes almost identical. Therefore, the combination of a high starting torque with the constant RPM capabilities of the resistance-start motor gives the capacitor-start motor the ability to maintain excellent speed regulation under a wide range of load conditions.

There are two ways in which the capacitor improves starting torque:

1. The capacitor causes the start winding current to lead the applied voltage. Because the run winding current lags the applied voltage the same way it does in the resistance-start motor, the phase shift between the currents of the two windings is nearly 90 degrees, as shown in Figure 5-10(b). Under this condition, the motor approaches two-phase operation.

2. In a resistance-start motor, the number of turns in the auxiliary winding must be kept low so that the current in the start winding is nearly in phase with the applied voltage. The result is that the starting surge current is very high. In the capacitor-start motor, the capacitor limits the initial current to a lower level while providing the proper phase shift. This condition enables the auxiliary winding to use coils with lower resistance and a greater number of turns. Therefore, the coil has a greater number of ampere-turns, which produces a larger rotating flux and an increase in starting torque.

It is very important that the centrifugal switch operates properly. If the capacitor is kept in the circuit too long it will be damaged or its life shortened appreciably. Also, it should not be used to start a motor more than eight times per hour; frequent starting can cause it to overheat. It is also important to use a replacement capacitor with a proper microfarad rating. If the capacitor is too small, the starting current will be less than 90 degrees out of phase with the run current. If the capacitor is too large, the starting current will be more than 90 degrees out of phase with the run current. In both cases the torque will be reduced.

A capacitor-start motor has a starting torque that ranges from 225 to 400 percent of its rated full-load running torque, or roughly 2.5 times greater than a resistance-start motor. Capacitor-start motors are manufactured in both fractional and integral sizes, up to 7.5hp. They are well suited for applications that require frequent starting and hard-to-start loads, such as pumps, conveyers, and compressors used by refrigeration and air conditioners. They also drive machine tools that require single-phase power.

The direction of rotation of a capacitor-start motor is changed by reversing the connections of either the main winding or the auxiliary winding, but not both.

Some larger capacitor-start motors use two capacitors in the start winding, as shown in Figure 5-11. With this type of motor, called a capacitor-start capacitor-run motor, its start winding is not disconnected from the line. One capacitor is larger than the other. The larger one is used when the motor starts. When the motor reaches about 75 percent of its operating speed, a centrifugal switch disconnects the start capacitor and connects the run capacitor. The purpose of keeping the start winding connected is to maintain split-phase power. This enables the motor to have excellent starting and running torque, and causes it to run quietly and efficiently.

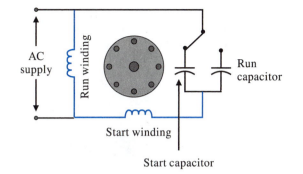

FIGURE 5-11 Capacitor-start, capacitor-
run single-phase AC motor

The capacitor-start capacitor-run motor is normally manufactured in sizes from 5 to 20hp. A practical application is driving the compressor of a central air conditioning unit that runs on single-phase power.

5-6 SHADED-POLE MOTOR

The shaded-pole motor is a type of induction motor that uses a squirrel cage rotor and a main stator winding. However, it differs from other types of induction motors in the manner in which it develops the required rotating field. Figure 5-12 shows that the stator poles are divided into two parts. The smaller segment is called the shaded pole and is surrounded by a metal ring called a *shading coil.* The larger segment is called the *unshaded pole* or *main pole.* The shading coil forms a complete circuit and operates in the same manner as a transformer with a shorted secondary winding. Its function is to delay the flux lines from passing through the shaded pole until they are about 90 degrees behind the applied voltage.

The movement of flux around the stator poles is described by the following explanations and diagrams in Figure 5-13:

1. Figure 5-13(a): When the current of the AC waveform increases from zero toward a positive peak, the flux builds up throughout the stator pole. As the flux lines cut through the shaded pole, an EMF is induced in the short-circuited ring which causes current to flow. The current develops flux lines around the conducting ring in the direction shown by the curved arrows. Note that the ring coil flux and main pole flux lines are in opposite directions within the inner area of the shaded pole. The effect is that they cancel each other, which weakens the main pole flux lines that pass through the shaded pole. Some of the shaded pole flux is diverted, which causes the bulk of the magnetism to pass through the unshaded side.

2. When the AC current reaches its peak value, the flux does not change. With a constant flux strength, the shading coils will not be cut by moving flux lines. There will be no flux lines developed around the ring to affect the main field because no EMF is induced. At this time the main pole flux lines will be distributed more uniformly over the entire stator pole piece.

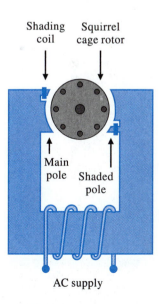

FIGURE 5-12
Shaded-pole motor

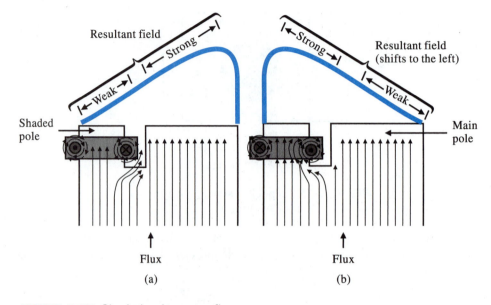

FIGURE 5-13 Shaded-pole motor flux

3. Figure 5-13(b): When the stator voltage decreases, the flux also decreases through the stator pole. The flux lines cut through the ring coil in the opposite direction. The induced EMF and resulting current reverse polarity and the flux lines around the ring coil change direction, as shown by the curved arrows. The ring coil and main pole flux lines are aligned in the same direction within the inner area of the shaded pole. The effect is that they reinforce each other and tend to oppose the decrease in flux. This delays the collapse of the stator pole flux lines that pass through the shaded poles. Some of the main pole flux is diverted and concentrated through the shaded pole.

4. When the AC current passes through zero and increases toward the negative peak, the flux lines in the stator pole change direction as the flux cycle is repeated. During either alternation, the resultant field moves in the direction from the unshaded pole toward the shaded pole.

Figure 5-14 shows the effect of the shifting flux lines between two field poles during one alternation of an AC cycle.

In the other AC motors described, the magnetic fields rotate. In the shaded pole motor, the field merely shifts across the pole face. As the flux lines shift, they cut the squirrel cage rotor bars and induce an EMF. The resulting current creates a rotor flux which interacts with the stator flux to develop the torque needed to turn the rotor.

The construction of shaded-pole induction motors is very simple. They have no auxiliary winding, no capacitor, and no centrifugal switch. They are mounted with cheap sleeve bearings and are designed to have air pass over them for cooling. Therefore, they are very inexpensive, rugged, require very little maintenance, and consume very little electricity.

(Time period 1)

(Time period 2)

(Time period 3)

(Time period 4)

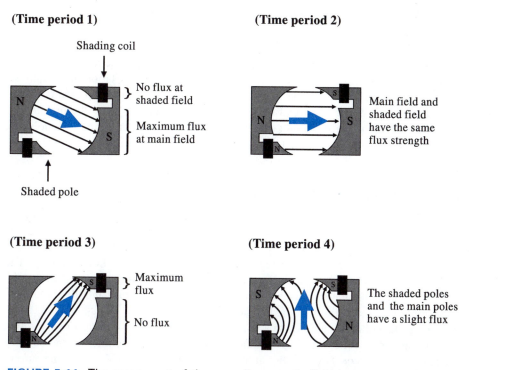

FIGURE 5-14 The movement of the overall magnetic field between two field poles during one alternation of an AC cycle in a shaded-pole motor

However, shaded-pole motors have several disadvantages. They are very small and inefficient. The smallest size, which produces $1/120$ hp, operates at an efficiency of only 5 percent. A motor of $1/20$ hp is 35 percent efficient, and the starting torque is only 40 to 50 percent of full-load torque though they handle overloading very well. Rotation is usually restricted to one direction.

These types of motors are used for light load applications that require small output horsepower, such as fans, blowers, pumps, toys, and other items that are inexpensive to make and operate. The most frequent cause of failure is dry bearings.

5-7 TROUBLESHOOTING SPLIT-PHASE AC MOTORS

The most frequent cause of malfunction in capacitor-start motors is a defective capacitor. If it opens, the start winding is disconnected, the motor begins to hum, and it will not start. If a short exists, the current will become high and will either blow a circuit breaker or burn out the start winding.

If a resistance-start motor does not turn when power is applied, give the rotor shaft a spin by hand. If it runs, either the centrifugal switch or the start winding is open. An ohmmeter can be used to check an open winding or a defective switch.

THERMAL PROTECTION

Some split-phase motors have built-in thermal overload protection. When a predetermined temperature is reached, a thermal switch made of a bimetal strip opens. It protects the run and start winding from overheating by removing them from the power source. The overheating can be caused by a lack of proper ventilation or an ambient temperature that is too high. It is also caused by excessive current that results from a motor load that is either too high or one that prevents rotation. Some thermal protection devices automatically reconnect the motor to the line after the motor cools off. Other devices are reactivated by a reset button mounted on the frame.

5-8   UNIVERSAL MOTORS

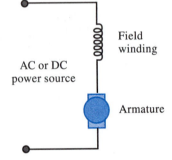

FIGURE 5-15 The circuit for a universal motor, similar to that of a series DC motor

A universal motor is usually categorized as an AC motor. It gets its name from its ability to operate on either AC or DC voltage. Its construction, shown in Figure 5-15, is very similar to a series wound DC motor. The rotor is made of laminated iron wound with loops of wire. A commutator segment is connected to the end of each loop. The stator is made of pole pieces that are wound with wire. Brushes that connect to the stator ride on the commutator as the rotor turns. The wound armature and the stator field are connected in series through the brushes and commutator.

When a DC voltage is applied to a universal motor, the same current flows through the stator and rotor coils. The magnetic fields around the winding interact and develop torque to turn the rotor. The direction it turns is determined by the direction the current flows through both sets of windings. When an AC voltage is applied, the direction of current will alternate. Since the current reverses in both the rotor and stator at the same time, the magnetic fields around both windings also change simultaneously. The result is that the interaction of the two fields causes the direction of the developed torque to remain the same. Therefore, the rotor turns one way, regardless of which direction the applied current flows.

There are two significant differences between the DC series motor and the universal motor:

1. In a DC motor, the iron cores are made of solid iron. The universal motor uses laminated iron to reduce energy loss from excessive heating due to eddy currents that are created from magnetic fields constantly reversing direction.

2. In a universal motor, the magnitude of the fields will fluctuate at twice the line frequency (120 times a second). This condition creates a reduction in output torque compared to a DC motor, which has constant field strength. This reduction of output power is partially compensated for by using more armature loops.

In a universal motor, a third set of coils called *compensating windings* is connected in series with the rotor and stator. These windings perform two functions: They correct the neutral plane position that is distorted by armature reaction, and they compensate for undesirable reactive voltages attributed to the inductance of the armature.

The speed of a universal motor is determined by its applied voltage, frequency, load, and physical design. Its speed increases as applied voltage increases, increases as frequency increases, and decreases as the load increases. At no-load, the speeds are the highest. Unlike DC series-wound motors, they do not encounter a runaway condition because they are usually connected to a gearbox that prevents them from reaching excessive speeds. Similar to series DC motors, they have a very large starting torque and very poor speed regulation.

Universal motors are used in many portable applications that require high horsepower for the size. A few examples are hand-drills, vacuum cleaners, sewing machines, routers, polishers, hedge trimmers, skill saws, blenders, and mixers. One of the most frequent malfunctions encountered with universal motors is worn-out brushes. They typically require replacement after 300 to 1000 hours of use. Another fault that develops is shorted armature windings, which occurs when excessive currents break down the insulation of the armature wire when the motor is overloaded.

5-9   THREE-PHASE MOTORS

Most of the motors used in industry operate directly on three-phase power. Also called *polyphase* motors, they have several advantages over single-phase motors. They are simpler in construction. Also, by using three phases, a more powerful machine can be built into a smaller frame, because maximum torque occurs once every 60 degrees during a revolution instead of once every 180 degrees.

There are three types of three-phase motors:

1. Induction motor
2. Wound-rotor motor
3. Synchronous motor

All three motors use the same basic design of the stator winding, but differ in the type of rotor used. They are discussed in sections 5-10, 5-11, and 5-12.

5-10   INDUCTION MOTOR

Most industrial machines are powered by three-phase squirrel cage induction motors. These motors are simple in construction and require very little maintenance. When compared with other types of motors, their physical size is small for a given horsepower rating. The wide usage of these motors results from their relatively low cost, rugged construction, and good performance.

The polyphase induction motor's basic construction is much like that of its single-phase counterpart. The rotor is a mass of laminated iron with embedded conductor bars and end rings called a squirrel cage. The stator consists of a group of coils wound on pole cores equally spaced inside the motor frame. A rotating field moves around the stator.

The concept of the rotating magnetic field is illustrated in Figure 5-16. There are six pole cores 60 degrees apart. There are three sets of paired coils labelled A, B, and C. Each set is connected to one phase of the three-phase power source. The coils in each pair are wound around cores that are located opposite each other. Since the coils in each pair are wound in the opposite direction, the magnetic fields developed by the poles across from each other are of different polarity. The diagram illustrates how a

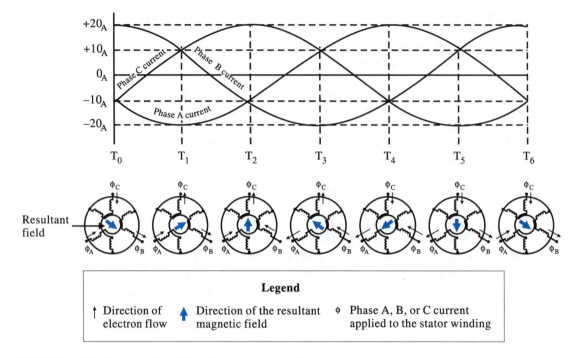

FIGURE 5-16 The rotating field of a three-phase induction motor

resultant magnetic field is created by current that flows through the three pairs of stator coils. Seven different time intervals are shown during one three-phase cycle.

T_0 At this time the Phase B is at the peak of the positive alternation and the current through the B coils is at a maximum value of 20 amps. Phases A and C are negative and at an amplitude of −10 amps. Most of the strength of this field is produced by the current flowing through the pair of B coils, which is at maximum strength. This field is aided by the adjacent A and C fields. The resultant field is in the direction shown by the inside arrow. It forms inside the pole pieces where the rotor is located.

T_1 At time period T_1 the current in Phase B reduces to +10 amps, Phase C reverses direction to +10 amps, and Phase A increases to a negative 20 amps. Most of the field is produced by current flowing through the pair of A coils, which is at maximum strength. The resultant field shown by the arrow rotates CCW.

T_2 At time period T_2, the current in Phase B reverses direction to −10 amps, Phase C increases to a maximum value of +20 amps, and Phase A changes to −10 amps. The resultant magnetic field rotates another 60 degrees CCW.

The remaining time periods illustrate that the resultant field continues to rotate CCW. Note that for every 60 degree change in the AC waveform, the resultant field also rotates 60 degrees. If the frequency of the line voltage is a typical 60Hz, the resultant field rotates 60 times a second, or 3600 revolutions per minute. By adding another pair of windings to each phase of the AC applied voltage, four poles are used. Since a four-pole field will rotate at half the speed of a two-pole field, the stator field will rotate at 1800 RPM. The same calculations are used as with single-phase induction motors to determine speed, slip, and speed regulation.

The revolving field produced by the stator currents cuts the squirrel cage conducting bars of the rotor. The induced voltages that cause currents to flow develop magnetic fields around the bars. The interaction between the rotor and stator fields produces a torque that causes the rotor to turn in the same direction as the stator field movement.

One favorable characteristic of a three-phase induction motor is that it is a constant speed machine with excellent speed regulation. Its behavior under load is similar to its single-phase counterparts. When the load is increased, speed slightly decreases, which causes the slip to increase. This action causes a greater EMF to be induced into the rotor. The higher current increases the torque to accommodate the new load requirement and prevent the speed from lowering further. If the load on the motor is increased beyond the full-load rating of the machine, it becomes overloaded. At a certain point the larger rotor current and torque cannot prevent the motor from stalling. This condition is called pull-out torque. The three-phase induction motor may be severely overloaded for short periods of time. However, a prolonged overload will increase the temperature and damage the motor. The pull-out torque of a three-phase motor is much greater than starting torque.

Sometimes, one of the motor's stator leads connected to the three-phase power line becomes open. This situation may result from a blown fuse or broken connection. A motor in this condition is called *single-phased*. A polyphase motor will not start when it is single-phased. Instead, it will hum because the remaining stator currents set up an alternating magnetic field instead of a rotating field. If the motor is running when the single-phase condition develops, it will continue to rotate as a single-phase motor, but its performance will not be normal.

5-11 WOUND-ROTOR MOTOR

The wound-rotor induction motor (or WRIM) is another type of polyphase motor. Its rotor consists of a set of three coils in place of the conducting bars of the squirrel cage rotor. The coils are preformed and are placed in the slots of a laminated iron core. It has many of the same characteristics as the squirrel cage motor. For example, the stator of a wound-rotor motor is the same as the stator of a squirrel cage motor. As the stator field rotates, it induces an alternating voltage into the rotor windings just as it would in the squirrel cage's shorting bars.

A squirrel cage motor's rotor resistance is fixed. Therefore, its speed-torque characteristics are fixed at full-load operation. Some motor applications require that the speed and torque of an induction motor be varied. This is possible with a wound-rotor induction motor. To perform these operations, certain design requirements must be met:

1. The rotor construction must be similar to that of the stator. It is wound for three-phase power in a wye-connected configuration and must have the same number of poles as the stator. The windings must be highly insulated. Since windings are used instead of copper or aluminum bars, there will be a

stronger interaction between the stator and rotor magnetic fields. The result is a greater amount of torque.

2. There must be a resistive network connected to the rotor windings made up of three rheostat resistors, one for each wye coil leg. Since the rotor will be spinning and the resistor network will be fixed, they must be connected by slip rings, as shown in Figure 5-17. By varying the amount of resistance, the induced rotor currents and the magnetic flux lines they create can be regulated. This feature allows for variable speed and torque control.

START-UP

During the start-up phase of the wound-rotor induction motor, the external resistance connected to the rotor is set to the maximum value. This resistance limits the amount of rotor current. It also causes the rotor to become more resistive and less inductive. Therefore, the stator flux and rotor flux are more closely in phase with each other. This causes the flux strengths of both fields to be at their maximum values during the peak of the alternation, producing a large amount of torque. This starting torque can be made equal to the maximum torque if a high starting torque is desired.

As the motor speed increases, the induced voltage will decrease because of less cutting action between the rotor coils and the rotating stator field. The decrease in induced voltage produces less current flow and a smaller torque. To provide maximum

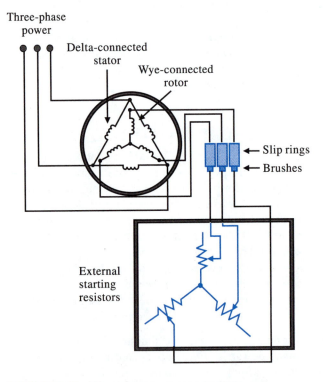

FIGURE 5-17 Wound-rotor motor system

torque throughout the acceleration range, the resistance is gradually reduced as the motor speeds up, either manually or automatically. Once the motor reaches operating speed, the external resistance is reduced to zero and each rotor coil is short circuited. The rotor windings are then electrically equivalent to a squirrel cage rotor. At full speed the two types of motors have similar characteristics. Since the maximum torque can be maintained throughout the acceleration period, the wound-rotor motor is desirable when starting high inertia loads.

Many power companies require that motors draw only a small amount of current when they are started, to avoid voltage fluctuations or the flickering of lights. For this reason, wound-rotor motors are often selected because they develop a starting torque of 150 percent of full-load torque with a starting current of 150 percent of full-load current. By comparison, a squirrel cage motor draws 600 percent of full-load current to develop a starting torque of 150 percent of full-load torque.

SPEED CONTROL

Even though the squirrel cage motor has excellent speed regulation, its speed cannot be varied without using complex electronic devices called AC drives. Wound-rotor motors also have excellent speed regulation, along with superior speed control capabilities. The speed cannot be made to run faster than synchronous speed, but can be slowed down by as much as 50 to 75 percent. The greater the resistance inserted into the rotor circuit, the slower it will turn.

Speed control is made possible by the external resistor controller. If the resistance is increased, the rotor current decreases. Since the stator current is proportional to rotor current because of transformer action, it also decreases. The magnetic field strength of both coils reduces, which causes the torque to decrease. As the rotor slows down, the slip increases, and the induced voltage increases. The rotor current will increase until it is sufficient to develop the torque necessary to turn the load at the slower speed.

The motor rotates at a slower speed, but with the same current and torque that it had before the resistance was increased. Because this slip is greater, the induced slip frequency is higher, which causes the rotor reactance to increase. The reactance is proportional to resistance, which causes the power factor to remain constant.

To change the direction of a wound-rotor motor, interchange any two of the three stator terminals. There will be no change in direction if the rotor terminals are interchanged because the rotor is connected only to external resistors.

The WRIM has been largely replaced by solid state AC drives for varying speed control of AC motors. However, there are some applications where the AC drives cannot be used. For example, WRIMs are used in applications where they are exposed to sudden loads that bog down during the starting period, such as in a rock crusher. If the motor is locked by being jammed against rock fragments, the slip shoots up, which causes a surge of high rotor current. However, the heat does not build up inside the motor and destroy the windings. Instead, most of the heat is dissipated by the external resistors.

Other applications of this motor are pulp chippers in paper mills, automobile crushers in junkyards, hammer mills, and printing presses, where frequent and smooth starting, stopping, and reversing of high-inertia loads and speed control are required.

The disadvantages of WRIMs compared to squirrel cage motors are as follows:

1. Decreased efficiency due to losses in the external resistance
2. Poorer speed regulation at low RPMs
3. Increased maintenance due to brushes and slip rings
4. Higher manufacturing cost because of insulated rotor windings, slip rings, and brushes.

5-12   SYNCHRONOUS MOTOR

The synchronous motor is the third type of polyphase motor. It gets its name from the term synchronous speed, which describes the rotating speed of the stator's magnetic field. Unlike induction motors, which have to run at less than their synchronous speed, synchronous motor rotors turn at the same RPM as the stator's magnetic field. The synchronous motor performs two primary functions: It converts AC electrical energy into mechanical power at accurate speeds and it performs power factor correction.

The stator windings of synchronous motors are excited with a three-phase voltage to establish a rotating magnetic field. The most common type of synchronous motor has two different rotor circuits. One circuit is a set of squirrel cage bars with shorting rings that resemble those used by induction motors. They are called *damper* or *amortisseur* windings. Instead of being embedded in the rotor core, the damper windings are locked on the outer periphery of the pole core called the *pole face*. The other circuit contains coils that are wound on laminated core bodies called *salient poles*. (The word *salient* means projecting out.) Therefore, as the name implies, the pole pieces project outward from the shaft. Figure 5-18 shows the rotor of a synchronous motor with damper windings and salient poles. When the motor approaches operating speed, direct current is fed to the coils through brushes and slip rings mounted on the motor shaft. Each coil makes the salient pole become electromagnetic. The number of rotor poles equal the number of stator coils.

Damper winding

Salient poles

FIGURE 5-18 The rotor of a synchronous motor

STARTING SYNCHRONOUS MOTORS

To start a three-phase synchronous motor, a three-phase voltage is applied to the stator, while no DC power is applied to the salient pole windings. A rotating magnetic field revolves around the stator and cuts across the rotor coils. Because the rotating magnetic field exerts opposing forces on each of the salient poles, the rotor does not turn.

The starting torque is provided by the amortisseur windings. Therefore, during the start-up period, the synchronous motor operates as a squirrel cage induction motor. At the beginning of the start-up period, the resultant rotor field lags behind the stator field by almost 180 degrees. As the motor speeds up, the field shifts until it is a little more than 90 degrees behind the stator field.

REACHING SYNCHRONOUS SPEED

When the rotor has accelerated to a speed close to the synchronous speed of the stator field, it is ready for synchronizing. Direct current is applied to the rotor and makes each salient pole an electromagnet. Since each coil is wound in the opposite direction from the adjacent coils, the adjacent poles around the rotor have different polarities.

Each rotor pole of fixed polarity is attracted to the rotating magnetic poles of the stator. At a certain point there is enough force to lock the rotor poles to the magnetic poles as they shift from one stator pole to the next. The amount of force needed to pull the rotor into synchronization is called *pull-in torque*. Because the rotor turns as fast as the field rotates, there is no cutting action between the stator field and amortisseur coils. This causes the current flow in the damper winding to cease so that it no longer operates as a squirrel cage motor.

It is critical to have the DC supply applied to the rotor at the precise moment that the rotor and stator fields are synchronized. If it is done too soon, the pull-in torque is too low. The rotor will slip back and be attracted to the previous stator field. If it continues to slip back, the motor jerks, which may shake it enough to cause damage. Programmable industrial controllers are used to apply excitation current to the rotor coils at the proper moment. Before programmable controllers existed, one person, usually a foreman, was trained to perform this function. Sometimes a Pony motor is used to help get the rotor up to synchronous speed, especially when the rotor is heavily loaded. The Pony motor is a small DC motor mounted on the same shaft as the synchronous motor, as shown in Figure 5-19. After the armature reaches synchronous speed, the motor is converted from a motor to a DC generator. The DC output voltage supplies the excitation current for the salient rotor poles through the brushes. Figure 5-20 shows the diagram of the motor/generator circuit. The DC motor/generator is sometimes called an *exciter*. Excitation current can also be supplied by a rectifier or a direct connection to a DC bus line.

When the motor runs at no-load, the center of the rotor field is aligned with the center of the stator field, as shown in Figure 5-21(a). This is known as *torque angle*. As the motor is loaded, the angle increases, producing more torque. At full load, the torque angle is around 30 degrees, as shown in Figure 5-21(b). The motor continues to turn at the same speed as before. At about 150 to 200 percent of full load, the motor becomes overloaded. The torque angle becomes too great and the motor pulls out of synchronization, which is called *pull-out torque*. In this condition the rotor slows down and lags behind the rotating stator field. A point is reached where the flux link between the rotor and stator is broken. This situation occurs when the rotor pole is

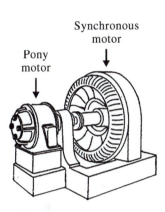

FIGURE 5-19 The Pony motor connected to the shaft of the synchronous motor

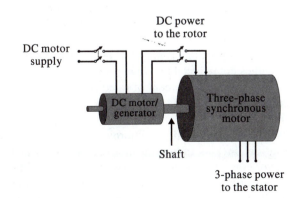

FIGURE 5-20 The schematic diagram of a motor/generator circuit

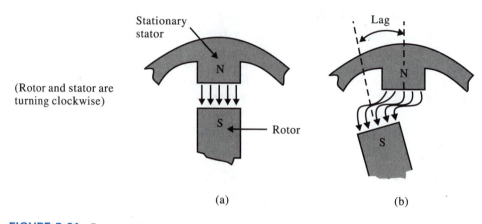

Stationary stator

(Rotor and stator are turning clockwise)

Lag

N

S — Rotor

N

S

(a)

(b)

FIGURE 5-21 Rotor-stator position

halfway between the stator pole to which it is attached and the one behind it. As the rotor speed slips behind the rotating stator, the rotor attempts to lock on with each stator field that passes. The motor may vibrate enough to become damaged. The motor must be shut down immediately and the load reduced before being restarted.

Synchronous motors are durable and dependable. They are used in applications that require constant speed, such as pumps. Because they do not have good starting torque, they should not be used to run equipment that frequently starts and stops, such as conveyers. Another important function of synchronous motors is power factor correction.

POWER FACTOR CORRECTION

Most motors used in industrial plants are induction motors. The total power supplied to induction motors consists of true power and reactive power. The actual work performed is produced by the true power supplied to the resistive components of the motor. The reactive power produces the magnetic fields in the stator, rotor, and air gap. Therefore, induction motors run with a lagging power factor. If the motors are lightly loaded, the power factor becomes low, which implies that the reactive component is large. The power factor of the entire power distribution system within the plant becomes low if several of these induction motors are connected to the same line.

There are several reasons why low power factors are undesirable. The voltage regulation of generators, transformers, and supply lines becomes low. The current-carrying capabilities of supply lines are reduced to a lower level than their rated values. For example, a system can only supply 60 percent of rated power at 0.6 power factor. Power companies may access a penalty charge for industrial sites that run at a power factor below a certain value.

The synchronous motor has the ability to improve the power factor when connected to the same distribution lines as induction motors, as shown in Figure 5-22. The amount of power factor correction is controlled by the amount of DC current applied to the rotor. An adjustment of the rotor current causes the stator current to lead, lag, or be equal in phase with the applied voltage. If the current supplied to the rotor is low, the motor is *underexcited,* and it will have a lagging power factor like an induction motor, as shown in Figure 5-23(a). As current to the rotor is increased, a CEMF is

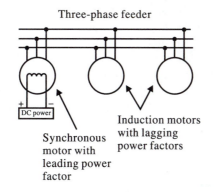

Three-phase feeder

DC power

Synchronous
motor with
leading power
factor

Induction motors
with lagging
power factors

FIGURE 5-22 Synchronous motor used to
correct the power factor of other motors

induced into the stator. The CEMF causes the applied voltage and incoming alternating current to be in phase. At a certain point the rotor is *normally excited,* and the
power factor of the motor is at unity, as shown in Figure 5-23(b). The current supplied
to the motor is at its lowest level. If the rotor current is further increased, the rotor becomes *overexcited.* The CEMF induced into the stator is so high that it causes the stator current to lead the applied voltage, just like a capacitor, as shown in Figure
5-23(c). In this condition the synchronous motor can supply reactive power to induction motors connected to the same line. The overall power factor in the distribution

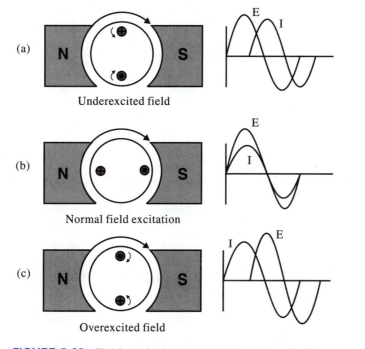

(a)

N   S

Underexcited field

(b)

N   S

Normal field excitation

(c)

N   S

Overexcited field

FIGURE 5-23 Field excitation in a synchronous motor

system then approaches unity and true power is consumed exclusively. To provide power factor correction, one synchronous motor is used for every six to ten induction motors. When the synchronous motor is used strictly for power factor correction without a load connected, it is called a *synchronous condenser.* Newer designs are efficient enough so that they can perform other work while power factor correcting.

Power factor is also corrected by using capacitor banks connected to the supply lines and by loading the induction motors as fully as possible.

CHAPTER PROBLEMS

1. Which members of a DC motor are not found in an AC motor? ____
   a. Armature
   b. Brushes and commutator
   c. Field windings
   d. Fan assembly

2. Which of the two types of motors requires less maintenance? ____
   a. DC
   b. AC

3. The stationary portion of the AC motor is called the _____ and the rotating part is called the _____.

4. The speed of an AC motor depends on the _____ of the power supply and the _____.

5. The RPM that the rotating field moves around the stator is called _____ _____.

6. What is the RPM speed of a single phase 12-pole AC motor operating from a 60Hz power source?

7. Calculate the percent slip of an induction motor that has a synchronous speed of 3600 RPM and a rotor speed of 3400 RPM.

8. A(n) ____ must be established in a split-phase motor to produce the starting torque.
   a. rotating field
   b. alternating field

9. To change the rotation of a split-phase AC motor, ____.
   a. reverse the power leads
   b. reverse the main and auxiliary windings with respect to each other
   c. both a and b

10. Once the split-phase motor reaches ____ percent of its synchronous speed, the centrifugal switch ____ to ____ the ____ windings.
    a. 45–50
    b. 75–80
    c. 95–100
    d. opens
    e. closes
    f. connect
    g. disconnect
    h. main
    i. auxiliary

11. In a resistance-start motor, current in the auxiliary winding always _____ (leads, lags) the current in the main winding.

12. The capacitor-start AC motor has a capacitor in ____ with the ____ winding.
    a. parallel/auxiliary   c. parallel/main
    b. series/auxiliary   d. series/main

13. The capacitor-start capacitor-run motor has how many capacitors? ____

14. The direction of the torque in a shaded-pole motor is toward the ____.
    a. main pole   b. shaded pole

15. Motors that operate from either AC or DC power are called _____ motors.

16. A three-phase induction motor uses a _____ (smaller, larger) frame than a single-phase induction motor of equal horsepower.

17. List two ways that a three-phase motor becomes single-phased.

18. __ T __ F   During operation, if the resistance of a wound-rotor motor is set to zero, it performs very much differently than a squirrel cage motor.

19. A(n) _____ (increase/decrease) in the external resistance of a wound-rotor motor causes an increase in the rotor resistance and therefore a reduction in the rotor current and speed.

20. __ T __ F   The amortisseur windings of a synchronous motor performs a similar function as the bars and shorting rings of a squirrel cage induction motor.

21. AC motors that turn at the same speed as the rotating magnetic field are called _____ motors.

22. In a synchronous motor, the DC excitation is applied to the _____ (rotor's, stator's) magnetic field.

23. In a synchronous motor, stator current will lead the applied voltage if the DC motor field is _____ (overexcited, underexcited).

24. List two favorable features of synchronous motors.

CHAPTER 6

Servo Motors

CHAPTER OBJECTIVES

At the conclusion of this chapter, you will be able to:

○ Describe the operation of the following servo motors:

Wound Armature PM Motor	PM Stepper Motor
Moving Coil Motor	VR Stepper Motor
Brushless DC Motor	AC Servo Motor

○ Define the following terms:

Servo Motor	Stepping Rate
Holding Torque	Step Angle

○ List a practical application of the different types of servo motors.

INTRODUCTION

The traditional motors described in the two previous chapters are used in applications requiring moderate to high power. Typically, they turn in one direction, and their speeds can be varied only within a very limited range. When traditional motors are stopped, they usually coast until they no longer turn.

For applications that require special performance characteristics such as precise speed or position control in both directions, specialty motors are often more suitable than traditional motors. These nontraditional motors are often used as **servo motors**. This term refers to any motor that uses a closed-loop feedback signal to monitor its velocity and position, or that uses open-loop digital equipment that provides precise input command signals.

The servo motors described in this chapter are bidirectional position devices that typically operate within the low to moderate power range.

6-1 DC SERVO MOTORS

DC servo motors are controlled by direct current command signals that are applied to coils, which become electromagnets. The magnetic fields that form around the coils

159

interact with permanent magnets (PMs) and cause the rotating member of the motor to turn. These DC servo motors are referred to as PM motors and are classified into two categories, depending on how the permanent magnet is used. One type of PM motor uses a wound armature and brushes like a conventional DC motor. The pole pieces, however, are permanent magnets instead of electromagnets. This category includes the wound armature motor and the moving coil motor. The other category of PM motors uses wound field coils and a permanent magnet rotor. This category includes brushless DC motors.

In the mid-1970s, the popularity of PM motors increased with the introduction of rare earth magnets. Rare earth magnets have greater flux strength than the ferrite magnets they replaced, which results in greater torque produced by the motor. PM motors are smaller and lighter than wound field DC motors that produce the same amount of torque. They are often used in applications that require portability and low maintenance requirements.

6-2   WOUND ARMATURE PM MOTOR

The wound armature PM motor is shown in Figure 6-1(a). It is similar in construction to a wound rotor motor. The armature contains wound coils that are placed in the slots of an iron core. Current is supplied to the rotor by using brushes, and a commutator switches the current as the armature turns, just as in a conventional direct current motor. The pole pieces are made of permanent magnets instead of series or shunt field winding. These motors typically use magnets to form either a two- or four-pole structure, although six or more poles have been used.

Since permanent magnets are used for the pole pieces, the field flux remains constant. There is no EMF induced into the field poles to cause the flux strength to vary. This gives the motor linear speed-torque characteristics similar to a conventional DC shunt motor, as shown in the chart in Figure 6-1(b). The speed is varied by changing the armature voltage. The direction is changed by reversing the current applied to the armature leads.

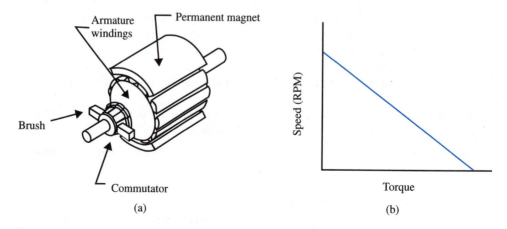

(a)   (b)

FIGURE 6-1 Wound armature permanent magnet

The wound armature PM motor is useful for applications that require small size and high torque. For example, they are used to drive forklifts, wheelchairs, and motorscooters.

6-3   MOVING COIL MOTOR

The moving coil motor (MCM) is designed very differently than other types of motors. The stator field is provided by eight pairs of permanent magnets that are on each side of the disc and parallel to the motor shaft. These magnets are placed around the perimeter of the motor housing. They are arranged so that they provide alternating magnetic fields as shown in Figure 6-2. By fitting as many magnets as possible around the circumference of the motor, the maximum number of stator flux lines is provided to produce the highest possible torque.

The armature, shown in Figure 6-3, is a thin disc made of fiberglass. Two layers of copper conductors are formed on each side of the fiberglass in much the same way as in a printed circuit. One layer, called the upper conductor, is placed on top of the other layer, called the lower conductor. The conductor paths of each layer are arranged at 30 degree angles to each other, as shown in Figure 6-4(a).

The ends of the conductors are located at the center and at one side of the disc in the shape of commutator segments. As the disc turns, brushes ride on the commutator to provide direct current to the conductors, as shown in Figure 6-5. As current flows through the upper and lower conductors, a resultant magnetic field is produced. The armature interacts with the permanent magnet field to produce a force tangent to each magnetic pole, as shown in Figure 6-4(b). Enough torque is provided to turn the armature. Because the armature is in the shape of a disc, it does not use iron. This provides two advantages. First, the disc is light, so it has low inertia. This enables the armature to accelerate rapidly (from 0 to 3000 RPM in $\frac{1}{6}$ of a revolution), stop quickly, and reverse direction easily. Second, the brush life is extended because its armature's low inductance does not cause arcing. Also, the large number of conductors enables the MCM to run smoothly at speeds as low as 1 RPM, unlike conventional DC motors, which tend to cog at low speeds.

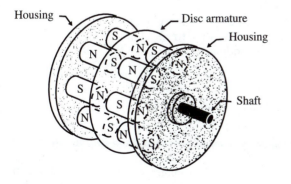

FIGURE 6-2 Permanent magnets are arranged to produce alternate magnetic polarities

FIGURE 6-3 A thin disc armature made of fiberglass with copper conductors placed on each side

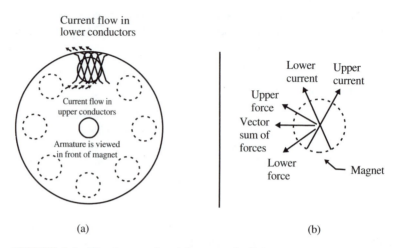

FIGURE 6-4 The interaction of magnetic fields creates the forces that turn the rotor

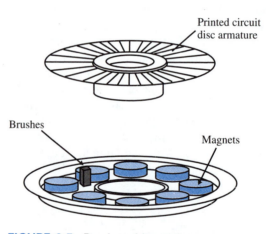

FIGURE 6-5 Brushes ride on the commutator located in the center of the disc to provide direct current to the conductors

The speed of the motor is varied by changing the amount of voltage supplied to the armature. The voltage is in the form of DC pulses at a frequency of about 20KHz. The average voltage varies by changing the width of the pulses. The ratio of time the pulses are on to the time they are off determines the amount of average voltage. For example, Figure 6-6 shows that when the pulses are turned on longer, the average voltage will be higher.

MCMs are used in applications that require high torque, fast acceleration, and small size, such as tape transport systems and computer peripheral devices.

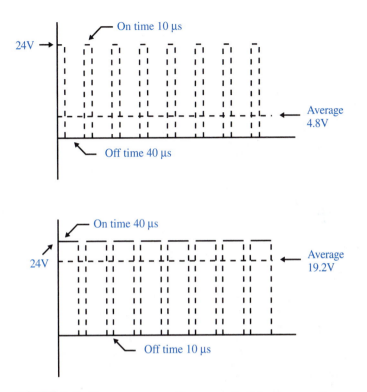

FIGURE 6-6 Square wave pulses supplied to the armature

6-4 BRUSHLESS DC MOTORS

Brushless DC motors (BDCM) contain a permanent magnet rotor and fixed stator windings. The stationary stator windings are usually three-phase, which means that three separate voltages are supplied to three different sets of windings.

The BDCM also contains a converter and a rotor position sensor. The converter is an electronic commutator that changes direct current into pulsating DC voltages. The pulses are applied to the stator windings to create a rotating magnetic field. This field attracts the permanent magnet rotor. As it follows the rotating field, the rotor turns. The rotor position sensor provides feedback signals to the converter so that it switches current pulses through the stator coils in the proper sequence and at the proper time.

The operation of a brushless DC motor is illustrated in Figure 6-7. Three transistor switching devices are connected to a DC power source. When a transistor is turned on, it supplies a phase current to a stator field coil. When current flows through a coil, a south pole is created at the end of the pole face. No more than one transistor can be turned on at any given moment. The north pole of the permanent magnet rotor aligns itself to a stator pole that is energized.

A rotor position sensor that consists of a round disc with 120 degrees cut away is mounted on the shaft. Three proximity detectors are mounted 120 degrees apart around the shaft, within sensing distance of the disc. Therefore, the disc is always

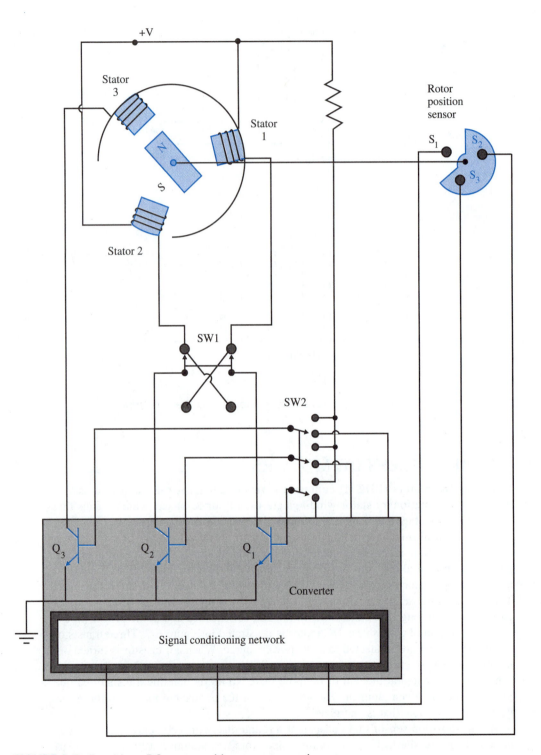

FIGURE 6-7 Brushless DC motor with converter and sensor

being sensed by two of the detectors. Each detector output is connected to a switching transistor through a signal conditioning network. A detector that does not sense the disc sends a positive voltage to the transistor and turns it on. Figure 6-7 shows the north pole of the rotor aligned with stator pole 3. With the disc in the position shown, sensor S_1 turns on transistor Q_1, causing current to flow through stator coil 1. The south pole created at the pole face will attract the rotor and cause it to turn 120 degrees clockwise. When the rotor aligns itself to stator pole 1, the disc will be turned so that it is sensed by detectors S_1 and S_3. Detector S_2 will then turn on transistor Q_2 and cause current to flow through stator coil 2. The rotor turns another 120 degrees clockwise as its north pole aligns itself with stator pole 2. The disc also rotates, causing detector 3 to turn on transistor Q_3. This switching sequence continues as the motor shaft turns in the clockwise direction. If the base connections of two transistors are reversed, by switch SW1, the motor will reverse direction. By moving switch SW2 upward, all three transistors turn on. The rotor stops and remains stationary as long as SW2 is in the upward position. The rotor remains aligned to one of the stator poles due to the magnetic attraction. The amount of force required to move the rotor away from this held position is called **holding torque**.

There are several advantages of a BDCM over a wound-field DC motor. In wound-field motors, the brushes wear out after only about 2000 hours. Also, as the surface contact between the brushes and commutator causes the electrical connections to open and close, sparks develop. The arcing that takes place creates magnetic fields called noise, which can cause interference problems in computer control equipment placed near the motor. Because the BDCM does not use brushes, it has lower maintenance requirements and no noise problem. Also, since it does not have a commutator and wound armature mounted to its shaft, the weight and size of the rotor is smaller. Therefore, its inertia is reduced, which allows the motor to accelerate or reverse its direction more quickly. Since higher supply voltages are used than those in conventional PM motors, they operate at higher speeds and greater torque.

BDCMs are used to drive equipment that requires high speeds, high reliability, and low maintenance. Applications are artificial heart pumps, aerospace gyroscopes, fans and blowers, and tape transport mechanisms for VCRs.

6-5 STEPPER MOTORS

All of the permanent magnet motors described so far in this chapter can be classified as continuous rotation motors. When power is applied to the motor, the armature turns. When power is removed, they coast to a stop and cannot be stopped at a desired position.

A stepper motor operates differently. In a stepper motor, the armature turns through a specific number of degrees and then stops. It converts electronic digital signals into mechanical motion in fixed increments. Each time an incoming pulse is applied to the motor, its shaft turns or steps a specific angular distance. The shaft can be driven in either direction and operated at low or very high stepping rates. Therefore, the stepper motor has the capability of controlling the velocity, distance, and direction of a mechanical load. It also produces a holding torque at standstill to prevent unwanted motion. A stepper motor is typically used as an actuator in motion control applications that require accurate positioning.

One attractive feature of the stepper motor is that it responds to digital signals. Therefore, it can be controlled by computers or a computer peripheral device. An interface device called *drive circuitry* is often connected between the computing device and the motor. This converts the data from the computer into a series of low and high voltage levels that are compatible with the motor.

Many types of stepper motors exist, of which the permanent magnet (PM) and the variable reluctance (VR) motors are the most popular.

6-6   PERMANENT MAGNET STEPPER MOTOR

The operation of a stepper motor is based on the magnetic principle that like poles repel and unlike poles attract. Its operation is illustrated by the simplified view of a stepper motor in Figure 6-8(a). It is constructed of four electromagnets that are energized by passing a current through their respective windings. A cylindrical rotor with two magnetic teeth is placed inside the four coils. Suppose that Coils A and A' become energized. Coil A becomes a south pole and Coil A' a north pole. They are of different polarities because the coils are wound in opposite directions. The rotor aligns itself vertically as shown. If the power applied to the coils is maintained, the rotor will be held in position by magnetic attraction. The amount of force needed to break the shaft away from the holding position is called holding torque.

Figure 6-8(b) shows another condition as Coils B and B' are energized with Coils A and A' remaining energized. If Coil B becomes a south pole and Coil B' becomes a north pole, the rotor turns 45 degrees CCW and aligns itself at position 2 between the poles. If the current flow through all four coils were reversed, the rotor would turn 180 degrees.

Table 6-1 shows how eight output signals of the drive circuitry control eight north pole positions of the rotor.

In practice, there are multiple teeth with alternating north and south poles on the rotor. There are also additional coil windings that alternate in the same sequence as

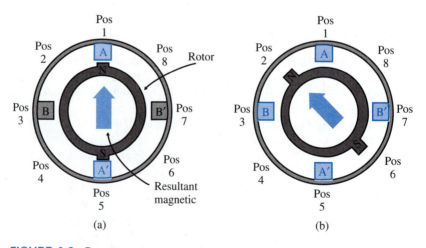

(a)   (b)

FIGURE 6-8 Permanent magnet stepper motor

TABLE 6-1   FUNCTION TABLE FOR EIGHT POSSIBLE POSITIONS OF PERMANENT MAGNET STEPPER MOTOR

Voltages				Motor (N-Pole)
A	A'	B	B'	Position
+5V		0V		1
+5V		+5V		2
0V		+5V		3
−5V		+5V		4
−5V		0V		5
−5V		−5V		6
0V		−5V		7
+5V		−5V		8

the four-pole stator shown in Figures 6-8(a) and (b). This allows for better stepping resolution.

The physical size and weight of a cylindrical-toothed rotor is relatively large; therefore it is more difficult to move, which causes the motor to react slowly. This limits the stepping rate of the motor. To overcome potential inertia problems, some PM stepper motors use a flat disc rotor instead, which weighs 60 percent less than the cylinder type.

The flat disc rotor is shown in Figure 6-9. It is supported on a non-magnetic hub and placed inside two C-shaped electromagnetic cores. The outer edge of the disc is composed of tiny individual magnets. The magnets are evenly spaced and are polarized with alternating north and south poles. Although the C-shaped electromagnets appear to be placed across from each other, they actually are offset from each other by half a rotor pole. Each electromagnet is energized by a different phase.

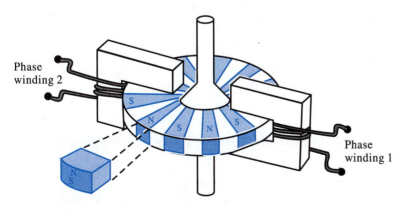

FIGURE 6-9 Disc-type permanent magnet stepper motor

As one electromagnet is energized, the rotor aligns itself to the magnetic field it produces. Next, the first phase is turned off and the second electromagnet is energized. The disc will turn one-half of a half rotor pole to align itself to the magnetic field produced by the second phase. The rotor continues to turn by simultaneously de-energizing one coil and energizing the other.

6-7 VARIABLE RELUCTANCE STEPPER MOTOR

The variable reluctance (VR) stepper motor uses electromagnetic stator poles. Its rotor is in the shape of a disc with teeth and slots around the outer edge. The soft-steel rotor is unmagnetized.

The operation of a VR stepper motor is illustrated in Figure 6-10. It uses four sets of stator windings to form four pole pairs. The poles in each pair are located directly across from each other and are energized at the same time because their coils are connected in series. Because the coils are wound in the opposite direction, the poles in each pair will always have different polarity. There are eight stator poles that are equally spaced apart by 30 degrees. The rotor has six teeth that are equally spaced at 45 degree intervals. Therefore, the alignment of the stator and rotor is different by 15 degrees.

Figure 6-10(a) shows that the teeth on the rotor will align themselves with the flux lines created by the north and south poles of coils A and A′ when they are energized. Note that the teeth are not lined up under the unenergized stator poles. To step the rotor 15 degrees counterclockwise, coils A and A′ are de-energized and coils B and B′ are energized. The teeth closest to the B coils will align themselves with the flux lines of both B poles. The rotor turns because it is made of soft iron which has a very low reluctance rating. The magnetic flux passes through the iron much more readily than through air. The difference of reluctance between iron and air creates a force that causes the rotor to turn so that the flux lines can pass through the iron teeth rather than the air slots between them. The next 15-degree step is made by de-energizing the B and B′ coils and energizing the C and C′ coils. This action continues by repeatedly energizing the coil pairs in the sequence of A, B, C, D, E, and F. Therefore it takes 24 steps to turn the rotor a full 360 degrees. The more rapid the sequence, the faster the rotor turns. The number of signal changes determines the distance the rotor travels. By reversing the sequence, the rotor will rotate in a clockwise direction.

In practice, there are multiple teeth with alternating north and south poles on the rotor. There are also eight coil windings that alternate in the same sequence as the four-pole stator, as shown in Figure 6-10(e). There are 50 teeth machined into the armature. Each pulse moves the armature a distance of $\frac{1}{4}$ tooth. Since it takes four steps to advance the width of one tooth, it takes 200 steps to complete one revolution. The step angle can be computed by dividing 360 degrees by 200 steps: 360/200 = 1.8 degrees.

Because the rotors of VR stepper motors do not have to be magnetized, they can be small and light. The rotor's small size gives it low inertia, so it can respond quickly to control signal changes.

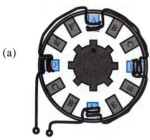

(a)

Phase A
energized

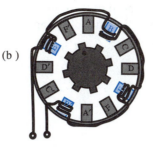

(b)

Phase B
energized

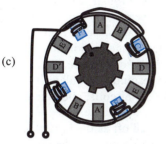

(c)

Phase C
energized

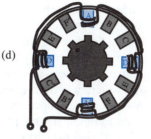

(d)

Phase A
energized

50 teeth on armature

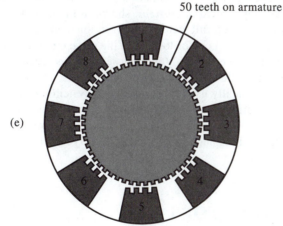

(e)

FIGURE 6-10 VR stepper motor

STEPPER MOTOR TERMINOLOGY

Two important terms relating to stepper motors are *stepping rate* and *step angle*. **Stepping rate** is the maximum number of steps the motor can make in a second. The number of degrees per arc the motor moves per step is called **step angle**. The step angle is determined by varying the number of rotor teeth and stator poles.

The actual speed of the rotor depends on the step angle and stepping rate. It can be calculated by using the following formula:

$$n = \frac{Y \times S}{6}$$

n = Speed in RPM
Y = Step Angles in degrees
S = Steps per second
6 = A Formula Constant

MICROSTEPPING

There are some undesirable characteristics associated with stepper motors. At low speeds the motor jerks as it steps, which can cause rough running in the mechanical mechanism to which it is attached. If a smooth operation is required at low speed, the motor speed is kept high and a gear reduction transmission is connected between the motor and load. Another drawback of the stepper motor is limited resolution. Some applications may require accurate positioning under one degree.

A technique that overcomes low speed and resolution problems is microstepping. Instead of using square waves to energize the stator coils, simulated sinewaves are used. Figure 6-11(a) shows conventional signals that cause a stepper motor to move from one position to the next. One phase is turned off as the next phase is turned on to produce motion. Figure 6-11(b) shows the microstepping technique using two sinewaves. The current is increased in one phase while the current in the other phase gradually decreases. The sinewaves are created by a varying voltage that increments or decrements in small steps called *microsteps,* shown in Figure 6-11(c). There are as many as 125 microsteps to each full step in a conventional stepper motor. These signals are developed by a microprocessor located in the drive circuitry. These sinusoidal signals cause the rotor to move smoothly without jerking, and accurate positioning is attained if the currents are held at intermittent values.

The stepper motor is typically used in an open-loop system. Position is determined by counting pulses. For critical applications, the system can be modified into a closed-loop configuration by using an encoder to verify the position. Stepper motors are used in many practical applications such as printers, CD players, floppy discs, and X-Y tables.

6-8   AC SERVO MOTORS

AC servo motors are controlled by alternating current command signals that are applied to coils which become electromagnets. The magnetic fields that form around the coils interact with other electromagnets and cause the rotating member of the motor to turn.

AC BRUSHLESS SERVO MOTOR

Recall that in a single-phase induction motor, two separate stator windings are energized by two AC voltages that are 90 degrees out of phase. One phase, called the main phase, is taken directly from the AC supply. The other phase, called the auxiliary phase, is tapped off of the supply source and is shifted 90 degrees by a capacitor or

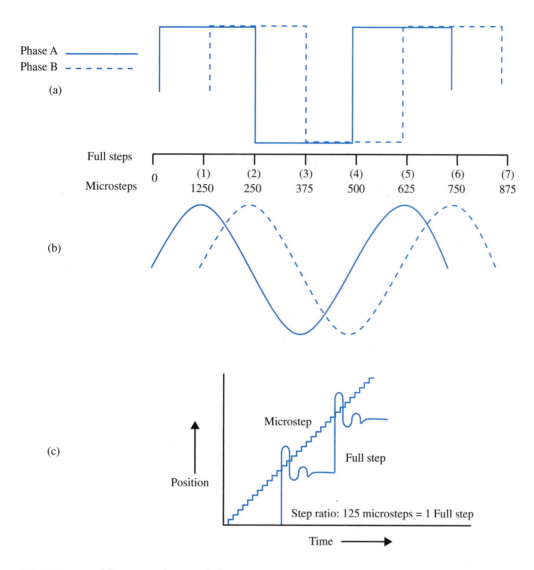

FIGURE 6-11 Microstepping vs. full step systems

inductor. These types of motors are sometimes referred to as split-phase motors. The purpose of the two phases is to create a resultant magnetic field that rotates around the rotor. As the field rotates, a voltage is induced into the rotor, causing it to turn due to the magnetic interaction between the rotor and stator fields.

The AC brushless servo motor operates on the same principle as the single-phase induction motor. Both motor types have squirrel cage rotors and two sets of stator windings that are energized by two AC voltages that are 90 degrees out of phase. Instead of using a capacitor or inductor to develop the auxiliary phase, the AC servo motor uses an electronic circuit to perform this function. This circuit, shown in Figure 6-12, is referred to as an AC servo drive amplifier.

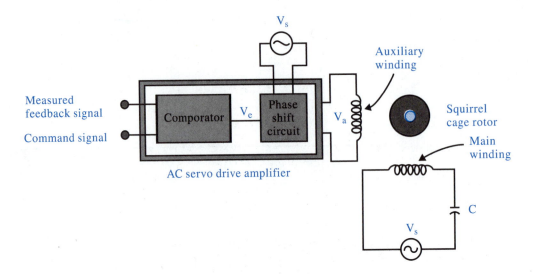

FIGURE 6-12 AC brushless servo motor

The AC line source supplies power to the main winding. It also provides power to the servo drive amplifier. A feedback signal from a transducer is used to indicate the actual position. Another input of the amplifier is a command signal that indicates the desired position. The difference between the measured feedback signal and the command signal creates an error signal. This error signal (V_e) is used to control the firing angles of phase shift circuits, such as those that are powered by a triac. By firing the triac at different points during each alternation, the RMS value supplied to the amplifier output, labeled V_a, can be varied.

When the difference between the actual and desired position is great, output V_a will be large. The strong magnetic field fed to the auxiliary winding will cause the servo motor to run at a high speed. As the difference between the actual position and desired position is reduced, the V_e also decreases. Therefore, the magnetic field in the auxiliary winding weakens and makes the motor run slower. When the object is in the desired position, V_a goes to zero, there is no magnetic field from the auxiliary winding, and the motor stops. Even though the voltage supplied to the main winding does not change, the motor will not turn because there is no rotating magnetic field due to the single phase condition.

Recall that it is possible for some split-phase motors to continue turning when single-phased. To prevent rotation under this condition, the rotor of the servo motor has high-resistance conducting bars.

CHAPTER PROBLEMS

1. __ T __ F   The difference between the two classifications of PM motors is whether the rotor and field are made of permanent magnets or coils.

2. List three applications of a wound armature PM motor.

3. List two applications of MCM motors.

4. __ T __ F   BDCM motors usually use three-phase power.

5. A _____ motor converts electronic digital signals into fixed increments of positional movement.

6. __ T __ F   The operation of a variable reluctance motor is based on the ease at which flux lines pass through soft steel.

7. If the stator's magnetic field of the VR stepper motor moves clockwise, the rotor turns _____.
   a. clockwise   b. counterclockwise

8. The stepper motor is typically used in a/n _____ (open, closed) -loop system.

9. A stepper motor that operates at a step angle of 12 degrees will rotate at what RPM if it has a stepping rate of 360?

10. In an AC servo motor, the phase-shifted voltage is supplied to the _____ winding.
    a. main   b. auxiliary

11. The speed of an AC servo motor is changed by varying the voltage applied to _____.
    a. the main winding   b. the auxiliary winding   c. none of the above.

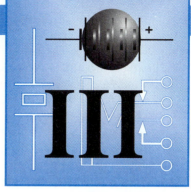

SECTION

ELECTRONIC MOTOR DRIVE SYSTEMS

Most of the machines used in industry are powered by motors. Some production process applications require that a motor maintain constant speed or torque regardless of the physical load placed on it. Other situations require the motor to change speeds either very quickly or very slowly. Mechanical mechanisms were the first devices used to control motors. For example, gear reducers were used to slow the speed of motors. When a speed alteration was required, the motor system was shut down while gears and belts were changed. When a gear mechanism was used for controlling position, backlash limited the accuracy.

Technological advancements in electronics have resulted in the development of the present electronic motor drive system to control electric motors. Commonly known as the *drive,* it uses integrated circuits and solid state devices in the circuitry that supplies power to the motor. The drive is capable of controlling the speed, torque, horsepower, direction, and position of motors with more precision than mechanical devices. The benefits of accurately controlling a motor are energy savings, increased productivity, and better quality. Electronic devices are typically categorized into two groups, DC and AC, according to the type of motor they control.

The DC industrial drive was the first type of electronic drive system. It controlled the speed of a DC motor by using variable resistors to regulate current flow through the armature. The resistors were replaced when high-current solid state devices such as diodes, SCRs, and power transistors were developed. AC drives were not used until a method was devised to vary the frequency of the voltage applied to the AC motor. Not until the advent of the microprocessor integrated circuit was this possible. Today, AC drive systems are becoming increasingly sophisticated. They now perform many of the operations that previously could be performed only by DC systems. However, DC drive systems are not being totally replaced because DC motors have some advantages over AC motors, such as higher starting torque and horsepower output.

The following two chapters discuss the most popular types of drives. Their basic parts are illustrated in block diagram form to describe the underlying theory of operation more clearly. Each block is further broken down into simplified discrete circuitry to show how the signals are developed, controlled, or processed. Typical applications of drives are given, as well as the drive's characteristics, strengths, and weaknesses.

The electronic motor drive primarily performs the function of the controller element of an industrial control system. Some drives also utilize the measurement feedback device and comparer elements of an industrial control system to perform their operation.

CHAPTER 7

DC Drives

CHAPTER OBJECTIVES

At the conclusion of this chapter, you will be able to:

○ List the reasons for using DC drives and describe their advantages over AC drives.

○ Identify the types of control functions performed by DC drives.

○ Describe the purposes of the following sections of a DC drive:

| Operator Control | DC Motor |
| Drive Controller | Speed Regulator |

○ Explain the operation of the variable voltage DC drive circuits that perform the following functions:

Motor Speed Control	Current Limiting
Speed Regulation	High/Low Speed Adjustment
IR Compensation	Field Current Speed Control

○ List the load characteristic requirements of several types of equipment controlled by DC drives.

○ Identify three classifications of motor braking techniques and describe the operation of each type.

INTRODUCTION

DC drives are used in both motion control and process control. They are ideal for material handling applications because they have high starting torque. They can start and stop DC motors quickly and efficiently. Additionally, they are capable of smoothly slowing down a motor to zero RPM and then immediately accelerating it in the opposite direction. Soft starts and stops by controlled acceleration and deceleration saves energy and reduces stress on expensive equipment.

The horsepower and constant torque characteristics of a DC drive and motor are higher than that of an AC drive system. Therefore, they are used in process control applications in which the physical load on the motor may be suddenly increased, such as extruders or conveyers.

7-1   DC DRIVE FUNDAMENTALS

DC drive systems are divided into three major sections, as shown in Figure 7-1: the operator controls, the drive controller, and the DC motor. A feedback configuration is also a part of the system.

OPERATOR CONTROL

The **operator control** section provides the operator a way to start, stop, and change the direction and speed of the motor. The controls used may be an integral part of the controller in the form of switches, push buttons, potentiometers, or a handheld keypad. They also may be remotely mounted in the form of programmable controllers or personal computers. Use of remote devices is increasing due to their greater flexibility in process or machine control. They also have diagnostic functions that provide information to identify a malfunction if it occurs.

DRIVE CONTROLLER

The **drive controller** converts an AC source voltage to an adjustable DC voltage which is then applied to a DC motor armature. Regulation characteristics of the controller will run the motor at a desired speed, torque, and horsepower set by a reference input. Additional circuits can be used to protect the drive system from overloads and various circuit faults.

DC MOTOR

The *DC motor* converts the adjustable DC voltage from the drive controller to rotating mechanical energy. The RPM and direction of the motor shaft are determined by the magnitude and polarity of the adjustable voltage applied to the motor. The motor shaft is usually coupled to a transmission device, which is then connected to the driven machine.

The DC motor in a typical drive control system is usually a shunt wound or permanent magnet type. An adjustable voltage supply is connected to the armature of both

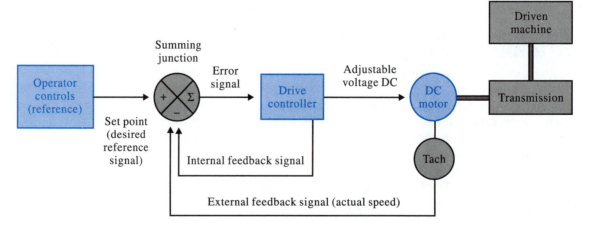

FIGURE 7-1 DC drive control system

types of motors to primarily control speed. A fixed or adjustable voltage is connected to the field coil of the shunt motor. An adjustable field control can be added to provide higher speed and better torque control of the motor.

SPEED REGULATOR (FEEDBACK METHOD)

The speed of the motor is regulated by a closed loop circuit. The desired speed of the motor is determined by a variable voltage from the operator control input device. The voltage produced, called the *set point,* is applied to a summing junction. The actual speed of the motor is converted to a proportional voltage. This voltage, called the *feedback signal,* is also applied to the summing junction. The set point and feedback signals are continuously compared. Any difference causes an error signal to form at the summing junction output. The result is that a speed correction is made, the set point and feedback signals become matched, and the error signal goes to zero volts.

Two methods are used to monitor the speed of the motor: internal and external. Each method produces a feedback voltage proportional to the motor velocity. The internal method uses armature current as the feedback signal. The external method uses a tachometer to produce a DC feedback voltage. Since the internal method monitors only the motor, the external method is more accurate because it monitors the entire closed-loop system.

The most popular DC drive is the variable voltage drive. It controls motors by regulating the amount of voltage applied to the armature winding, shunt field winding, or both.

7-2   VARIABLE VOLTAGE DC DRIVE

A simple speed control circuit is shown in Figure 7-2. It consists of a motor armature and an SCR. The SCR performs two functions. First, it rectifies the AC supply into a pulsating DC current, causing current to flow through the DC armature in one direction. Second, since it can be turned on at any time during the alternation, it controls the amount of current that flows through the DC motor armature. Therefore, the speed of the motor can be controlled, since it is directly proportional to the current that flows through the armature. The 120VAC source is converted into 0 to 55 Average DC volts.

The half-wave rectifier is rarely used because of motor noise and excessive heating. By supplying power to the motor armature with full wave rectification, the overall

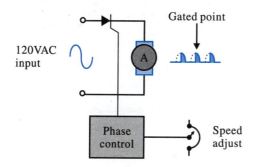

FIGURE 7-2 Half-wave speed control

performance of the motor is significantly improved. Figure 7-3(a) shows a full-wave rectifier that can supply current to the armature during each alternation. It consists of two rectifier diodes and two SCRs. During each alternation, current flows through one of the SCRs and the diode connected to it in series. Current that flows through the diode and armature is regulated by the SCR.

DC motor drives control either permanent magnet or shunt field wound motors. If the motor is shunt wound, a field supply is also required. Figure 7-3(b) shows two diodes, D_5 and D_6, and a field coil added to the circuit. Full wave rectification is provided to the field coil by diodes D_5 and D_6, and D_3 and D_4 from the armature bridge rectifier. The field coil also functions as an inductor to provide filtering for the pulsating DC output of the rectifier.

A firing circuit, shown in Figure 7-4, is used to turn the SCRs on during the alternations. It consists of two parallel branches, one with an RC network and the other with a UJT pulse configuration. The capacitor and resistor in the RC branch remain constant, and the resistance of the transistor varies depending on how hard it is turned on.

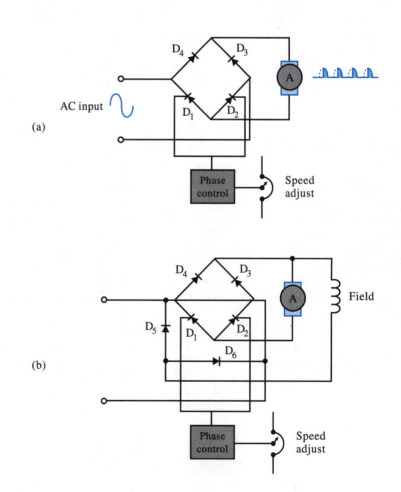

FIGURE 7-3 Full-wave rectifier

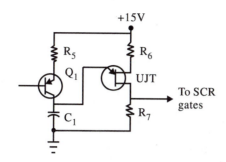

FIGURE 7-4 Firing circuit

When the charge on the capacitor reaches a certain voltage, the depletion region of the UJT collapses and the capacitor discharges through it. As the pulse of current flows through R_7, a momentary voltage develops across it. The voltage formed at the top of R_7 is fed to the gate leads of SCRs D_1 and D_2 of Figure 7-3. The SCR that is anode/cathode forward biased will fire when the gate lead receives the pulse voltage. The SCR turns off when the voltage across the anode and cathode approaches zero volts.

When the voltage at the transistor base of Q_1 in Figure 7-4 becomes more negative, it turns on harder and its resistance decreases. The time constant of the branch decreases, which causes the capacitor to charge more quickly. The result is that the firing delay angle decreases, which allows more current to flow through the armature.

Figure 7-5 shows the entire variable voltage DC drive. Figures 7-6 through 7-8 focus on parts of the drive shown in Figure 7-5. Figure 7-9 and 7-10 are optional circuits that can be added to the drive for greater control capabilities.

MOTOR SPEED CONTROL

The speed of the motor is operator controlled by adjusting a knob connected to potentiometer R_1 in Figure 7-5.

The fixed terminals of the potentiometer are connected to the positive 15-volt supply. As the wiper arm is moved upward, the input lead connected to R_2 of the summing op amp goes more positive. The result is that the op amp output goes negative, turns Q_1 on harder, and causes the capacitor to charge more quickly. As it does, the UJT turns on sooner during the alternation and fires the SCR earlier. This action causes the current to increase through the armature. The rise in armature current increases the speed of the motor.

SPEED REGULATION

When the operator sets the speed control potentiometer, the motor rotates at the desired RPM. However, if the physical load that the motor is turning happens to change, the motor speed will change as well. For example, suppose that the motor is turning a paper machine roll. The more paper that is wound on the roll, the heavier it gets, slowing the motor down. To compensate for the change in speed due to a changing physical load, an armature voltage feedback network is added to the circuit. This provides **speed regulation**.

DC Drive System

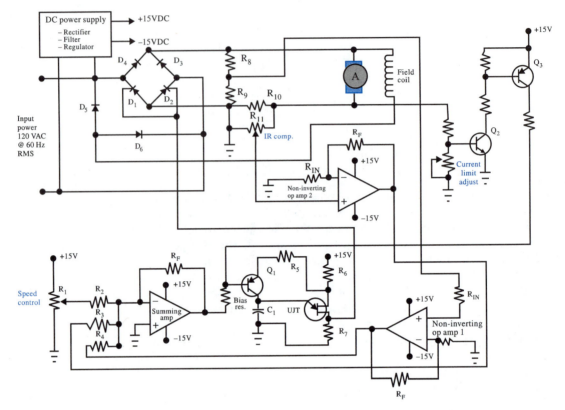

FIGURE 7-5 DC drive system

The armature voltage feedback network, shown in Figure 7-6, consists of two resistors, R_8 and R_9, that are connected across the output of the bridge rectifier. The feedback signal develops at the junction between the resistors and is amplified by non-inverting op amp 1 before it is applied to summing op amp input resistor R_4. Suppose that the motor RPM decreases. The CEMF decreases, causing the net voltage applied across the armature to increase. The voltage at the feedback junction between R_8 and R_9 also increases and is amplified before being fed to the summing op amp. The result is that the summing amp output goes more negative, turns Q_1 on harder, and causes the SCR to fire sooner. More current flows through the armature, which speeds the motor back up to its desired RPM.

IR COMPENSATION

An **IR compensation** circuit, shown in Figure 7-7, further improves the speed regulation of the DC drive. The IR circuit, which refers to *internal resistance* of the motor, is used to compensate for the inefficiency and power losses in the motor. The IR comp signal is obtained at R_{11}, which is the IR comp adjustment potentiometer. It divides the voltage developed across resistor R_{10}. The voltage across R_{10} is proportional to the

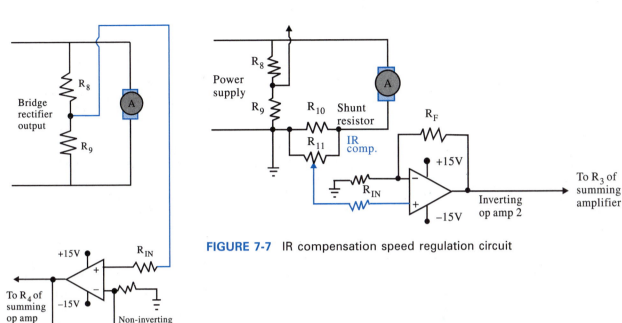

FIGURE 7-7   IR compensation speed regulation circuit

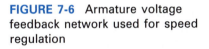

FIGURE 7-6   Armature voltage feedback network used for speed regulation

motor current and the load. In the event that the motor slows down, the CEMF decreases, causing the current through the armature and R_{10} to increase. The IR voltage at R_{11} increases and is amplified by non-inverting op amp 2 before being fed back to summing op amp input resistor R_3. The effect is that the timing circuit fires the SCRs sooner, which speeds the motor back to the desired RPM. Potentiometer R_{11} is used to adjust the feedback voltage at a level that properly compensates for any load variances.

CURRENT LIMITING

If a malfunction develops that causes the motor to stall or slow down far below its rated speed, the current through the armature will become excessive. These large currents occur because the slow rotation of the motor reduces the CEMF so that the net voltage at the armature becomes very high. The current levels may rise above the maximum forward current rating of the semiconductors in the bridge assembly. Either the semiconductors or the armature windings are destroyed. To limit the current to a safe amount, a **current limiting** network, shown in Figure 7-8, is added to the circuit. The current through the armature is monitored by connecting an input lead to the bottom terminal of the armature. If the current becomes too high, a greater voltage forms across R_{10}. The voltage at the top of the current limit adjust rheostat goes more

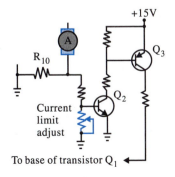

FIGURE 7-8   Current limiting circuit

positive. Transistor Q_2 turns on harder and causes the voltage at the base of transistor Q_3 to go in a negative direction. As PNP transistor Q_3 turns on harder, the potential at the base of Q_1 goes in the positive direction. The effect is that as transistor Q_1 turns on less, it prevents the timing circuit from firing any sooner, and therefore limits the current flow through the armature. The current limit adjust rheostat is used to adjust the network so that the current is limited to the proper level.

HIGH SPEED/LOW SPEED ADJUSTMENT

There are some applications that should avoid maximum motor speed when the speed control is adjusted to the maximum motor speed setting. Some applications should avoid 0 RPM when the speed control is set at minimum speed. Figure 7-9 shows **High/Low Speed Adjustment**. A rheostat is connected between the speed control potentiometer and the +15-volt supply. Another rheostat is connected between the potentiometer and the ground. If the wiper arms of both rheostats are set all the way to the bottom, they are shorted out. The speed control setting varies from 0 to 15 volts which enables the motor to rotate from 0 RPM to maximum speed. If the wiper arm at the top rheostat is moved slightly upward, the voltage divider changes. Therefore, when the speed control is set at maximum speed, the voltage at the wiper arm may be only +12 volts instead of +15 volts. Likewise, when the bottom rheostat wiper arm is moved slightly upward, the voltage at the potentiometer's wiper arm may be +3 volts instead of zero volts when the speed control setting is adjusted to minimum speed.

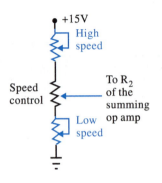

FIGURE 7-9 High speed and low speed adjustments

ACCELERATION/DECELERATION ADJUSTMENT

Suppose that a paper machine is at 0 RPM when the speed control is suddenly adjusted from the minimum to the maximum speed setting. Maximum current will flow through the armature. If the adjustment is made too quickly, something may become damaged. For example, the surge of current may overheat the armature windings, since the CEMF is minimal because of the low motor RPM. Even if the armature windings withstand the large momentary current, the motor shaft coupled to the paper machine may sheer if the inertia of the paper roll is too great to overcome.

By placing a capacitor and rheostat between the speed adjust wiper arm and ground, as shown in Figure 7-10, both problems can be avoided. If the speed control is suddenly changed from the minimum to maximum setting, the voltage of the wiper

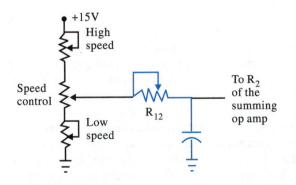

FIGURE 7-10 Acceleration/deceleration control

arm will not be felt immediately at the summing input. Instead, the voltage forms at the top plate of the capacitor at a rate determined by the setting of rheostat R_{12}. The charging time delays application of the voltage to the motor, causing it to accelerate at a slowed rate. This action is referred to as a *soft start*. If the speed control is suddenly adjusted to a lower voltage, the potential at the top plate of the capacitor decreases at a slower rate. The voltage on the capacitor lowers at a rate determined by the setting of rheostat R_{12}. As the capacitor discharges, the motor decelerates at a slowed rate.

FIELD CURRENT SPEED CONTROL

The speed of a shunt motor is controlled primarily by varying the voltage at the armature and keeping the field voltage constant. When the maximum rated armature and field voltages are applied to the motor, it runs at *base speed*. It is possible to further control motor speed by changing the field current, called **field current speed control**. The following formula reinforces this statement:

$$N = \frac{V - IR}{K \cdot \theta}$$

where,   N = Speed
V = Applied Armature Voltage
IR = Internal Motor Resistance Drop
K = Motor Design Constant
θ = Motor Field Flux

By reducing the field current to half its original value, the armature CEMF will decrease and allow more current flow through the armature windings. With the additional current, speed increases and CEMF builds back up to its original level. When equilibrium is reached, the motor RPM is doubled. It is possible to increase the motor speed up to five times the base speed. However, weakening the field by reducing the field voltage also reduces motor torque.

The field current may be varied manually or automatically. The control circuit uses a rheostat or an SCR bridge to regulate the amount of current flow. DC drives with this capability are frequently called *voltage drives* and *field drives*.

LOAD CHARACTERISTICS

One of the primary factors used to select an adjustable drive is what type of load the motor will be driving. Therefore it is important to understand both the speed and torque characteristics and the horsepower requirements of the load being considered. Motor loads are classified into three main categories, depending on how the torque and horsepower requirements vary with changing operating speed. These three groups are *constant torque load, constant horsepower load,* and *variable torque load.*

Constant Torque Load

In this category, the torque required by the load is constant. An example of this type of load is friction. The constant torque overcomes the friction at low speeds. Applications that provide this type of load are conveyers, extruders, and hoists. Motors that provide constant torque are also used when shock loads, overloads, or high inertia loads are encountered.

A DC drive that controls a permanent magnet or shunt field motor is used for this type of load requirement. Speed is adjustable from 0 to 100 percent by controlling the armature voltage as the field strength is held constant.

Constant Horsepower Load

In this category, the horsepower required by the load is constant within the speed range of the motor. When the motor is starting at a low speed, the load requires high torque for quick acceleration. According to the following formula, the torque will decrease as the speed increases to maintain a constant horsepower:

$$HP = \frac{Torque \times Speed}{5252}$$

where,   Torque is measured in lb-ft
            Speed is measured in RPM
            5252 is a proportional constant

An application of this load type is a machine tool lathe. It requires slow speeds and high torque for rough cuts, and high speed with less torque for fine cuts where little material is removed. Another application is a mixer that operates at a low continuous speed.

Like the constant torque load, DC drives that control shunt field or permanent magnet motors are used for this type of load requirement.

Variable Torque Load

In this category, the torque required by the load varies proportionally to the mathematical square of the speed. Such applications often require that the load be added as the speed of the motor is increased. Large centrifugal fans and pumps demand variable torque from a motor, as do punch presses, which are high inertia machines that use flywheels to supply most of the operating energy.

7-3   MOTOR BRAKING

Some industrial applications require that a rotary mechanism be stopped quickly. However, because of inertia, it may be difficult to stop a mechanism abruptly. Since DC electric motors are used as the prime mover for some types of rotating machines, certain types of motor braking techniques have been developed. Two popular methods are dynamic braking and regenerative braking.

DYNAMIC BRAKING

Dynamic braking uses a resistor to absorb the kinetic energy from the rotating armature and the load to which it is connected. Figure 7-11 shows the circuit configuration for dynamic braking. When the start button is closed, current flows through the control relay (CR) coil. All normally open (N.O.) contacts close and all normally closed (N.C.) contacts open. Current continues to energize the CR coil when the start button is released, as current flows through the contact that is parallel to it. Current flows through the motor armature from the negative to the positive terminal of the motor control circuitry. When the stop button is pressed, current no longer flows through the

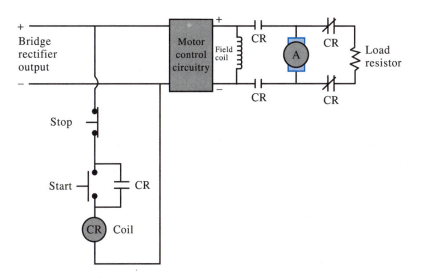

FIGURE 7-11 Dynamic braking circuit

CR coil. The N.O. contacts open and the N.C. contacts close. The field coil remains energized, but current no longer flows through the armature from the motor control circuitry. The armature becomes connected in series with the load resistor. As it spins due to inertia, the armature coil cuts the shunt field flux lines and consequently becomes a generator. Current induced into the armature now flows in the opposite direction than the original current. The result is that a reverse torque is developed. At first, the current and torque are high, so the motor speed drops quickly. As the motor slows down, less current is induced into the armature coil. The reverse torque progressively gets smaller, which causes the motor to gradually slow to a halt. This braking action is the result of kinetic energy being extracted from the rotating machinery as dissipated heat by the resistor.

The smaller the load resistance, the more current is drawn from the armature and the faster it slows down. Dynamic braking resistors of small ratings are typically constructed of wire wound on ceramic. Resistors with greater ratings are usually made of fabricated steel stampings, a convoluted strip, cast iron, or liquid, all of which provide the large surface areas required to dissipate the heat produced during braking.

When selecting a load resistor, it is necessary to choose one with a rating based on the duty it is required to perform, taking account of load inertia and the number of stops per hour. In practice, a load resistor value is chosen so that the initial braking current is about twice the rated motor current. The initial braking torque is then twice the normal torque of the motor.

REGENERATIVE BRAKING

In regenerative braking, power from the kinetic energy is returned to the power supply. Figure 7-12(a) shows the circuit that causes the motor to stop quickly. One bridge supplies current to the motor armature in one direction; the other bridge supplies current

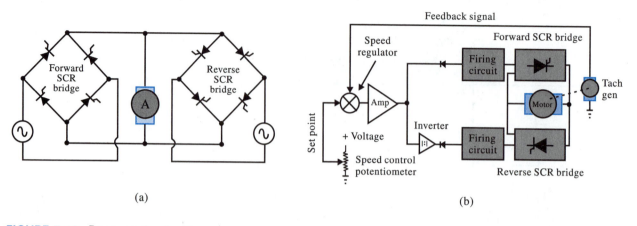

(a)   (b)

FIGURE 7-12 Regenerative braking

in the opposite direction. Suppose the motor is running in the forward direction. The braking action takes place by shutting the forward bridge off and then turning the reverse bridge on, causing the direction of the current flow through the armature to change. This action causes the torque of the armature to develop in the opposite direction. As the torque applied to the motor shaft reverses, it counters the forward inertia of the motor, causing it to decelerate quickly. When the motor rotation stops, the reverse SCR bridge turns off. If the reverse bridge is kept on, the motor will begin rotating in the opposite direction. This system is ideal for applications that require rapid starts and stops or frequent direction reversals.

In addition to stopping a motor quickly, the regenerative motor braking circuit can increase the deceleration rate. Suppose that the speed adjust knob on a DC drive without the regenerative circuit is suddenly turned to a lower RPM setting. The firing circuit will turn off the signals applied to the SCR gates in the bridge circuit. No voltage will be applied to the motor armature as it coasts to its new speed setting. However, if a regenerative circuit is present, the motor decelerates quickly as it is braking. When the new speed is reached, the reverse bridge turns off and the forward bridge turns back on.

Regeneration also provides a precise method of speed regulation. Figure 7-12(b) shows a block diagram of the circuit that performs this function. A speed control potentiometer provides a setpoint voltage to a speed regulator to obtain the desired RPM. The output voltage of a DC generator/tachometer reliably indicates both the magnitude and direction of motor speed. This voltage is also applied to the speed regulator. The setpoint and feedback signals are opposite polarity. Assume that the forward SCR bridge is on and the motor is rotating as desired, when the inertia of the rotating load causes the motor to accelerate to a rate that is higher than the desired RPM. The feedback signal becomes greater than the set point. The output of the speed regulator changes polarity, turns the forward bridge off, and turns the reverse bridge on. The current change through the armature reverses the motor torque and the motor slows down until the set point is greater than the feedback voltage. The motor then runs at that speed.

CHAPTER PROBLEMS

1. List two advantages of DC drives over AC drives.

2. List two advantages of soft starts and stops controlled by acceleration and deceleration.

3. Which of the following names are major sections of a DC drive system? ____
   a. Operating Controls
   b. Drive Controller
   c. DC Motor
   d. Feedback
   e. All of the above

4. Which of the following motor functions can be manipulated by the operator control section? ____
   a. Starting
   b. Stopping
   c. Change of direction
   d. Speed
   e. All of the above

5. Which of the following types of circuits are used by the Drive Controller section? ____
   a. AC to DC Converter
   b. Regulation
   c. Protection
   d. All of the above

6. __ T __ F   The armature is the primary part of the motor where an adjustable voltage is connected for speed control.

7. What are the two functions performed by the SCRs in a variable voltage drive?

8. The function of Q_1 in Figure 7-4 is ____.
   a. to operate as a switching transistor
   b. to operate as a variable resistance device

9. Referring to Figure 7-5, if the motor slows down, the voltage at the bottom input of the summing amplifier _____ (increases, decreases), causing the summing amplifier to become more _____ (negative, positive), which turns Q_1 on _____ (less, harder), causing the SCR to fire _____ (sooner, later), allowing _____ (less, more) current to flow through the armature.

10. Referring to Figure 7-7, if the motor speed slows down, the IR voltage at R_{11} _____ (increases, decreases), causing the SCRs to fire _____ (sooner, later) in the alternation.

11. As the motor runs at base speed, the voltage applied to the armature and field is ____.
    a. low   b. moderate   c. maximum

12. By adjusting the DC drive so that the field current is reduced to half, the motor speed ____.
    a. is cut off   b. doubles   c. stays the same

13. Match the application example with the appropriate motor load classifications.

    Constant Torque Load   ____
    Constant Horsepower Load   ____
    Variable Torque Load   ____

    a. Machine Tool Lathe   d. Centrifugal Fan
    b. Pump   e. Extruder
    c. Hoist

14. Match the proper term with the description that explains the method used to perform motor braking.

    ____ Uses a resistor to absorb kinetic energy from the rotating armature and load.

    ____ Returns power from kinetic energy to the power supply.

    a. Dynamic   b. Regenerative

CHAPTER 8

AC Drives

CHAPTER OBJECTIVES

At the conclusion of this chapter, you will be able to:

○ List the reasons for using AC drives to control the speed of AC motors.

○ Explain the methods of controlling the speed of an AC motor and identify which one is most effective.

○ Describe the purpose of the following sections of an AC drive:

    Operator Control   AC Motor

    Drive Controller

○ Explain the differences among VVI, PWM, and Vector AC Drives.

○ Describe the operation of the following major sections of the VVI drive:

    Converter   Inverter

    Intermediate Circuit

○ Determine the V/H ratio when frequency changes and the voltage is constant.

○ Describe the operation of the following VVI circuits:

    Control Circuit   Overcurrent Protection

    Base Driver   Auto Boost

    Phase Control Rectifier   Chopper Control

○ Describe the operation of the PWM drive and list its advantages over a VVI drive.

○ Describe the operation of the Flux Vector drive and list its advantages over the VVI and PWM drives.

○ List the types of parameters that are preset by a programmable drive unit.

INTRODUCTION

Much of the electricity produced by electric power utility companies is consumed by AC motors that run fans, blowers, and pumps. A 1989 United States Department of Commerce report stated that approximately 50 percent of AC motors operate these load devices. Such loads typically operate inefficiently. For example, fans and pumps are designed to provide the maximum air or liquids for the system in which they are installed. To do so, they run at the maximum rated speed. However, the actual demand is often less than the designed capacity. To control flow, outlet dampers are used for fans and throttling valves are used for pumps. Even though these are effective control methods, up to 30 percent of the power consumed by the motors is not applied to the work they are meant to perform.

A much more efficient method to control flow is a system which monitors the load and adjusts the speed of the motor to satisfy the demand. This function is accomplished by AC adjustable frequency drives. Because a motor may consume as much as twenty times its acquisition cost in electricity every year, it is financially and ecologically advantageous to improve the efficiency of the motor. Also, by maintaining only the flow needed to satisfy system requirements, the pump or fan is not subjected to any higher pressure than necessary. Therefore the components last longer.

8-1 AC DRIVE FUNDAMENTALS

Unlike the DC motor, the speed of an AC motor is not controlled exclusively by varying the applied voltage. It is true that reducing the AC supply voltage will lower motor RPM. However, it will also reduce the torque, so that any changes in torque demand imposed by the mechanical load will cause the motor to run at erratic speeds.

The speed of an AC motor typically is determined by how many cycles of alternating current it receives each second. In the United States, the frequency applied is 60Hz. Speed is also determined by the number of poles. Electric currents in the motor pole windings create electromagnets. Each magnet must have a north and a south pole. Therefore, magnetic poles always come in pairs. Motors are commonly wired to have two, four, six, or eight poles. The speed of a given AC motor is derived from the following equation:

$$N = \frac{60f}{P}$$

where, N = speed (RPM)
            f = frequency (Hz)
            P = Number of Pole Pairs
           60 = To determine RPM, the frequency given in seconds
                  is converted to the number of cycles in a minute by
                  multiplying 60 times f.

Since the number of poles (P) of a motor remains constant, the speed (N) of the motor is more easily controlled by varying the applied frequency (f).

AC adjustable frequency drives take standard commercial AC power and convert it to a variable frequency and voltage, which is delivered to single-phase or three-phase motors. The output frequency of some drives ranges from 0 to several thousand hertz, which smoothly controls the motor speed from a low RPM to the base speed and above.

8-2   AC DRIVE SYSTEM

The block diagram of a typical AC adjustable drive system is shown in Figure 8-1. It has three basic sections: the *operator control,* the *drive controller,* and an *AC motor.*

> *Operator Control.*   The **operator control** provides the operator a way to control the AC motor. Its features are similar to those of the operator control used by the DC Drive System.

> *Drive Controller.*   The **drive controller** converts the fixed voltage and frequency of an AC power line to an adjustable AC voltage and frequency.

> *AC Motor.*   The **AC motor** converts the adjustable AC frequency and voltage from the drive controller output to rotating mechanical energy. Three-phase squirrel cage induction motors are primarily used. AC drives can also control synchronous motors.

The part of the system that is actually the adjustable frequency drive is the *drive controller.* Also referred to as an *inverter,* a drive controller can operate motors ranging in size from a fraction of a horsepower to several thousand horsepower. Inverters can be sold separately because the operating control and the AC motor may already be in place.

The objective of most AC drive systems is to control an induction motor so that it meets the following criteria:

1. Provide full load torque from zero RPM to full speed.
2. Prevent torque fluctuations at low speed.
3. Maintain a set speed when the load torque varies.
4. Provide a starting torque of at least 150 percent of the rated full-load torque.
5. Control the rate of acceleration and deceleration.

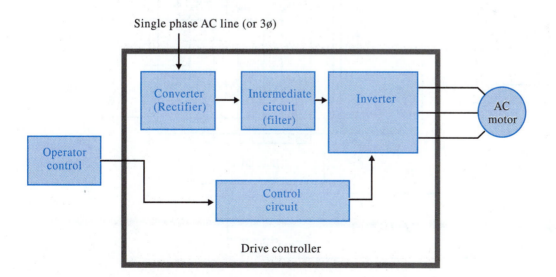

FIGURE 8-1 Block diagram of an AC drive system

The most common types of AC adjustable frequency drives used in industry are *variable voltage inverter* (VVI), *pulse width modulated* (PWM), and *flux vector.* They differ in the switching patterns each uses to approximate a sinewave, and in the method each uses for controlling the motor voltage in proportion to frequency.

8-3   VARIABLE VOLTAGE INVERTER (VVI)

The variable voltage inverter adjustable frequency drive is also known as a six-step voltage source inverter, or six-pack. Its name is derived from the six distinct output states of the DC to AC inverter during each cycle.

A VVI drive is shown in Figure 8-2. The function of each section is described as follows:

> *Converter.*   A **converter** is a rectifier that converts AC line voltage to a pulsating DC voltage. If the drive runs a low horsepower motor, a single-phase bridge circuit can be used for rectification. A three-phase rectifier is used to supply extra current if a high horsepower motor is used. VVI drives use either uncontrolled or controlled rectifiers. A rectifier that has only diodes is uncontrolled because it is capable of producing one voltage. A controlled rectifier uses thyristors. The control function is achieved by regulating the

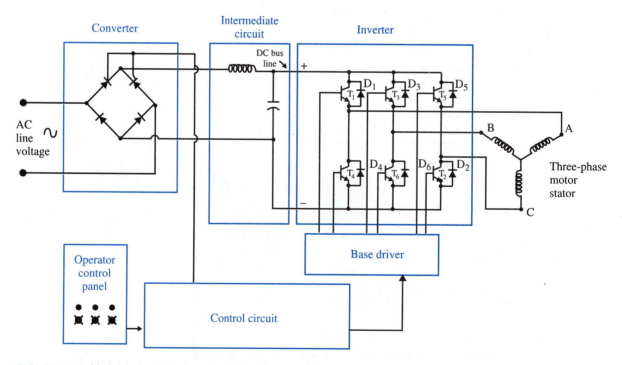

FIGURE 8-2   Variable voltage inverter AC drive

point during the alternation at which the thyristor conducts to vary the DC voltage.

Intermediate Circuit.   The **intermediate circuit** transforms the pulsating DC signal from the converter to a smooth DC waveform. Figure 8-2 shows a low pass LC circuit used to perform the filtering action for the controlled converter in the diagram. An intermediate circuit called a *chopper* is used to vary the DC voltage if the converter section is uncontrolled. The output of the intermediate circuit is called the **DC bus line**.

Inverter.   The function of the **inverter** is to "invert" the DC bus-line voltage back to simulated AC power at the desired frequency. Some types of inverters also convert a constant DC bus voltage into a variable AC voltage. An inverter consists of six electronic switching devices to develop an approximate three-phase AC signal. In the diagram, switching transistors are used. The transistors turn on and off in the proper sequence to create three sets of voltage waveforms between A-B, B-C, and C-A.

Control Circuit.   The **control circuit** transfers signals to the inverter to cause its semiconductor device to switch on or off. In some types of drives, the control circuits send signals to either the converter or the intermediate circuit, which will cause the DC bus-line voltage to vary.

THE INVERTER OPERATION

The VVI drive inverter controls nothing except frequency. The operation of the inverter section can be illustrated by using the switch model in Figure 8-3(a). Three SPDT switches are connected in parallel across a DC supply. Depending on the switch position, either a positive voltage or a zero volt ground connects to the swivel terminal. The voltage is positive in the up position and zero in the down position. The wires connected to each swivel terminal are labeled A, B, and C, to identify each of the three-phase inverter output lines. Figure 8-3(b) shows how the settings of switches A and B create a resultant AC square wave line voltage. Consider output line B as the reference and compare output line A to it. During time period 1, both lines are the same ground potential, so the line voltage AB is zero. During time period 2, reference line B is zero volts and A is positive, which produces a positive line voltage. During time period 3 both lines are the same positive potential, so the line voltage is zero. During time period 4 reference line B is positive and A is zero volts, which produces a negative line voltage.

The transistors in the inverter perform the function of SPDT switches. For example, compare switch A of the switch model to the left parallel branch of the inverter section in Figure 8-2. The swivel terminal of the switch represents the junction between transistors T_1 and T_4. When transistor T_1 is on and T_4 is off, the circuit operates as if the SPDT switch is in the up position. Therefore, a positive voltage exists at junction A. If transistor T_1 is off and T_4 is on, the circuit operates as if the SPDT switch is in the down position and zero voltage is felt at junction A. When all six switching transistors are sequenced properly, they develop square wave signals to simulate three-phase voltage patterns. The line voltages form across terminals AB, BC, and AC, and

(a)

(b)

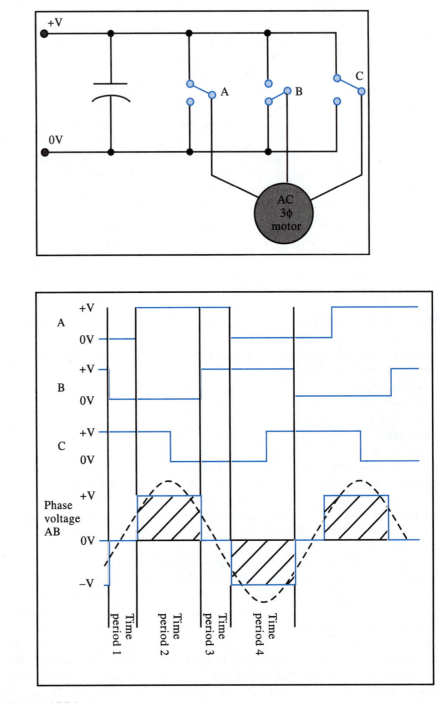

FIGURE 8-3 VVI Inverter

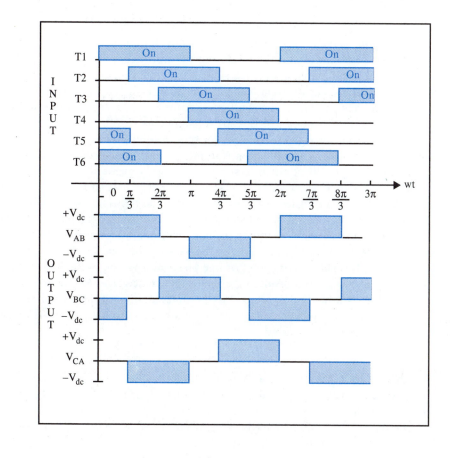

(c)

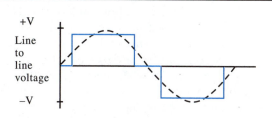

(d)

FIGURE 8-3   *(continued)*

are 120 degrees out of phase with one another. The frequency is controlled by the speed at which the inverter transistors develop the six-step waveform.

To further illustrate the operation, the input and output waveforms are shown in Figure 8-3(c). The signals represented by the input waveforms are applied to the NPN transistors. A positive voltage applied to the base of a transistor turns it on and causes it to operate like a closed switch. A zero volt potential turns the transistor off and causes it to act as an open switch.

During the first step, which lasts a sixth of a cycle, transistors T_1, T_5, and T_6 turn on. During the second step (second sixth of a cycle) transistors T_1 and T_6 remain on, but T_5 turns off and T_2 turns on. During the third step, T_6 turns off and T_3 turns on, while T_1 and T_2 remain on. Each subsequent sixth of a cycle performs in a similar manner; one transistor turns on while another transistor turns off. During the six-step cycle, each transistor turns on and off once. After the sixth step is completed, the cycle is back to its starting point and is ready to begin a new cycle.

The diodes D_1 through D_6 are placed across the collector and emitter leads of each switching transistor. Their function is to protect each transistor from being damaged by a current surge. When a transistor turns off, the magnetic field of the stator coil to which it supplies current will collapse. The result is that an induced current is fed to the collector lead. Instead of flowing through the transistor to the emitter, current flows through the forward biased shorting diode and back to the bus line. In addition to permitting the return of energy to the bus line from the reactive load, the diode also provides a path for regenerative energy fed back by the load.

Figure 8-3(d) shows the output of one phase during one cycle. The smoothing of the rectangular wave is a result of the motor inductance. The AC frequency of the inverter output is directly proportional to the time duration of one complete cycle. The input signals to the power transistors that cause them to turn on or off originate in the *control circuit* of the drive.

CONTROL CIRCUIT OPERATION

The six square wave signals fed into the inverter originate in the control circuit, which is shown in the block diagram in Figure 8-4(a).

The speed control potentiometer at the operator control panel feeds a voltage to the control circuit input terminal. This terminal could be the input of an astable multivibrator. Its function is to generate a continuous square wave signal. The output of the multivibrator is called a clock signal. It is applied to a digital ring counter that consists of six flip-flops (FFs). This type of digital circuit is actually a shift register that first moves bit patterns to the right and then recirculates them from the left.

Observe the waveform diagram in Figure 8-4(b). It is identical to the input waveforms of Figure 8-3(c). During the first sixth of a cycle, FF1, FF5, and FF6 contain a 1 state at their Q outputs. The other flip-flops have 0 states. The bit pattern is 100011. As a clock pulse from the multivibrator occurs, the bits in FF1 through FF5 are shifted one position to the right. Also, the bit from FF6 is recirculated by shifting to FF1. The bit pattern is now 110001. When the next clock pulse arrives, the bits are shifted another place to the right and the FF6 bit is recirculated to FF1. The bit pattern is now 111000. The bits are shifted one position every one-sixth of a cycle, as each additional clock pulse occurs.

When the operator increases the speed, the wiper arm of the potentiometer goes upward. By reducing the resistance, the RC time constant decreases, which causes the output frequency to increase. The result is that the bits circulate faster, which causes the inverter's three-phase output frequency to increase.

AC drives do not use a multivibrator and discrete flip-flops in the control circuit. Instead, they utilize a microprocessor or programmable controller to form the bit pattern fed to the inverter.

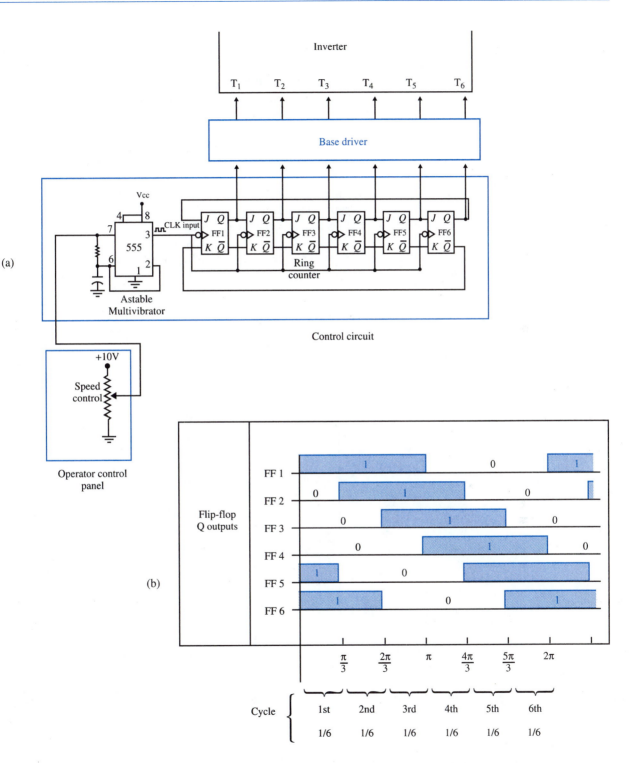

FIGURE 8-4   The control circuit in the drive controller section

BASE DRIVER

The circuitry of flip-flops, microprocessors, and programmable controllers is manufactured on integrated circuit (IC) chips. The output current of ICs is relatively small. If the ICs were connected directly to the inverter power transistors, there would not be enough base current to turn the transistors on. Therefore the IC output leads are connected to a **base driver**. This section amplifies and isolates control circuit signals to make the inverter transistors operate in the appropriate switching scheme.

The circuitry of the base driver is illustrated in Figure 8-5, which shows one of the driver's six NPN opto transistors. When the output lead of the control circuit is in a "0" state, the LED is forward biased. The LED lights as current flows through it. Light shines on the base of the transistor and turns it fully on. Because the transistor acts as a closed switch when it saturates, the base driver output lead tied to the collector will be a low state potential. When the control circuit output lead is high, the LED is not forward biased. Therefore the transistor is in the cutoff mode and acts like an open switch. A high voltage one state potential is then formed at the collector output lead.

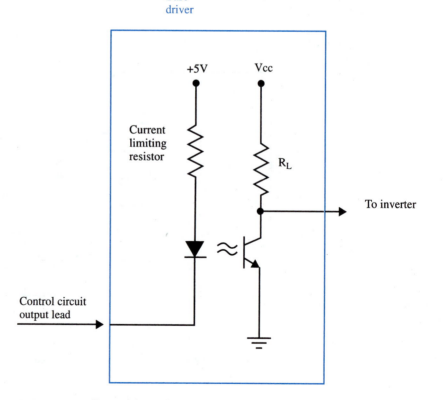

FIGURE 8-5 Base driver circuitry

VOLTAGE/Hz RATIO

As the speed of the motor is changed, a constant torque is required. To achieve this, current (I) that flows through the stator windings of the motor must be kept at a constant level.

It is evident that the speed of the motor can be changed by simply varying the frequency. However, if the frequency alone is varied without changing the motor voltage, the current and resulting torque will change, as is proven by the following formulas:

$$\textit{Inductive Reactance:} \quad X_L = 2\pi fL$$

$$\textit{Ohm's Law:} \quad I = \frac{V}{X_L}$$

If the frequency is lowered and the voltage remains constant, the current will rise dramatically. The result is that the torque will rise by the same proportion. However, the motor may overheat due to overexcitation from the excessive current. To maintain a constant current and torque, the bus voltage must be varied in direct proportion to the inverter output frequency. This variation is called voltage-to-frequency (V/Hz) ratio.

The V/Hz ratio in a VVI drive is controlled in one of two ways: by a phase-controlled rectifier or by a diode rectifer and chopper configuration.

Phase Control

A variable voltage source drive with phase control is illustrated in Figure 8-6. Two of the diodes used by the converter's bridge rectifier network are replaced by SCRs. The SCRs—or *phase control rectifiers*—perform two functions. First, they

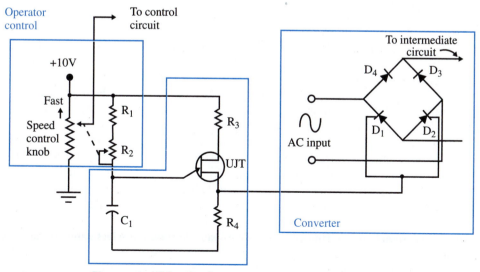

FIGURE 8-6   The phase control circuitry of a VVI AC drive

provide full-wave rectification, rectifying AC power with the diodes. Second, the SCRs control the amount of average DC voltage applied to the converter output. By turning the SCRs on early in the alternations, the average voltage produced at the bridge output is high. The later the SCRs fire in the alternations, the smaller the voltage generated. By controlling the amount of time that elapses before the SCRs are turned on, the output voltage of the converter can be adjusted to any value desired. The instant at which the SCRs fire during an alternation is controlled by the firing circuit in Figure 8-6. The RC network uses a rheostat to supply resistance. It is mechanically coupled to the speed control potentiometer.

Recall that to maintain a constant V/Hz ratio, the inverter's voltage and frequency must increase or decrease together. Suppose that the speed control knob is adjusted to a lower speed setting. The ring counter shifts more slowly, which causes the switching sequence to the inverter to slow down, decreasing the output frequency. The reason the output voltage also decreases is that the resistance of rheostat R_2 increases. As it does, the capacitor takes more time to charge, which causes the UJT to fire later. Since the SCRs turn on late during the alternations, the firing delay angle increases. The result is that the converter's average voltage decreases. The filter components in the intermediate circuit are required for smoothing the pulses.

Overcurrent Protection   If the motor current in the output lines of the inverter rises above the rated value of the drive, it is likely that a malfunction has developed. As a precautionary measure, the drive is designed to keep the current from exceeding a certain level. An overcurrent condition can result when:

1. The physical load that the motor is driving becomes excessive.
2. The starting inertia of the motor load is excessive and the acceleration time is too short.
3. A short circuit exists in the inverter output leads or motor windings.
4. A device in the drive inverter has short circuited.

A protective circuit in the control section monitors the load current. When it becomes too high, the circuit shuts the inverter down by turning off the transistors.

Figure 8-7 shows an overcurrent protection circuit. A current-sensing coil is wound around one of the three-phase output lines of the inverter section. The changing lines of flux around the wire induce a voltage into the coil, which is connected to the primary input of a step-up transformer. The secondary voltage is rectified, filtered, and applied to the inverting input of comparator op amp 1. A reference voltage from the current limit adjust potentiometer is connected to the non-inverting input of op amp 1.

If the output line current is within the rated level, the voltage at the inverting input is less than the non-inverting input. The output of the op amp goes high. This enables the output signals of the ring counter FFs to pass through the AND gates to the base driver. Therefore the inverter transistors go through their switching sequence. When the output line current becomes too high, the inverting voltage becomes greater than the non-inverting voltage. The op amp output goes low and disables the AND gates. This blocks the 1 state signals from passing from the ring counter FFs to the base driver. The drive shuts down because the inverter transistors cannot turn on.

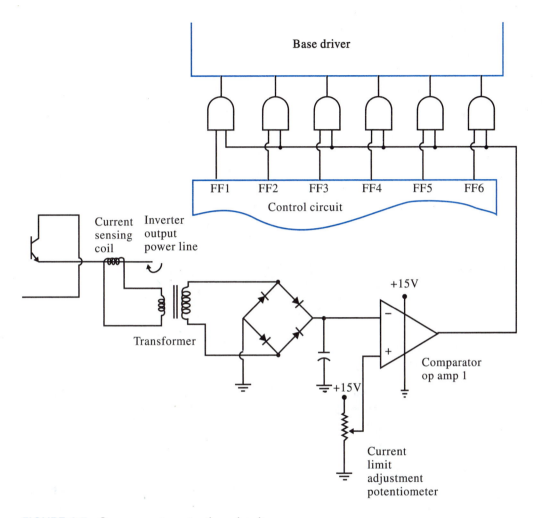

FIGURE 8-7   Overcurrent protection circuit

The current limit potentiometer is used to adjust the circuit to shut down at different current levels.

Auto Boost Circuit   When a motor is being started or is running at a slow RPM, the drive must provide enough current to produce sufficient torque to overcome inertia.

Figure 8-8 shows an **auto boost** circuit that will cause the current to increase when the motor is running at a low RPM. Within the normal operating speed, Q_1 is biased within its active region. As the speed control knob is adjusted to a slower speed setting, the potentiometer voltage at the comparator's inverting input reduces. When it becomes lower than the non-inverting input voltage, the comparator output goes high. The result is that transistor Q_1 turns on harder, charges C_1 more quickly, and fires the SCRs in the converter earlier. The average voltage increases and provides additional current for the motor.

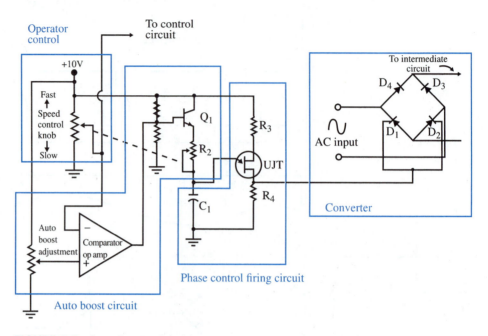

FIGURE 8-8 Auto boost circuitry

The auto boost circuit is usually located in the control section of the drive. The auto boost adjustment potentiometer is used to set the speed at which the added current will be supplied.

Chopper Control

A VVI with **chopper control** is shown in Figure 8-9. It uses a single-phase bridge rectifier in the converter section to produce a pulsating DC voltage. The intermediate circuit has two capacitors and a chopper switch. The capacitor placed before the chopper smooths out the DC pulsations of the converter output. The chopper switches on and off very rapidly. As it does, the ratio of on-time to off-time is changed to vary the bus voltage applied to the inverter section. Turning the chopper off for a large portion of the alternation produces a small average voltage at the bus line; turning the chopper on for a large portion of the alternation produces a large average voltage at the bus line. By varying this ratio, a precise voltage can be produced. A second capacitor is placed after the chopper for smoothing.

The circuit that controls the chopper function is shown in Figure 8-10. An astable multivibrator is used to produce a square wave. Because the current from the 555 IC is relatively small, it cannot provide enough base current to control the standard power transistors in the inverter section. The current limitations of the 555 IC can be overcome by connecting its output line to an opto-transistor. When the square wave output of the IC goes high, it forward biases the LED. The opto-transistor turns on and acts as a closed switch. A low at the IC output turns the LED off. The opto-transistor also turns off and acts as an open switch.

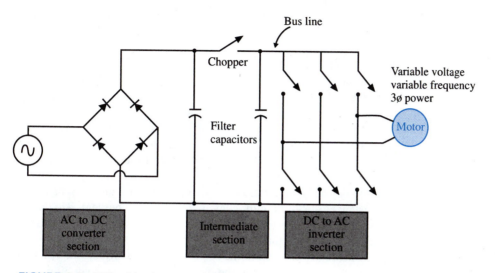

FIGURE 8-9 VVI with chopper control

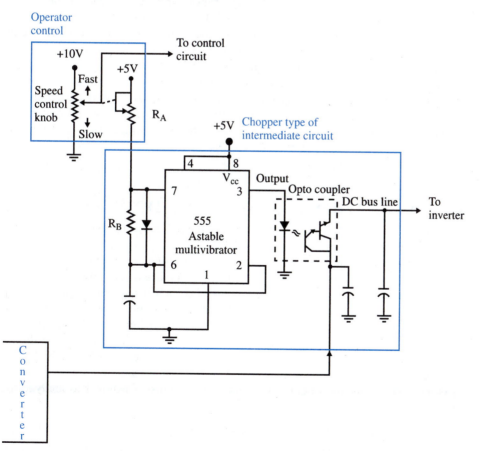

FIGURE 8-10 The chopper control cicuitry of a VVI AC drive

The on-time/off-time ratio of the chopper is controlled by the multivibrator's duty cycle. The duty cycle is the ratio of time the output terminal is high to the total duration of one cycle. The duty cycle is determined precisely by the ratio of resistors R_A and R_B. If the diode was not connected in parallel with R_B, the capacitor would charge up through R_A and R_B. It would then discharge only through R_B (and the IC internally from pin 1 to pin 7). Therefore the duty cycle would be less than 50 percent. With the diode connected, R_B is bypassed and the capacitor charges only through R_A. When the capacitor discharges, its current path is blocked by the reverse biased diode, and therefore only flows through R_B. Depending on the resistance ratio of R_A and R_B, this configuration allows the duty cycle to operate over a range of 5 percent to 95 percent.

One way to change the resistance ratio is by using a rheostat at R_A. When R_A is smaller than the resistance of R_B, the duty cycle is less than 50 percent. This means that the IC output is high for less time than it is low. When R_A is larger than the resistance of R_B, the duty cycle is more than 50 percent.

The rheostat is mechanically coupled to the speed control potentiometer. When the speed control is adjusted to a faster setting, the duty cycle is greater than 50 percent. The result is that the chopper supplies a greater average voltage to the DC bus line.

The VVI drive is usually used in industrial applications where low to medium power is required. Although the motor operates more effectively when it is powered by a sinewave from the supply mains, the square wave output of a VVI drive is usually satisfactory. However, at low speeds, such as below 10Hz, a stepping or cogging motion of the output shaft may occur as torque pulsations develop each time an output transistor is switched. At higher speeds the cogging disappears, but the pulsations may cause vibration or audible noise problems. Also, the non-sinusoidal waveforms shown in Figure 8-3(d) may cause extra heating in the motor windings. This situation may require a derating of the motor by up to 25 percent.

8-4   PULSE WIDTH MODULATION DRIVES

The VVI drive has been largely superseded by the pulse width modulation drive, or PWM. Its circuitry is shown in Figure 8-11.

Since most drives run high horsepower motors, the single-phase bridge used to describe the VVI's operation may not generate enough current. The PWM drive's converter uses a three-phase rectifier to supply the extra power required by large motors. A three-phase rectifier supplies more power than a single-phase rectifier because its pulsations are closer together and therefore, the filtering required is minimized.

A full-wave three-phase rectifier circuit is shown in Figure 8-12. Its six diodes either block or pass current. A wye transformer is the source that supplies both positive and negative alternations of three-phase current to the rectifier.

Three facts about the operation of the rectifier must be known to analyze the circuit:

1. At any given time, two diodes are on and the remaining four are off.

2. One odd-numbered diode (D_1, D_3, or D_5) and one even-numbered diode (D_2, D_4, or D_6) is always on.

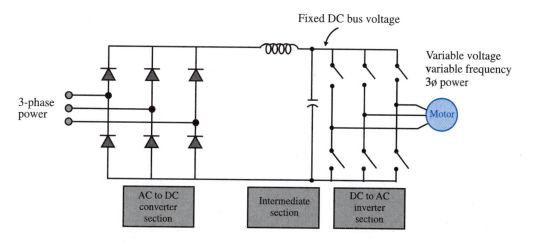

FIGURE 8-11   A pulse width modulated (PWM) drive circuit

3. Electron current flows out of the A, B, or C phase transformer terminals from the most negative voltage, through an odd-numbered diode, through the load resistor, and through an even-numbered diode, to whichever phase terminal—A, B, or C—has the highest positive voltage.

Figure 8-13 graphically illustrates the rectification process in detail. Figure 8-13(a) shows the 208V RMS three-phase voltage supplied from the source terminals of the transformer to the rectifier. The timing chart in Figure 8-13(b) shows intervals when the corresponding diodes are on. Figure 8-13(c) uses closed switches to show which diodes are forward biased during 30- or 60-degree intervals.

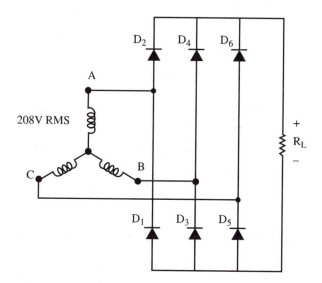

FIGURE 8-12   A wye source supplies power to a three-phase full-wave rectifier

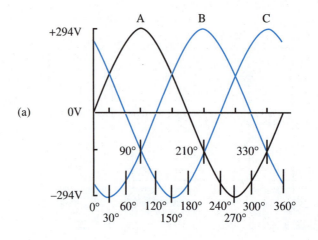

(a)

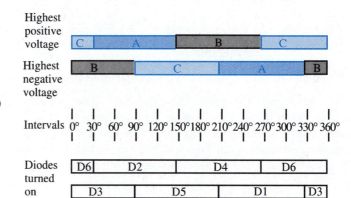

(b)

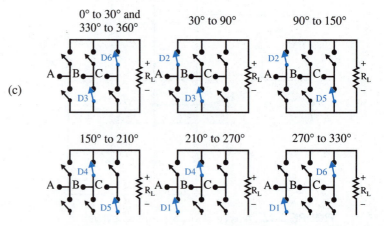

(c)

FIGURE 8-13 A graphical illustration of the three-phase rectification process

Between 0 and 30 degrees, terminal C is most positive, which forward biases diode D_6, and terminal B is most negative, which forward biases D_3. The resulting 30-degree alternation is shown in Figure 8-14. From 30 to 90 degrees, terminal A is most positive, which turns D_2 on, and terminal B is most negative, which turns D_3 on. The resulting voltage pulses form across the load resistor during this, and every, 60-degree period. Since the magnitude of the pulses fluctuates approximately 59 volts (235V to 294V), the rectifier output requires only minimal filtering.

The pulsating waveform from the converter is filtered by the intermediate section. Unlike the VVI drive, which varies the bus-line voltage with frequency, the PWM bus voltage is constant. Therefore it is required that the inverter section provide motor voltage control as well as frequency control.

The inverter section consists of six transistor switches. The output switching scheme is similar to the six-step inversion strategy. The difference is that the transistors are pulsed many times rather than remaining either on or off during the entire alternation. By varying the widths of each individual pulse, the RMS voltage is adjusted. Figure 8-15(a) shows the waveform voltages developed at the inverter A, B, and C output lines. These signals are formed by the transistors switching in a prescribed sequence.

Figure 8-15(b) uses SPDT switches to show how the transistors create the voltage patterns. All of the switches are connected in parallel across a DC supply. Depending on the switch position, either a positive voltage or a zero-volt ground potential connects to the swivel output terminal. In Figure 8-15(a), observe the output waveforms at terminals A and B, and the resultant line voltage that forms across them. Assume that output line B is the reference and compare line A to it:

○ During time periods a, g, i, m, o, and u, both switches are in the down position. Since both lines are the same ground potential, line voltage AB is zero.

○ During time periods b, d, f, h, and j, switch A is up and switch B is down. Output A is positive in respect to B. The result is line voltage AB becomes positive.

○ During time periods c, e, k, q, and s, both switches are up. Because outputs A and B are the same positive potential, line voltage AB is zero.

○ During time periods l, n, p, r, and t, switch A is down and switch B is up. Output A is negative with respect to B. The result is line voltage AB becomes negative.

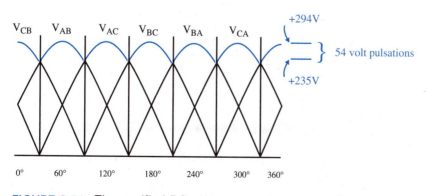

FIGURE 8-14   The rectified DC voltage waveform across R_L

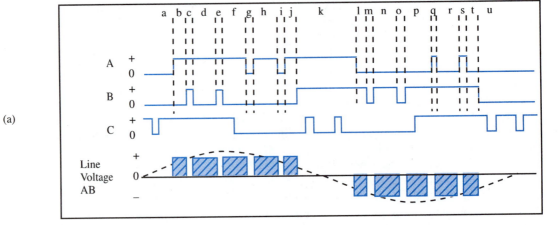

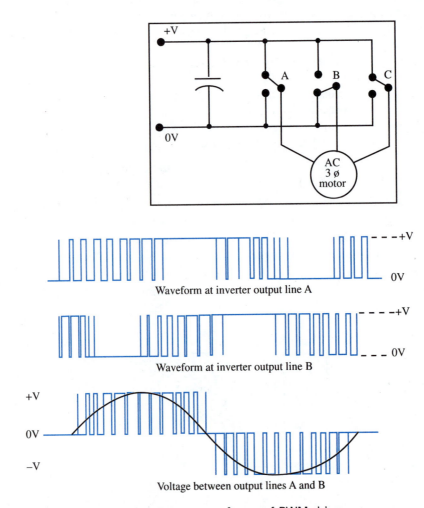

FIGURE 8-15 Inverter section output waveforms of PWM drive

The switching sequence forms a voltage pattern that consists of five pulses during each alternation. The dashed line is the RMS voltage created. The longer the pulse duration, the higher the RMS voltage. The way in which the signals are produced is called *modulation*.

In a typical PWM inverter circuit, more than five pulses are developed during an alternation. Figure 8-15(c) shows an actual square wave pattern on two inverter output terminals and the resultant line voltage waveforms they develop. The shape of the pulses form an RMS voltage that becomes sinusoidal, which causes a current, approximate to sinusoidal, to flow through the motor. The signals that activate the switching transistors originate in the control section.

There are several advantages of the PWM drive over the six-step VVI drive. As previously stated, when the VVI drive is running at a low frequency, the speed of the motor may fluctuate or cog due to the square wave-shaped output of its inverter. A PWM drive will run a motor smoothly at any RPM within its speed range because its inverter can produce a sinusoidal-shaped waveform by generating pulses.

Another disadvantage of the VVI is that its output voltage is controlled by the intermediate circuit. The result is poor speed control because the large filtering capacitor has to be charged or discharged to change the voltage. The PWM can vary the output voltage instantly by changing the number of pulses—and their widths—during each alternation.

The pulses generated from the control circuit originate in a modulator logic board. The board consists of low power signal generating circuits and microprocessor-based LSI logic circuits, which control the performance characteristics of the drive. Pulse width is decreased for lower RMS voltages and increased for higher voltages. Lower frequencies have a greater number of pulses in one alternation. As the frequency increases, the circuits on the logic board select the number of pulses per cycle.

Because the semiconductors in the inverter must switch at extremely high speeds to produce a PWM output, insulated gate bipolar transistors (IGBT) are frequently used. The IGBT transistor is a combination of the bipolar transistor and the MOS-FET transistor. It utilizes the desirable characteristics of each one, including high power capacity, good conductivity, and simple gate control, to provide a high switching frequency.

The standard induction motor is designed for a fixed speed determined by its rated operating frequency—usually 60Hz—and its number of poles. However, the AC drive is capable of achieving motor performance beyond its nameplate rating. Therefore, to prevent motor damage, it is necessary to take precautions when its velocity is above or below its rated speed. Normally the motor is cooled by a fan which is attached to the rotor shaft. When the drive causes the motor to run at a velocity slower than its base speed, the fan will not produce sufficient cooling. As the temperature rises, the insulation temperature ratings are exceeded, which can shorten the life of the motor's insulation or cause the motor to fail. To protect the motor from overheating, a thermal protection device can be connected to the motor winding to stop the motor if its temperature gets too high. Another solution is to add a constant speed, separately-powered blower to the motor to ensure adequate cooling at all speeds. Preventing vibration becomes especially important in higher-speed motors. If the motor is expected to run at speeds greater than 3000 RPM, accurate coupling and shaft alignment are important to reduce vibration, and the rotor must be precision balanced. Special bearings must also be fitted if the motor speed exceeds 7500 RPM.

The motor temperature may also become very high when a drive operates at full speed and full load. Some of the power applied to the motor is lost because it generates heat when operated by the drive's simulated waveform, which is not an exact duplicate of a sinewave. By using a motor with a derated power factor, the excessive heat that is generated will be reduced.

8-5 FLUX VECTOR DRIVES

The operation of VVI and standard PWM drives is adequate for steady-state conditions, or in applications where a slow response speed change is adequate when load variations are encountered. However, there are some specialized applications where abrupt load, speed, and position changes are likely to occur. A flux vector drive is better suited to handle these conditions.

The vector drive is one of the most recent and significant developments in drive technology. It is capable of achieving control over an asynchronous squirrel cage motor so that its performance is similar to that of a DC motor and advanced DC drive. For example, maximum torque is available at all speeds and the dynamic response is quick, thereby precisely maintaining a constant velocity under varying load conditions. In these respects a vector drive far exceeds other AC drives.

The term *vector* is derived from the method in which the drive controls the magnetic flux vector. A vector in mathematics indicates amplitude and phase angle. The vector drive controls the motor by varying the magnitude of the stator flux and adjusts the angle of the stator flux in reference to the rotor.

The primary reason for a DC motor's effective operating characteristics is its physical construction. The way in which the brushes make contact with the commutator causes the resultant flux of the rotor to be perpendicular to the stator field. Because the armature's magnetic field and the stator field are always quadrature (90 degrees to one another), maximum torque is always produced. The drawback of a DC motor is that it is a high-maintenance machine; the brushes require frequent replacement due to their physical contact with the commutator. The AC asynchronous motor is preferrable to the DC motor for most industrial applications because of the low capital and maintenance cost, high efficiency, and long-term dependability.

The fixed 90-degree relationship between the rotor and stator in a DC motor does not naturally occur in an AC induction motor. The induction motor rotor is basically two rings with bars placed between them like a wheel for a squirrel, hence the name squirrel cage. Unlike the DC motor, rotor flux is produced by having a rotating stator field induce current into the bars of the cage. As the stator and rotor fields interact, torque is produced, which causes the rotor to turn. However, the rotor lags behind, and rotates slower than, the stator field. This difference in rotational speed is referred to as *slip*. Slip is measured as an angular velocity expressed as a frequency. Under normal operating conditions, the (slip) frequency is about 3Hz. When a greater load is placed on the rotor, slip increases, more current is induced into the armature, and torque increases. The maximum motor torque is produced when the rotor flux relative to the stator field is at 90 degrees. The drawback of the AC induction motor is that it responds slowly to load changes. It is acceptable for driving such applications as pumps, fans, and compressors. However, when rapid dynamic control of speed, torque, and response to a change in load is necessary, its operation is too crude.

The fundamental problem with controlling an AC induction motor is that the stator and rotor flux cannot be regulated separately. Instead, current induced into the rotor is only responsive to the variation of the frequency and current amplitude supplied to the stator. Recent developments in digital signal processing (DSP) have overcome the difficulty of providing accurate control of the rotor, by manipulating the stator flux. The control scheme of digital signal processing is to constantly monitor and control the amplitude of the stator MMF flux, as well as its position relative to the rotor. A high-performance microprocessor repeats 2000 or more computations per second by using a look-up table to make readjustments if they are required. This method is referred to as *flux vector control.*

A vector drive is similar in construction to a PWM drive. It consists of a power conversion section which uses six input diodes that rectify the three-phase power to a fixed pulsating DC voltage. A capacitor filters the pulsating ripple from the rectifier to a smooth DC bus-line voltage. A set of six transistors in the inverter section convert the DC bus line voltage into a synthesized three-phase PWM voltage to power the motor. The vector drive differs from the PWM drive in using two independent feedback control loops, as shown in Figure 8-16(a). The loop that controls amplitude is referred to as the *current feedback.* It uses a sensor such as a clamp-on ammeter or Hall effect transducer to detect the magnitude of stator current at each of the three stator lines. When an increase in load occurs, the vector drive will not allow the slip to increase. Therefore, the current reduces. The vector drive responds by increasing the amplitude of the current to the stator, as shown in Figure 8-16(b). The result is that the torque increases to meet the demand of the load. The loop that controls relative positioning is referred to as the *speed feedback.* It uses a tachometer, resolver, or encoder to sense the velocity of the rotor. The vector drive's control technique of positioning causes a slip frequency to occur where the flux of the stator and rotor are always perpendicular, thus creating maximum torque. If the load changes, the rotor velocity is altered and the slip varies. When the new rotor RPM is sensed by the feedback detector, the vector drive changes the stator frequency until the slip causes the stator and rotor flux lines to return to the desired 90-degree angle. This method is referred to as *field orientation.*

Figure 8-17 illustrates how field orientation is achieved. Figure 8-17(a) shows half of one sinusoidal three-phase sinewave supplied to the stator coils. Figure 8-17(b) represents the 120-degree spacing of the stator windings, and the direction of current flowing through them at successive 60-degree intervals. For example, at zero degrees, current from lines B and C flows to neutral and combines to flow through line A. The direction of these currents corresponds to the polarity of the instantaneous current. Figure 8-17(c) shows vectors positioned head-to-tail that represent current flow during each interval. Vectors B and C at zero degrees are shown with arrows pointing toward the neutral, while in vector A the arrow points away from neutral. Also, the length of each vector varies to represent the magnitude of current. Figure 8-17(c) also shows the resulting magnetic field at the zero degree interval. Because the stator coils are primarily inductive, the magnetic field lags the current by 90 degrees. As the diagram advances through each interval from zero to 360 degrees, the resulting flux makes one revolution in a counterclockwise direction. Because the stator current shown in the diagram is supplied by a three-phase commercial line, the amplitude and frequency are constant. In a vector drive where the inverter stage artificially produces a synthesized

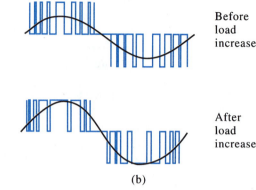

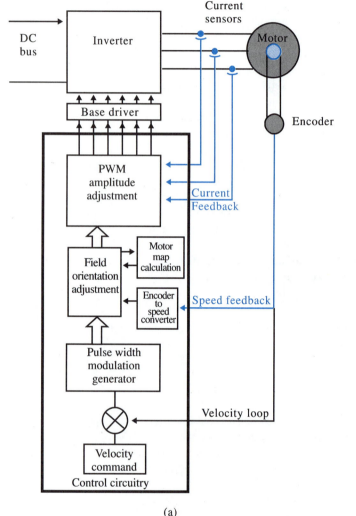

(a)

FIGURE 8-16 Block diagram of the vector drive control

waveform, the sinewave can be disrupted. For example, Figure 8-18(a) shows the stator current suddenly advanced by 30 degrees. The effect is that the stator flux also is instantaneously advanced through 30 degrees. This jump is referred to as a *step change*. Figure 8-18(b) shows the stator current held constant for 30 degrees. The result is that the stator flux is held stationary for one 30-degree interval. The ability to advance or delay the stator currents makes it possible to adjust the stator flux to maintain a 90-degree angle with the rotor, even when a change in the load causes the rotor speed to vary, and disrupts its positional orientation to the stator field.

The feedback signal from the speed loop is used by the DSP operation to cause the stator currents to make an advanced or delayed step change so that the stator and rotor fluxes return back into quadrature. To cause a rotor to vary its speed, a proportional

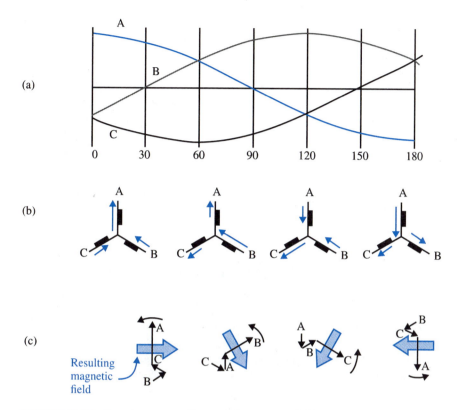

(a)

(b)

(c)

Resulting magnetic field

FIGURE 8-17   Development of a rotating stator field

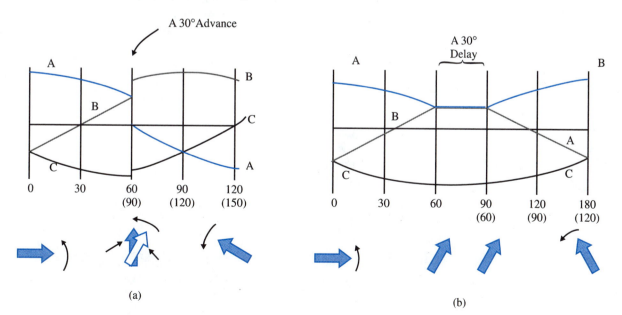

A 30°Advance

A 30° Delay

(a)

(b)

FIGURE 8-18   Affects of stator flux step changes

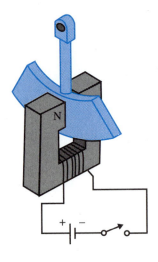

output frequency change from the inverter section is produced. The direction of the motor's rotation is reversed by producing an inverter output signal with the opposite sequence of instantaneous voltage steps. A similar switching scheme is used to apply reverse torque when the motor needs to be stopped quickly. There is, therefore, no special need for independent braking. To hold a motor in a stationary position, the phase current and frequency are held constant. Figure 8-19 illustrates the concept of how the rotor can be held in a fixed position. An axe head made of a conducting metal is suspended within the flux lines of an electromagnet. Any movement results in voltage being induced into the conductor. The current that is produced develops a magnetic field which is opposite to the electromagnet, thus exerting a force to resist the movement. If the current supplied to the electromagnet is removed, the axe will move more freely. Therefore, the rotor, which is a conductor, will also oppose movement if it is placed within the fixed field of the stator.

FIGURE 8-19
Conductor movement
resisted by induction

8-6   PROGRAMMABLE MOTOR DRIVES

Technological improvements are constantly being made in the design of electronic motor drives. The most advanced drives have a functional operator station called a *control panel* to set up, operate, and control the drive. To optimize the operation of the driven load, the control panel replaces potentiometers that adjust parameters and test equipment used to calibrate the drive. These "smart drives" are programmable and use a microprocessor and EEPROM memory chips.

Figure 8-20 shows the control panel for the Model 1352 Allen-Bradley AC drive. All parameters are entered on a keypad simply by pressing buttons. The data are shown on a display and are stored in memory. Fault conditions are also displayed if a malfunction develops. The control panel has input and output terminals for making

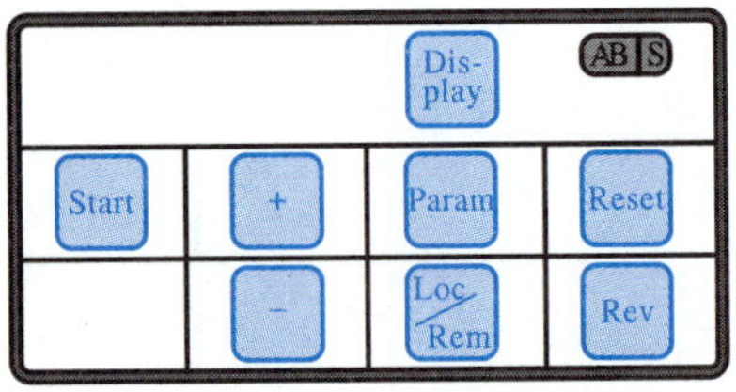

To program the drive parameters, follow the sequence:

1. Push [Param] to select the parameter mode

2. Push [+] or [–] to select the parameter to be changed

3. Push [Display] to show the present value of the parameter

4. Push [+] or [–] to change the value of the parameter

5. Push [Reset] to enter the new value

Pushing the display button allows the operator to monitor:

☐ Drive frequency reference(s)
☐ Output frequency
☐ Motor load
☐ Output voltage

A diagnostic code is displayed if one of the following conditions exits:

☐ Overcurrent
☐ Short circuit
☐ Undervoltage
☐ Overvoltage
☐ Overtemperature
☐ Semiconductor fault
☐ Processor fault

FIGURE 8-20 Control panel operation for the Allen-Bradley Model 1352 programmable AC drive

hardwire connections to external devices or for interfacing to other equipment. To further enhance the troubleshooting capabilities of the drive, the input and output points are monitored and shown on the control panel display.

Some of the standard controls and adjustments typically on the control panel are:

Minimum and Maximum Frequency Adjust.   These two controls set the drive's overall frequency.

Volts per Hertz.   This control sets the voltage level for every cycle per second. The V/Hz ratio is affected by this setting.

Operator Speed Control.   The actual speed of the motor is adjusted by this control. Minimum and maximum speed settings can also be programmed if a desired speed range is required.

Acceleration and Deceleration Rate.   These controls are used for starting and stopping loads gently. They adjust the output frequency rate of change that is applied to the inverter switching transistors. The minimum amount of acceleration or deceleration time is less than one second. The maximum time is as high as several minutes. Minimum settings are usually for lighter loads and maximum settings are for heavy loads.

Auto Boost Adjustment.   This control sets how much current is supplied to the motor as it starts or when it is running at low frequencies. All these parameters can be protected by a security code to prevent unauthorized or accidental interference. Data can also be entered from a remote location through a serial communication link by a programmable controller or computer.

CHAPTER PROBLEMS

1. __ T __ F   One common method of slowing the speed of an AC motor is to reduce the applied voltage.

2. Which of the following conditions may cause an AC motor that is controlled by an AC drive to overheat? ____
    a. The RPM are lower than the rated speed of the motor.
    b. A non-sinusoidal six-step output is applied to the motor.
    c. Both a and b.
    d. None of the above.

3. Determine the RPM of an eight-pole AC motor when 60Hz is applied.

4. AC drives change the speed of an AC induction motor by which of the following methods: ____
    a. Varying the bus voltage.   c. Varying the applied frequency.
    b. Varying the armature current.   d. Changing the number of poles.

5. AC drives are also commonly referred to as _____.

6. List the two major differences between the VVI drive and the PWM drive.

7. Match the following names of the VVI section with the appropriate circuit functions that they perform:
   a. Converter   b. Intermediate Circuit   c. Inverter
   ____ Rectifies AC into pulsating DC.
   ____ Filters pulsating DC signals into a pure DC voltage.
   ____ Converts a DC voltage into a simulating AC waveform.

8. How many step voltage levels are produced by the inverter section of a VVI AC drive? ____

9. How many times does each transistor in the inverter section of a VVI drive turn on and off during the production of one cycle? ____
   a. 1   b. 2   c. 3   d. 6

10. List two types of control circuit devices that feed switching signals to the transistors in the inverter section of a VVI AC drive.

11. List two circuits that maintain a constant V/Hz ratio as frequency is varied in a VVI drive.

12. List four malfunctions in a VVI AC drive that cause an overcurrent condition.

13. __ T __ F   The auto boost circuit is used to help accelerate an AC motor when it is starting or running at a low speed.

14. The duty cycle is the ratio of time the output terminal is at a _____ (low, high) logic level compared to the total time period of one cycle.

15. The motor voltage and frequency are controlled exclusively by the inverter section of which of the following types of AC drives? ____
   a. VVI   b. PWM   c. Vector Drive

16. As the pulse width of a PWM signal decreases, the RMS voltage _____ (increases, decreases).

17. The inverter section of the Vector drive produces a ____ output.
   a. Six-step voltage   b. PWM

18. Which feedback loop in a Vector Drive performs the field orientation function? ____
   a. Speed   b. Current

19. List a device that produces the feedback signal for the following control loops in a vector drive: ____
   a. Speed   b. Current

20. List five parameters that are typically adjusted by programmable AC motor drives.

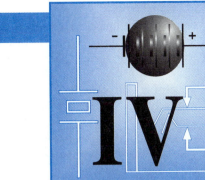
PROCESS CONTROL

Process control is an industrial field that involves the control of variables that affect products as they are being manufactured. Specifically, these variables, called *process variables,* include pressure, temperature, flow, and levels of gases, liquids, and solids. During the manufacturing process, variables are manipulated by equipment to alter a raw material until it reaches the desired condition. The number and types of variables being manipulated depend upon the product being manufactured. Products produced in process industries include such things as chemicals, refined petroleum, treated foodstuffs, paper, plastics, and metals.

During the early years of industrial manufacturing, the process variables were controlled manually. Today, the same operations are performed by automated systems that require only minimal human intervention. The hardware used in an automated system is called *instrumentation* equipment.

An instrument used in process control directly performs one or more of the following three functions:

Measurement During the manufacturing process, it is often necessary to monitor the existence or magnitude of a process variable. Measuring instruments perform this function. The measurements may be read, used for control purposes, or stored as data for use at a later time.

Control Another function of an instrumentation device is to ensure that a process variable is maintained at a specific value or within specific limits. For example, if a disturbance causes the controlled variable to deviate from the set point, the controller must call for a corrective action. The control section is the brain of the automated system, which compares the set point to a feedback signal from the measuring device. In response it provides an appropriate signal to the output actuator device, which makes any necessary changes.

Manipulation   The final control element is the actuator, or muscle of an automated system. Its function is to manipulate energy or flow of materials at a desired rate.

An example of an automated system is a closed-loop pneumatic system widely used in industry to power handtools and machines. As the pressure of compressed air in a storage tank decreases, less force is applied to the sensing element. The sensing element performs the measurement function. At a given pressure, the element causes an electrical switch in the control section to trip. The switch is used to activate or deactivate an actuating device that performs the manipulation function. In this case, a compressor used as a final control element turns on or off to maintain a pressure within a predetermined range.

Through the continuous control of systems and processes, automated production provides the advantages of better product consistency, more precise tolerances, and reduced waste. The improved efficiency it offers cuts energy cost, which increases profit and minimizes conservation concerns.

This section of the book consists of five chapters that provide information on the operation of process control and instrumentation equipment. The first four of the five chapters define the properties and characteristics of one of four process variables: pressure, temperature, level, and flow. They also describe various types of instrumentation equipment used for measuring and manipulating the variable to maintain a desired condition. The measuring instruments are both mechanical and electronic. The electronic devices are more commonly referred to as *sensors*. Technological advancements enable these devices to be more accurate, more reliable, and less expensive than most of the mechanical measuring instruments that they are replacing. Sensors are designed to produce analog or discrete output signals that are compatible with the computer-based equipment used to perform the control function of most closed-loop systems. The fifth chapter of this section describes the operation of specific types of sensors that detect the presence or absence of an object. Some manufacturers call these sensors *switches,* referring to the on or off voltages that their outputs produce in response to the objects they detect. These sensors are used in both process and motion control applications.

The selection of which measuring instrument to use is often based on such considerations as the type of material being measured, the instrument's cost, its accuracy, its range, its maintenance requirements, and its life expectancy.

CHAPTER 9

Pressure Systems

CHAPTER OBJECTIVES

At the conclusion of this chapter, you will be able to:

○ Define pressure.

○ Define fluid.

○ Given force and area, calculate the pressure exerted by a fluid.

○ Identify five factors that affect the pressure exerted by a liquid.

○ Calculate pressure by using specific gravity and depth values for a liquid in a container.

○ Identify three factors that affect the pressure exerted by a gas.

○ List the reference value for gage, absolute, and vacuum pressures.

○ Convert psia to psig, and psig to psia.

○ Calculate differential pressure.

○ Identify the difference between direct and indirect measurements.

○ Describe the operation of the following non-electrical measuring devices:

Barometer	Diaphragm
Manometer	Bellows
Bourdon tube	Capsular

○ Describe the operation of the following electronic pressure sensors:

Semiconductor Strain Gauge	Traverse Voltage Strain Gauge

○ Explain the operation of the following pressure systems:

Hydroelectric	Steam
Pneumatic	Static distribution
Vacuum	

INTRODUCTION

Many products manufactured in industry result from a process that involves **pressure**. Pressure is defined as force exerted by a **fluid** over a unit of surface area. Different forms of fluids include liquids, gases, steam, or air. They are the medium used to transfer energy from one location to another within a confined network of pipes, tubing, and vessels. Various mechanical devices control the fluids that ultimately provide the power to perform some type of work. Pressures that fluids exert in industrial systems range from a near-vacuum level up to 10,000 psig and above.

Various industrial applications involve increasing or decreasing the pressure exerted by a fluid. Reasons for increasing pressure in a process system are:

1. To perform work with a machine such as a hydraulic-powered press or a pneumatic air drill.

2. To move fluids through pipes or to a higher elevation by overcoming friction and gravity.

3. To transform a fluid into a desired physical state. For example, a nitrogen gas will become a liquid when the pressure is high enough.

Reasons for decreasing pressure in a process system are:

1. To perform work, such as lifting and moving sheets of paper in a printing press operation.

2. To draw liquid into a vessel. Lowering the pressure inside the container helps to fill it with a liquid.

3. Changing the physical state of a fluid, such as causing a liquid to vaporize.

To perform these various operations, it is often necessary to control the pressure accurately. Closed-loop systems that perform this function may maintain the pressure at some specified value. They can also be designed to ensure that pressures are kept within a required range. For example, if a sensor detects that the pressure has moved out of a safe range by exceeding the upper limit, the system will either set off an alarm or shut down the equipment responsible for the deviation.

9-1 PRESSURE LAWS

Pressure is measured as force per unit area. In the English system, force is measured in pounds, and unit area for pressure measurements is the square inch. Therefore, pressure is commonly expressed in terms of pounds per square inch, or *psi*. The formula for pressure is:

$$\text{Pressure (P)} = \frac{\text{Force (F)}}{\text{Area (A)}}$$

9-2 PROPERTIES OF A LIQUID

The molecules of a liquid are closely attracted to each other, which is why substances in this state are incompressible. Its molecules are constantly moving, slipping and sliding past one another. This enables liquids to flow and take the shape of a container. The ability of a liquid to accomplish this action is related to a property called **viscosity**.

Liquids exert a pressure, which is affected by several factors: height, weight, temperature, atmospheric pressure, and mechanical machines.

HEIGHT

The height of the liquid affects the pressure it exerts. The term *head* is commonly used to describe the height of a liquid above the measurement point. Head is given in inches, feet, or other units of distance.

WEIGHT

Another factor that affects liquid pressure is its weight. Different liquids have different weights due to their **density**. Density is defined as the weight of a certain volume of liquid, expressed in pounds per unit volume.

Hydrostatic Pressure

By multiplying the height of the liquid by its density, the pressure, called **hydrostatic pressure**, can be calculated.

$$\text{Pressure} = \text{Height} \times \text{Density}$$

For example, the head of water inside a tank can be determined by measuring the height and obtaining the weight density of water. Figure 9-1 shows a tank with a water level of 15 feet. The weight of one square inch of water one foot high is 0.433psi. The hydrostatic pressure is computed by the following calculation:

$$\text{Hydrostatic Pressure (head)} = 15 \text{ feet} \times 0.433\text{psi} = 6.495\text{psi}$$

Specific Gravity

If a liquid other than water is in the container, allowances must be made for its **specific gravity** (S.G.). Specific gravity of a liquid indicates how much lighter or heavier it is compared to water at 60 degrees Fahrenheit. This relationship is calculated by dividing the weight of a specific volume of liquid into the same volume of water. Because water is used as a standard, its specific gravity is 1.0 If a liquid is lighter than water, its specific gravity will be less than 1.0. An example is ethyl alcohol, with an S.G. of 0.79. A liquid heavier than water has a specific gravity number greater than 1.0. An example is mercury with an S.G. value of 13.57. Therefore, the hydrostatic pressure of 10 feet of mercury is determined by the following calculations:

Step 1: Determine the weight of a one-foot column of mercury using its specific gravity value and the weight of an equivalent volume of water.

$$\text{Mercury Weight (Density)} = 13.57 \text{ (S.G.)} \times 0.433\text{psi (Water)}$$
$$= 5.876\text{psi}$$

Step 2: Multiply the density of a one-foot column of mercury times the level of the mercury.

$$\text{Pressure} = 5.876 \times 10\text{ft}$$
$$= 58.76\text{psi}$$

Hydrostatic Pressure

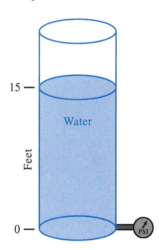

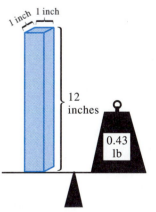

FIGURE 9-1 Hydrostatic pressure is developed by the weight of a column of liquid

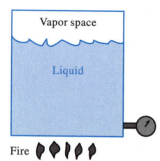

FIGURE 9-2 As heat expands the liquid, the vapor in the enclosed space compresses, creating a larger pressure at the bottom of the vessel

TEMPERATURE

The temperature of a liquid affects the pressure it exerts. Increasing the temperature expands the liquid (i.e., its molecules move farther apart) and reduces its density. If the liquid is in an open vessel, the hydrostatic pressure remains the same, because the density reduces at the same rate at which the level rises; the decrease in density and the increase in volume cancel each other to keep the weight of the container constant. However, if the liquid is confined in a closed vessel, as shown in Figure 9-2, the pressure rises. As the liquid expands, it compresses the air in the vapor space above the liquid. The compressed air applies a larger force on the liquid, which adds weight to the liquid, causing its psi pressure to increase at the bottom of the vessel.

ATMOSPHERIC PRESSURE

The earth is surrounded by its atmosphere, a layer of air approximately 100 miles thick that is pressed against its surface by gravity. Under normal conditions, the weight of a one-square-inch column of air from the top of the layer to sea level is 14.7psi. At an elevation higher than sea level, such as Mexico City (altitude 7,800 feet), the atmospheric pressure is only 11.1 psi, because the column of air above the earth's surface at that location is shorter. Figure 9-3 illustrates the difference in hydrostatic pressure between liquids in Boston and Mexico City.

The pressure from the atmosphere also will exert a force on a liquid in an open vessel.

Atmospheric weather conditions also affect the hydrostatic reading of the liquid in an open vessel. The reading will decrease if a low pressure front has moved into the region.

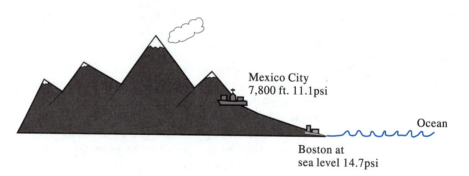

FIGURE 9-3 The higher elevations produce less atmospheric pressure

MECHANICAL MACHINES

The pressure of a liquid can be changed by a mechanical machine. For example, pumps are used to move liquids. If the liquid is moved out of an enclosed space, the pressure reduces. When the liquid is pumped into a confined area, the pressure increases.

9-3   PROPERTIES OF A GAS

Gases are another type of fluid used in industrial process applications. A gas can be air from the atmosphere, vapor, or steam.

Unlike molecules of solids and liquids, which remain attracted to one another, gas molecules remain separate. They are also constantly moving at a very rapid speed, crashing both into each other and into nearby surfaces. The molecular energy of the gas causes it to take the shape and fill the volume of its container. Unless the gas is confined in a container, it will disperse. Most of the volume occupied by a gas is space. If the space was eliminated, the gas molecules would be in contact with one another and become a solid or liquid.

Because of their high degree of molecular activity, gases take the shape of their container, and exert equal pressure in all directions on its walls. The factors that affect this pressure are temperature of the gas, volume of the container, and air removal from the container.

TEMPERATURE OF THE GAS

In a gas, the molecules are constantly moving, crashing into each other and the walls of the container. In a confined vessel, billions of molecules collide each second. A pressure gage placed in a gas container will interpret these collisions as a single pressure. As the temperature of the gas increases, its molecules move faster and collide more frequently in a given span of time. The result is that the pressure increases proportionately with the rise in temperature.

If the temperature of a gas is increased without its being confined in a sealed container, it will expand proportionately and the pressure will remain constant.

VOLUME OF THE GAS CONTAINER

When the area of the enclosed container is decreased, the space between the gas molecules is reduced. This action is called **compression**. By decreasing the space between the molecules, a proportionately greater number of collisions will occur in a given span of time, resulting in a higher pressure. The temperature of the gas will also increase. Compressed gas is stored in a confined tank at a higher pressure than atmosphere. The pressure also increases as an additional amount of gas is pumped into the system. Compression can be accomplished by a mechanical compressor, as shown in Figure 9-4. This device is a cylindrical container with a close-fitting piston and a seal. This piston is driven in a reciprocating manner by a crankshaft connected to a motor. An inlet port allows air in from the atmosphere. The piston pushes the air out of the exhaust port into an enclosed system with a fixed volume.

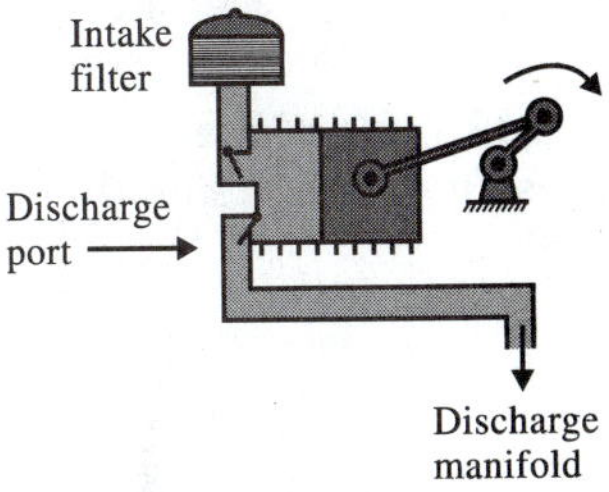

GAS REMOVAL FROM A CONTAINER

If any gas is removed from a sealed container, its pressure will become less than the atmospheric pressure surrounding the vessel. Any reduction of pressure compared to the atmospheric pressure is a partial **vacuum**. If the gas is completely removed, a full vacuum exists.

The vacuum is created by a pump, as shown in Figure 9-5. It operates on the same principle of the compressor. The difference is that the intake port is connected to the enclosed system, and the discharged air is pumped out of the exhausted port into the atmosphere. Most systems require only a partial vacuum to perform the desired operation.

FIGURE 9-4 Single stage one-cylinder compressor

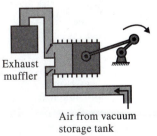

FIGURE 9-5 Single stage one-cylinder vacuum pump

(a)

Gage
pressure
(reference is
atmospheric
pressure)

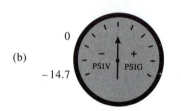

(b)

Compound gage
pressure
(reference is
atmospheric
pressure)

FIGURE 9-6 Gage
pressure measurement
scales

Absolute
pressure
[reference is
a vacuum
(absolute zero)]

FIGURE 9-7 Absolute
pressure measurement
scale with a vacuum
reading of zero as the
reference

9-4   PRESSURE MEASUREMENT SCALES

Instruments that measure pressure use one of four pressure scales: *gage pressure, absolute pressure, differential pressure,* or *vacuum pressure.* What distinguishes each from the other is the reference pressure it uses.

GAGE PRESSURE SCALE

Instruments that utilize the gage pressure scale use atmospheric pressure as the reference point. If the sensing element is exposed to the atmosphere, the measurement scale records zero. The units of measurement are recorded in *psig* (pounds per square inch, gage).

Gage pressure is either positive or negative, depending upon its level above or below the atmospheric pressure reference. A gage pressure instrument will read +30psi when measuring an inflated tire with 30 pounds per square inch of air pressure. This value indicates a positive pressure 30psi above atmospheric pressure. A negative gage pressure indicates a pressure in pounds per square inch below atmospheric pressure. A negative pressure of -14.7psi indicates a full vacuum. A gage pressure scale that makes a positive reading only is shown in Figure 9-6(a). A compound gage pressure scale, which makes both positive and negative readings, is shown in Figure 9-6(b).

ABSOLUTE PRESSURE SCALE

Instruments that use the absolute pressure scale are referenced to absolute zero, or the complete absence of pressure. Since it is not possible to have a pressure less than a vacuum, absolute pressure readings are only positive values. If the sensing element is exposed to the atmosphere at sea level, the measurement scale will read 14.7 pounds per square inch. The units of measurement are recorded in *psia* (pounds per square inch, absolute).

Absolute pressure readings are generally more accurate than gage readings. The reference of gage pressure instruments is not consistent: atmospheric pressure fluctuates with weather changes and altitude. With absolute pressure instruments, the reference point is consistent. A pure vacuum is the same at sea level as it is on top of Mt. Everest. An absolute pressure scale is shown in Figure 9-7.

To convert gage to absolute pressure, simply add atmospheric pressure to the psig pressure value. To convert absolute to gage, subtract atmospheric pressure from the psia measurement. The diagram in Figure 9-8 shows a comparison of both gauges when measuring pressures under various conditions.

PROBLEM

A gage reading of 30psig is taken. What will an absolute pressure instrument read under the same conditions?

SOLUTION

Absolute Pressure = Gage Reading + 14.7

= 30psig + 14.7

= 44.7psia

PROBLEM

An absolute reading of 60psia is taken. What will a gage pressure instrument read under the same conditions?

SOLUTION

$$\text{Gage Pressure} = \text{AbsoluteReading} - 14.7$$

$$= 60 - 14.7$$

$$= 45.3 \text{psig}$$

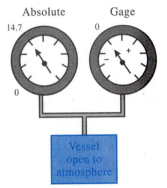

DIFFERENTIAL PRESSURE

Pressure readings are also measured in units of differential pressure, given in *psid* (pounds per square inch, differential), or Δ P. Differential pressure is used to express the difference in pressure between two measured pressures. It can be determined by subtracting the lower reading from the higher reading. The calculation must be made by using values from the same type of measurement scale.

VACUUM PRESSURE

Instruments that measure a vacuum use a scale that begins at atmospheric pressure, just as gage pressure, but works its way down to a vacuum. In the United States, the most common vacuum scale is listed in units of inches of mercury *(in Hg),* as shown in Figure 9-9. The gauge reads zero when measuring atmospheric pressure, and 29.92 in Hg when measuring a complete vacuum. This unit of measurement is based on a barometer tube that uses mercury to indicate atmospheric pressure. The operation of this device is described in the next section.

FIGURE 9-8 Comparison of readings by absolute and gage measurement scales at a vacuum and **at** atmospheric pressure

9-5   PRESSURE MEASUREMENT INSTRUMENTS

Many manufacturing operations require that specific conditions exist. For example, to combine a powder and a liquid in preparation for a food blending operation, a pressure condition within a few pounds may be required. Pressure above or below the required range can ruin the food. Excessive pressure in a system can rupture equipment or cause an explosion.

Measuring instruments are used to monitor pressure conditions so that corrective action can be taken if necessary. The forces created by the pressure produce a deflection, a distortion or some change in volume or dimension of the instrument's sensing element. The physical alteration of the element provides the movement that is necessary to indicate the pressure reading.

Measuring instruments that detect pressure, temperature, level, and flow conditions are classified by whether they make the measurements *directly* or *indirectly*. For example, the level of liquid in a tank can be measured directly with a dipstick. When measurements are taken indirectly, some variable other than the liquid is read. For example, by making an indirect measurement of a tank's weight, the level of its contents can be

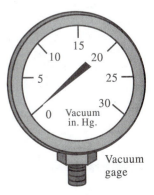

FIGURE 9-9 A vacuum pressure measurement scale

determined. This method is also referred to as an **inferred measurement.** The weight is used to *infer* a level measurement.

Both non-electrical measuring devices and electronic sensors used to measure pressure are described below.

9-6   NON-ELECTRICAL PRESSURE SENSORS

Non-electrical pressure sensors fall into two categories: liquid column gauges and mechanical gauges.

LIQUID COLUMN GAUGES

As previously stated, it is possible to measure pressure by monitoring the height of liquid in a column. For example, a pressure of 0.433psi will support a twelve-inch column of water. This method, called head pressure, is very simple and accurate. For this reason, it is often used to calibrate other types of pressure gauges or instruments.

The operation of this measuring device is based on the same principle as the barometer that measures atmospheric pressure, shown in Figure 9-10. It consists of a glass tube filled with mercury that is sealed at one end and open at the other end. Mercury starts to drain out of the tube when it is positioned vertically with its open end inserted into a dish of mercury. When there is enough atmospheric pressure exerted on the mercury in the dish to support the mercury in the tube, it will stop draining. The gap that forms above the mercury is a vacuum. Changes in pressure cause the mercury in the tube to rise or fall. The larger the pressure, the higher the mercury will rise. A scale on the side of the glass tube allows the height of the liquid to be read directly. At sea level, 14.7psi of atmospheric pressure will support 29.92 inches of mercury.

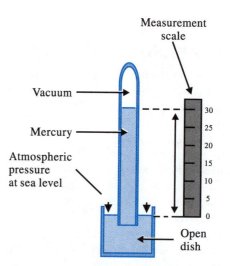

FIGURE 9-10 A barometer reading of 29.92 inches of mercury at sea level

In addition to being affected by pressure, the level of the liquid is also influenced by its density. Because it takes less pressure to move liquids lighter than mercury, water is often used in liquid column gauges. It is often used to measure very low pressures, vapors, pressures below atmospheric, or a vacuum. Mercury is typically used to detect and indicate higher pressures. Liquid column devices that use water as the medium should never be exposed to temperatures below freezing.

Manometer

The most common liquid column device used to measure pressures is the manometer. It consists of a glass tube bent in the shape of a U so that there are two columns. Each column is exposed to a different pressure source. To determine the amount of pressure, read the rise of liquid in one column and the drop in the other, and add them together. There are several types of manometers used to measure various types of pressure.

Gage Pressure   A manometer that measures gage pressure uses one column as a reference. The end of the reference tube is open so that it is exposed to atmospheric pressure. The other column is connected to the process being measured. The pressure measurement is taken by reading the difference in height of the two columns on the scale. If the level in the reference column is higher than the liquid in the measurement column, a positive value above atmospheric pressure is read. The opposite situation indicates a negative value below atmospheric pressure. The maximum negative pressure reading indicates a vacuum. Figure 9-11 shows a U-shaped manometer and how a gage pressure manometer measures both positive and negative readings.

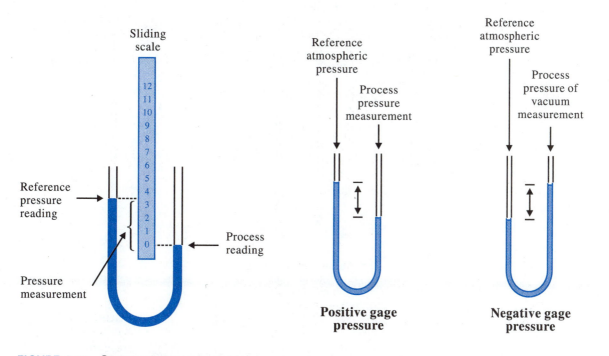

FIGURE 9-11   Gage pressure manometer

Absolute Pressure   A manometer that measures absolute pressure of a vacuum has a reference column with the end of the tube sealed. The closed end creates a vacuum and allows an evacuation space to form. The other column is connected to the measured pressure. If a high pressure exists, it will push the level in the measured column downward and the level in the reference column upward, thus creating a large level difference between both columns. If a vacuum exists, the levels in both columns will be the same. Figure 9-12 shows how an absolute pressure manometer makes measurements of both a vacuum (Figure 9-12(a)) and a positive pressure (Figure 9-12(b)).

Differential Pressure   The manometer can also be used to measure differential pressure. Simply connect the open ends of each column to the two pressure sources being compared, as shown in Figure 9-13. The level differences of the liquid provide the pressure reading.

MECHANICAL GAUGES

Because of their durability, mechanical pressure gauges are often preferred over liquid-filled glass gauges. They are also inexpensive and reliably accurate. Mechanical gauges are constructed by using two major parts, a sensing element and an indicator. The element is elastic and changes form when exposed to pressure. The element is mechanically connected to an indicator device, such as a needle-shaped pointer. As the pressure changes, the element's shape changes and moves the pointer over a scale to indicate a reading.

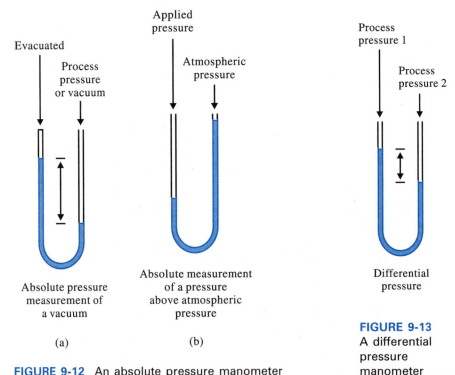

Evacuated

Process pressure or vacuum

Absolute pressure measurement of a vacuum

(a)

Applied pressure

Atmospheric pressure

Absolute measurement of a pressure above atmospheric pressure

(b)

FIGURE 9-12 An absolute pressure manometer

Process pressure 1

Process pressure 2

Differential pressure

FIGURE 9-13
A differential pressure manometer

There are several types of mechanical gauges. They differ primarily by the type of device each uses as the sensing element.

Bourdon Tube Gauge

The Bourdon tube gauge illustrated in Figure 9-14 shows a C-shaped metal tube as the element. The tube is hollow, sealed at one end, and open at the other end. The open end is exposed to the pressure being measured. An increase in process pressure creates a higher pressure inside the tube, while the pressure outside the tube remains the same. This increase in differential pressure causes the coil to unwind. As it becomes straighter, the needle coupled to the tube will move toward the right of the scale to indicate a higher pressure reading. A decrease in measured pressure reduces the differential pressure between the inside and outside of the tube. The tube returns toward its original shape, and the needle moves to the left.

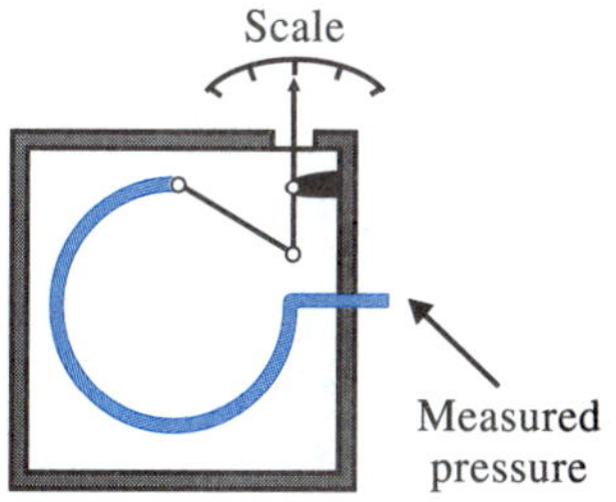

FIGURE 9-14   A Bourdon tube gauge

Diaphragm Gauge

The diaphragm gauge in Figure 9-15(a) measures absolute pressure. It uses a flat flexible material that bends or flexes when exposed to pressure. One side of the diaphragm is connected to the process being measured and the other side is exposed to a vacuum, which is the reference pressure. The diaphragm element bends toward the side that has

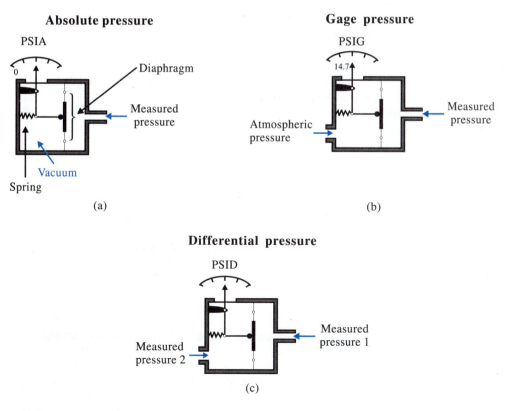

FIGURE 9-15   The diaphragm mechanical pressure gauge

the lowest pressure and pushes against an opposing spring. Also referred to as a *load spring,* its strength determines the range and sensitivity of the instrument. The spring also helps to return the element to its original shape as the pressure is reduced. The diaphragm element is mechanically connected to a pointer that indicates the pressure reading.

Figure 9-15(b) shows a gage pressure measuring device. One side of the diaphragm is exposed to atmospheric pressure as a reference. The other side of the element senses the process pressure. The element is pushed toward the side with the least pressure. Figure 9-15(c) shows a differential pressure measuring device. The differential pressure gauge does not have a reference. It only measures the difference between two pressures. The construction of the differential gauge is identical to the gage device. Each side of the element is connected to one of the two pressures being measured. Instead of having a meter face with a gage scale, a differential scale is used.

For low pressure readings, a non-metallic element such as teflon is used, often without an opposing spring. Because of its resiliency, it is able to distort when lower pressures are applied. Metal elements are not as flexible, but bend enough when measuring high pressure. Because metals tend to withstand exposure to harmful elements, they are used in environments that are corrosive and the temperatures are high. Diaphragm gauges sense pressure from 30 inches Hg vacuum to 6000psig.

Bellows Gauge

Flat diaphragm elements have a limited range of motion and produce non-linear readings. By using sensing devices with pleated walls, the flexibility and movement increase. An elastic element that resembles an accordion bellows is shown in Figure 9-16. When applying the measurement pressure to the inside of the element, it expands and causes a larger needle deflection than the flat element. The bellows material may be brass, phosphor bronze, or stainless steel, depending on required range, sensitivity, corrosion resistance, and cost. An opposing spring is employed to control range and sensitivity. Bellows instruments measure pressures that range from 30 inches Hg vacuum up to 500psig.

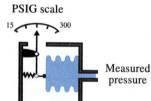

FIGURE 9-16 A bellows pressure gauge

9-7   ELECTRONIC PRESSURE SENSORS

New advancements in electronic technology have resulted in the development of electronic sensors to measure pressure. These devices are more accurate, more reliable, and less expensive than many of the mechanical measuring instruments they are replacing.

SEMICONDUCTOR STRAIN GAUGES

The strain gauge is used to detect the strain on a body caused by forces due to pressure. This device is typically constructed with a Wheatstone bridge arrangement. It is shown in both schematic and exploded views in Figure 9-17. The four resistive elements that make up the bridge network are made of piezoelectric semiconductors. The elements are bonded to a pressure-sensitive diaphragm. A pressure change causes the elements to expand. A compressive strain on the diaphragm will cause the element to contract. The distortion of the elements produces a differential resistance change which is measured by applying a constant excitation voltage to the bridge. The diaphragm deflection is an analog output voltage up to 250 millivolts proportional to pressure. Without pressure the bridge output is 0 volts because the four elements are balanced.

The formula $R + \Delta R$ and $R - \Delta R$ represents the elements' actual resistance value without pressure applied to the diaphragm. The R represents the resistor value without pressure applied to the diaphragm. The ΔR represents the change in resistance due to an applied pressure. When pressure is applied, all four resistors' elements change the

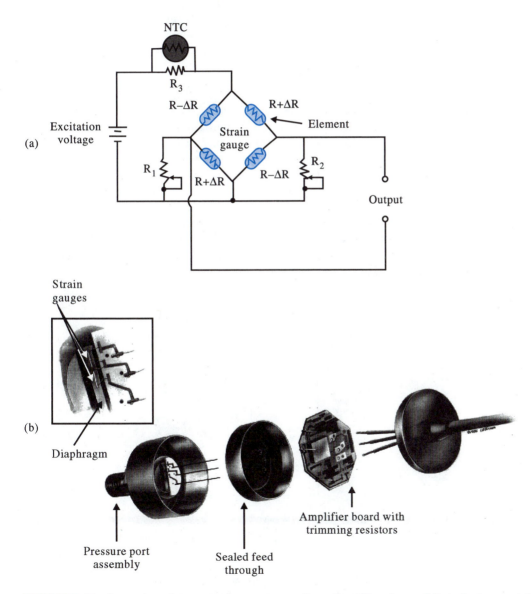

FIGURE 9-17 A semiconductor strain gauge configuration (Courtesy of Data Instruments)

same amount. The elements are mechanically connected to the diaphragm so that two of them compress while the other two expand. The result is that the resistance of two elements increases, and the resistance of the other two decreases.

Variable resistors R_1 and R_2 are trim pots. They are used to externally balance the resistance of any element changes due to component aging. The RTD connected across R_3 is used to temperature compensate the bridge network when the ambient temperature changes. The output voltage is applied to an amplifier and to the transducer's internal or external conditioning circuits. The sensitivity of semiconductor strain gauges is 100 times that of wire strain gauges. Silicon has very good elasticity, which makes it an ideal material for receiving an applied force. Since it is a perfect crystal, it does not become permanently stretched. Instead, it returns back to its original shape after the force is removed.

For better temperature ranges and stability, semiconductor elements are made of epoxy bonded on stainless steel diaphragms.

TRANSVERSE VOLTAGE STRAIN GAUGE

The transverse voltage strain gauge is a new configuration developed by Motorola. Figure 9-18 shows the top view of the sensor element. Current passes through a semiconductor piezoresistor from pins 1 to 3. Pressure that stresses the diaphragm is applied at a right angle to current flow. The deflection of the diaphragm causes the resistor element to bend. As it does, a transverse electric field is developed that is sensed as voltages at pins 2 and 4, which are located at the midpoint of the resistor.

The advantage of this type of strain gauge is that it uses only one element. A single element eliminates the need to closely match the four stress- and temperature-sensitive resistors of a Wheatstone bridge design. It also simplifies the additional circuitry necessary to accomplish calibration and temperature compensation.

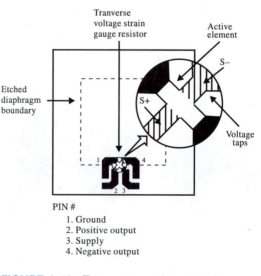

PIN #
1. Ground
2. Positive output
3. Supply
4. Negative output

FIGURE 9-18 Transverse voltage strain gauge (Courtesy of Motorola)

9-8   PRESSURE CONTROL SYSTEMS

The most common types of industrial process systems that employ pressure are hydraulic, pneumatic, vacuum, static, and steam pressure distribution systems.

HYDRAULIC SYSTEMS

Many types of machinery used in the manufacturing industry are powered by hydraulic pressure. Most types of hydraulic systems recirculate an oil-based fluid, as shown in Figure 9-19. The illustration shows a double-acting hydraulic system that controls a punch press.

The figure shows a motor-driven rotary pump as the energy source of the system. It converts electrical energy into mechanical energy. With each revolution of the pump, a fixed amount of hydraulic fluid enters the inlet port from the reservoir. The liquid is set into motion by being forced through the outlet port into the system through transmission pipes or flexible tubing. This outlet port is called the pressure line. The hydraulic fluid is in the state of ready operation as it circulates throughout the system.

A filter in the feedline is placed between the reservoir and pump. In this position, it removes any dirt particles or contaminants from the oil before it enters the system. Its purpose is to extend the life of the system.

As fluid flows through the system, it encounters resistance due to friction from surface areas of the transmission lines, and from various components. The pressure is developed as fluid is forced against the surface areas. Indicator instruments such as pressure gauges and flowmeters are placed throughout the system to show different operating conditions, or to monitor the components in order to show they are functioning properly.

The mechanical control of the hydraulic system is achieved by a number of components, such as directional valves, flow control valves, and regulators. They provide either full or partial control of system fluid.

The directional control valve, or four-way valve, alters the directional flow path of the fluid. It consists of a valve body with four internal passages and a sliding spool

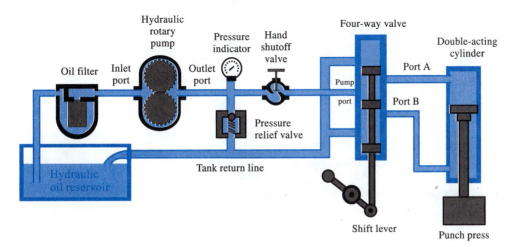

FIGURE 9-19 Hydraulic system

that connects and disconnects the passages. When the spool is moved to the extreme bottom position, the pump port is connected to Port A, and the tank return line is connected to Port B. Pressurized fluid enters the cylinder at Port A and fluid is forced out of Port B into the return lines, causing the rod to extend. When the spool is in the extreme top position, the pump port is connected to Port B and the tank return line to Port A. Fluid enters port B and exits port A, causing the rod to retract.

The double-acting cylinder is used to control a punch-press and it serves as the load device of the system. It performs work by changing the mechanical energy of hydraulic fluid into linear motion that moves the ram of the press. Ports are located at each end of the cylinder body through which fluid can enter and exit, thus allowing the piston rod to move in two directions.

The hand shutoff is a flow control valve. By turning the knob, the amount of fluid flow in both directions can be adjusted between maximum flow and no flow. When the valve is fully open, maximum fluid will flow; fully closed, no fluid will flow.

A pressure relief valve is connected between the pump output port and the reservoir. It consists of a valve body and a spool that is biased by a spring. When the pressure at the pump end of the spool opposite the spring is high enough, the spool is pushed open. A path is created for flow between the pump and tank. The purpose of the relief valve is to prevent excessive pressure from building up in the system. This function is accomplished by providing a route for the fluid between the pump and tank when the flow paths downstream are blocked. This situation would occur under the following conditions: the cylinder is fully extended or retracted, the flow control valve is fully closed, or the four-way directional control valve is in a position that blocks flow to and from the cylinder.

PNEUMATIC SYSTEMS

One of the major applications of pneumatic pressure systems is on mass production assembly lines. The compressed air of these systems provides the power for industrial processes that require high forces or high-impact blows to produce products.

Figure 9-20 illustrates a pneumatic system that operates a handtool. An air compressor unit serves as the energy source of the system. It forces surrounding air into a tank under pressure. The tank where the compressed air is stored serves as a reservoir until it is eventually distributed into the system when needed. In an industrial setting, the compressor is driven by an electric motor. Portable pneumatic systems use internal combustion engines. The compressor forces the air into the tank, stopping when the

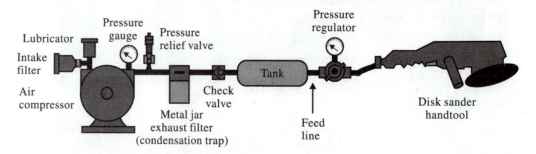

FIGURE 9-20 A pneumatic system that drives a rotary actuator

pressure reaches a certain level. As the air is used, the pressure must be maintained during system operation. As the air is used, the pressure drops. If it falls below a predetermined level, the compressor turns on until enough air is replaced.

A feedline from the tank provides air for distribution throughout the plant. Solid pipes, tubing, and flexible hoses are used to transmit pressure in the system. The air is not recirculated and returned back to the tank. Instead, it is released from the system back into the atmosphere.

Before the air enters the compressor, it must first be conditioned. Conditioning involves the removal of dirt by an air intake filter, the removal of moisture by a condensation trap, and the injection of a fine oil mist to provide lubrication for moving parts. Figure 9-20 shows the filter and lubricator at the inlet of the compressor and a metal jar exhaust that collects moisture from oil vapors produced by the lubricator. The pressure gauge monitors system pressure. The pressure relief valve vents air into the atmosphere if the pressure in the system becomes excessive. The check valve prevents high air pressure from returning to the compressor from the tank due to back-flow. The tank is a reservoir that holds pressurized air for intermittent usage. The pressure regulator is a variable pressure valve that operates by restricting and blocking flow to the working portion of the circuit. The actuating speed of the load can be regulated to different speeds by adjusting the flow control setting of the valve.

The load of the system that performs work is a rotary actuator that is capable of variable speed control. This type of load device would be used in industrial applications such as buffing, drilling, grinding, and mixing. Pneumatic load devices are also designed to produce linear motion to perform work. For example, a double-acting cylinder is used in industrial applications such as hoisting, elevating, pile driving, and clamping.

The pneumatic system operation is monitored by pressure indicators placed at strategic locations. Their readings provide information on system operation and troubleshooting.

VACUUM SYSTEMS

Any enclosed space containing air or other gas at a pressure lower than atmospheric pressure is defined as a *vacuum*. Just as with compressed air, a vacuum condition can be utilized to perform various types of work applications.

Figure 9-21 shows a diagram of an On-Off cycling vacuum system. All of the components and pipes are enclosed and isolated from the outside atmosphere. A vacuum pump removes air from the system. A storage tank is used to accumulate vacuum for on-demand power needs. The vacuum pressure in the tank is monitored by a measuring sensor. If the vacuum pressure is not great enough, the sensor develops a signal which turns the pump on. The vacuum pressure lowers when actuator devices pull air into the system as they perform work, or from any leaks that may develop in the system. When the sensor detects that the vacuum pressure has reached a preset level, its output signal changes and turns off the pump. The usual range between the turn-on and turn-off points is about 5 to 15 Hg inches. The reason for using a reservoir tank is that it permits On-Off operation, which allows the vacuum pump time to cool down.

A vacuum relief valve provides protection from an excessive vacuum condition. The check valve installed between the vacuum pump and tank allows air flow out of the tank. This one-directional component prevents the back-flow of air into the tank. An intake filter is used to prevent foreign particles such as dust or sand from entering

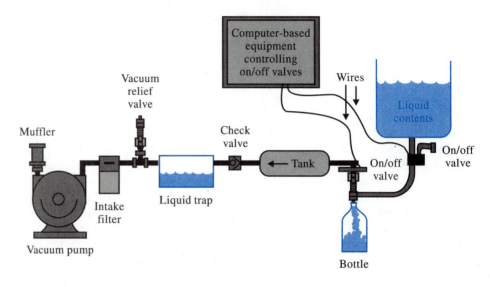

FIGURE 9-21 A vacuum system used to fill bottles with liquid

the pump mechanism. A bottle-like tank called a liquid trap uses gravity to prevent liquid materials from being sucked into the pump.

The vacuum system performs work by creating a pressure differential between the air-tight surfaces of the equipment. As controlled actuators open vents, the outside air creates a mechanical force as it is sucked into the system. This force is able to produce different types of work operation.

The actuator in this system is an injector device that fills a bottle with liquid. A vacuum tube is placed in the center of a drain tap. At the moment liquid is dispensed into the container, the vacuum tube is activated to suck air out of the bottle. The filling operation is much faster because the liquid is literally forced into the bottle, rather than filling because of gravity.

STATIC PRESSURE SYSTEMS

Static pressure systems are also referred to as *hydrostatic*. They are used for industrial applications where fluids are distributed during the manufacturing process. These fluids can include liquids, gases, or solids (such as powders) that flow. Specifically, hydrostatic systems are used for batch processing applications such as mixing or blending operations that occur for a limited period of time.

The pressure developed is the result of the fluid source elevated above the working section of the system. The fluid is usually held in a storage tank where it is stored until it is needed. The depth and density of the fluid develop a force at the bottom of the tank called *static head*. When the fluid is released, this pressure is required for the distribution of fluids to the required location throughout the system.

The variety of control components used in the static system is limited to flow control/shutoff valves. Since cylinders or other types of actuators are not used, there are no load components that perform work. Instead, the load is the resistance of the entire system and the only work function that exists is the result of heat generated due to friction. The operation of the system is monitored by strategically placed pressure and flow measurement instruments.

A batch process static pressure system that manufactures soft drinks of different flavors is illustrated in Figure 11-1 in the chapter on flow control. Ingredients are stored in elevated containers. As the ingredients are needed, valves open to allow drainage into a mixing tank.

STEAM PRESSURE SYSTEMS

Steam pressure is used in industry for a variety of purposes. It is used as a heat source for food processing, chemical processing, refining, or simply for warming the plant facility. Steam pressure is also used to perform some type of work, such as driving a turbine to generate electricity. Steam pressure is developed by applying heat to water. Water is transformed into a vapor, which creates a pressure as it expands throughout the system.

The energy source of the steam pressure system is a boiler. The water tube boiler design illustrated in Figure 9-22 is a very popular method of developing steam pressure. Tubes that contain water are placed inside a sealed metal chamber that is heated

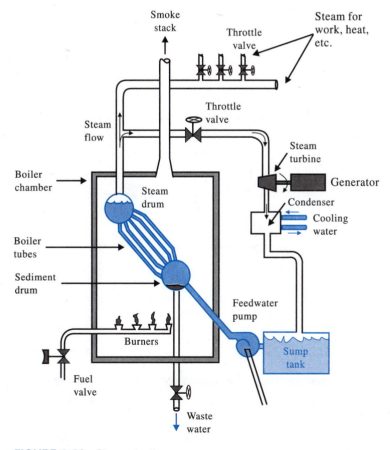

FIGURE 9-22 Steam boiler

by a furnace unit. The tubes are exposed to high temperatures by the surrounding air inside the chamber. Electricity or fossil fuel such as coal, oil, or natural gas is the energy source for the furnace.

As the water inside the tubes is heated, it changes into pressurized steam which is then forced through pipes and tubes that serve as transmission lines. As the steam flows throughout the system during the production process, its pressure and flow are controlled by directional and flow control valves. Pressure gauges monitor the steam at strategic locations, and may provide proportional electrical signals as input data to control equipment for automation purposes. The pressure in a steam system can be described as *head pressure*.

In some systems, the steam is recirculated. After passing through the actuator section, it is condensed back into water before returning to the boiler to be used again.

CHAPTER PROBLEMS

1. Which of the following are defined as fluid? ____
   a. Liquid   d. Air
   b. Gas   e. All of the above
   c. Steam

2. To perform work such as a hydraulic press or pneumatic air drill, pressure must be _____ (increased, decreased).

3. To lift individual sheets from a stack of papers, pressure is _____.
   a. increased   b. decreased

4. Liquids _____ (are, are not) compressible.

5. Water has _____ (more, less) density than an equal volume of mercury.

6. The hydrostatic pressure of one square inch of water 100 feet deep is how many psi? If the liquid is ethyl alcohol instead of water, the hydrostatic pressure is how many psi?

7. A decrease in temperature causes the head pressure of a liquid in a confined container to ____.
   a. increase   b. decrease

8. The head pressure of one liquid gallon in Mexico City is ____ the pressure of an identical open vessel in Miami.
   a. less than   b. greater than   c. the same as

9. The pressure exerted by a gas in an open container will ____ when its temperature rises.
   a. increase   b. decrease   c. stay the same

10. A condition where air is forced into a confined container is called ____.
    a. compression   b. a vacuum

11. As air is removed from a confined container, the inside pressure _____ (increases, decreases).

12. List the reference points of the following pressure scales:
    _____ Gage
    _____ Absolute

13. Convert an absolute pressure of 64.7 psia to gage pressure.

14. At sea level, 14.7psi of atmospheric pressure will support _____ inches of mercury in a glass tube.

15. Of the two columns in a U-shaped manometer, the one with the lowest level of liquid has the ____ pressure applied to it.
    a. lowest   b. highest

16. A (lower, higher) _____ pressure detected by a Bourdon tube causes the coil to unwind.

17. Describe the purpose of variable resistors R_1 and R_2 in Figure 9-17.

18. Identify the output terminals of the transverse voltage strain gauge in Figure 9-18.

19. What is the power source of a hydraulic system?

20. What is the power source of a pneumatic system?

21. Air is pumped _____ (out of, into) the storage tank of a vacuum system.

22. What is the power source of a static pressure system?

23. What is the energy source of a steam pressure system?

CHAPTER 10

Temperature Control

CHAPTER OBJECTIVES

At the conclusion of this chapter, you will be able to:

○ Define thermal energy.

○ Explain the law of thermodynamics.

○ List the three category types of heat transfer.

○ Describe the operation of the following heat sources of thermal energy:

Blast Furnace	Arc
Electronic Heat Element	Resistance Induction

○ Describe the operation of a cold thermal energy source (refrigeration system).

○ Define temperature.

○ Identify the Fahrenheit and Celsius scales and convert specific values from one scale to another.

○ List several reasons for monitoring temperatures in process control applications.

○ Define BTUs.

○ Describe the principle of operation of the following temperature indicator devices:

Crayons	Labels
Paints	Liquid Crystals
Pellets	

○ Describe the principle of operation for the following mechanical temperature measurement instruments:

Bulb Thermometer	Bimetallic Thermometer

○ Describe the principle of operation for the following electrical measurement instruments:

Thermocouple	Thermistor
Resistance Temperature Detectors (RTD)	Radiation Thermometry

243

INTRODUCTION

Many products manufactured today are the result of a process that involves temperature control. High temperatures may be used to soften metals before they are formed into a desired shape or to melt plastic in an injection molding machine. Low temperatures are necessary to preserve dairy products in the food processing industry. Manufacturing processes that are affected by temperature are referred to as *thermal systems*.

10-1 FUNDAMENTALS OF TEMPERATURE

Scientific theory states that molecules of matter are in continuous motion due to kinetic energy. Molecular movement creates heat known as **thermal energy**. Thermal energy is measured in temperature. Suppose one end of an object is exposed to the elevated temperature of a flame. The object's molecules in contact with the flame will move faster and create heat. The heat transfers from the heated area to the cooler areas throughout the object. Thermal equilibrium is attained when the object's temperature reaches the elevated temperature. Thermal movement from hot to cold is called *thermodynamics*. Each type of matter has an ignition point, at which a chemical reaction (called a fire) causes it to burn. When a cold theoretical temperature (called absolute zero) is reached, the molecular movement stops and no heat is generated.

10-2 THERMAL CONTROL SYSTEMS

Thermal systems supply thermal energy from a source, provide a path for its distribution, and convert the energy into some kind of work.

Temperature control is maintained at a desired level either manually by a human operator or automatically. In an automated system, the thermal energy is regulated by a controller. The control function is accomplished by altering the flow of thermal energy from the energy source to the load device that performs the work. There are two types of control methods: On-Off and proportional. The On-Off controller directs energy from the source when the load device's temperature falls below a certain level and stops the flow of energy to the load device when the desired temperature has been reached. An example is a home heating system. The proportional control method provides only the amount of energy from the source that the load needs. Both types of controllers respond to the temperature difference between the set point and the measured value.

10-3 THERMODYNAMIC TRANSFER

The transmission path of thermodynamic transfer can be through materials that are solids, liquids, gases, or a vacuum. The process by which heat is transferred by a solid is called **conductance**. Figure 10-1 shows an example of conductance. One end of a metal bar is placed over an open flame heat source. The molecules over the flame move more rapidly. The increased molecular velocity causes collisions with neighboring molecules, which in turn causes them to move faster. This action continues until

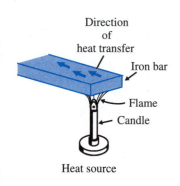

Direction
of
heat transfer

Iron bar

Flame

Candle

Heat source

FIGURE 10-1 The heat conduction principle

the molecular movement throughout the bar has increased. The best solid material thermal conductors are metals. Some types of non-metal solids are insulators.

Since most fluids are poor conductors of heat, very little thermodynamic transfer takes place through the process of conductance. The transfer of heat through fluids such as liquids and gases takes place through a process called **convection**. When a container of fluid is placed above a heat source, the bottom layer begins to expand. The warmer fluid becomes less dense than the fluid above it and therefore moves to the top of the container. The cooler fluid—which is heavier—goes to the bottom. In this manner, heat is transferred through the constant movement of circulating currents, as shown in Figure 10-2.

Thermal energy can also be transferred through a vacuum by a process called **radiation**. In theory, bundles of energy are radiated away from atoms in the heat source as wavelike patterns that travel at the speed of light. Radiated heat transfer also takes place through air. A prime example is the sun, which radiates heat through the vacuum of space and through the earth's atmosphere, as shown in Figure 10-3.

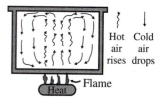

FIGURE 10-2 The heat convection principle

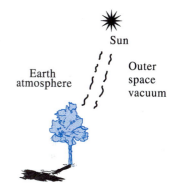

FIGURE 10-3 The heat radiation principle

10-4   THERMAL ENERGY SOURCE

Thermal energy is primarily produced by a change in the state of matter. For example, when a fuel such as coal is combined with oxygen at a high enough temperature, it ignites into a chemical reaction called a fire. As it burns, it changes into another material called carbon dioxide. The combining of carbon and oxygen causes heat to be released. Thermal energy is also released by matter changing form. For example, when a liquid changes to a gas, evaporation occurs and lowers the temperature. When a gas is converted to a liquid, heat is released.

This section describes several common types of thermal energy sources used in industrial applications. This study will include hot and cold temperature systems.

INDUSTRIAL FURNACES

The most common heat source for many manufacturing operations is the furnace. It is usually built of metal and brick, because these materials will withstand high temperatures. The shapes of industrial furnaces vary, but all of them confine the heat generated inside their chambers. Industrial furnaces provide heat to heat-treat materials, whether by cooking food or hardening metals. They are also designed to melt materials, such as butter to pour over popcorn or iron for foundry use.

Combustion Furnace

The most common heat source in industry is the combustion furnace. It usually combines fossil fuel and oxygen at a high temperature to produce heat. Smaller furnaces are ideal for heat treating machined gears to make their metal harder. For a higher temperature, the blast furnace is used. The high temperature is achieved by forcing large amounts of air into the burning chamber. The blast furnace is used to manufacture products such as glass, steel, and cement.

Systems that use steam from boilers to develop pressure are also used to supply heat. Fossil fuels are often burned to convert the water into steam. Boiler systems provide very safe and accurate temperature control.

Electric Furnaces

Furnaces can also be heated by electricity. There are three types of electrical heating systems commonly used in industry: arc, resistance, and induction.

Arc Furnaces   The arc furnace is used in the process of smelting steel in a foundry, as illustrated in Figure 10-4. The voltage is applied by connecting one power supply terminal to an electrode and the other terminal to a crucible made of a conductive material such as graphite. When the voltage is applied, the electrode is in contact with the metal to be melted. Current begins to flow and an arc is formed at ignition. The electrode is placed a small distance from the metal. To keep the arc burning, the heat from the arc melts the metal to a liquid state.

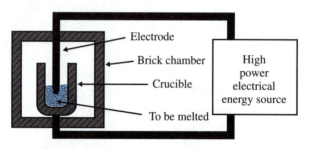

FIGURE 10-4 An arc furnace

Resistance Furnaces   Resistance furnaces use heating elements similar to a kitchen oven. The element has large resistance, which produces heat when current flows through. The elements are placed inside an insulated chamber where all surfaces of the object or material being heated are uniformly exposed to heat. This type of furnace is used in batch processing for heat treating purposes. It is also used in burn-in tests where integrated circuits are exposed to high temperatures for several hours to test their reliability.

Induction Furnaces   Induction furnaces are also used to melt metals. Again, insulated chambers are used to hold the work to be heated. Coils of wire are wrapped around the chamber and AC current is applied. A magnetic field constantly expands and contracts around the coils. As the field sweeps across the iron, a current is induced, which causes its molecules to move around. As the molecules shift positions, heat is created due to friction. An intense heat results in a very short period of time.

COOLING SYSTEMS

The most common type of refrigeration system is the household refrigerator. Refrigeration systems are used in the food processing industry to keep perishable products at low temperatures. They are also used in other manufacturing fields, such as chemical plants, to cool liquids to required levels before they are used in a blending operation.

The principles of refrigeration can best be explained by describing an old icebox used in the home 75 years ago. This unit was a wooden box insulated with cork or sawdust. Chunks of ice were placed in an upper compartment and the perishables underneath. The cooling process works on the principle that heat energy from warmer objects transfers to cooler objects nearby. In the icebox, heat from the food moves to the ice. The food cools as it receives the energy because it has lost energy, and the ice warms up. The transfer of heat continues until everything inside reaches the same temperature.

The modern refrigerator operates on the principle that evaporation loses heat energy, and condensation gains heat energy. By controlling these two reactions mechanically, refrigerators are able to perform the same functions as the iceboxes they replaced. Freon is evaporated and condensed in the refrigerator system. This refrigerant is used because it evaporates quickly when exposed to room temperature.

The refrigeration system is illustrated in Figure 10-5(a). Its components include a compressor, a condenser, a capillary tube, an evaporator, and two fans. The compressor is the device that pumps the refrigerant throughout the system. The freon in liquid form enters the capillary tube, shown in Figure 10-5(b), through a port with a small diameter. As the freon leaves the small capillary tube, it enters a larger-sized tube called the evaporator. The sudden increase in tube size creates an abrupt drop in pressure which causes the liquid freon to evaporate into a gas and cool down. As the cold gas flows through the evaporation coils, it absorbs heat from the contents in the refrigerator compartment. The heat energy transfer is assisted by an evaporator fan. As the

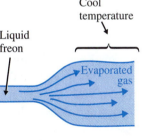

(b)

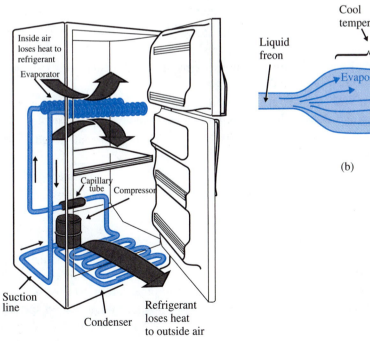

(a)

FIGURE 10-5 A refrigeration system

compressor runs, it creates a suction and draws the warmer gas into its inlet port. Inside the compressor, the gas is compressed into a liquid. The conversion from a gas to a liquid state also generates heat. As the liquid freon is pumped from the compressor outlet port, it flows through a series of folded tubes called a condenser. The condenser is a radiator that releases heat from the warm freon flowing through the tubes to the outside. The heat transfer is assisted by the fan that blows air over the condenser coils.

If the refrigerator compressor were to continue running, the food would freeze. To keep the temperature within a desired range, a thermostat is used to turn the compressor on and off. When the temperature rises to a certain level, the thermostat activates a switch to turn the compressor on. The thermostat turns the compressor off when the temperature reaches the desired lower level.

10-5   TEMPERATURE MEASUREMENTS

There are many manufacturing applications that require precise measurements of temperature. This function is performed by instruments placed at the energy source, the controller, or the system load. These devices provide visual indication, or a mechanical or electrical feedback signal in a closed-loop system for automated control.

The components of non-electrical instruments are physically altered as they respond to temperature changes. Electronic instruments are designed to produce electronic signals proportional to variations of temperature.

To provide good control of industrial processes, accurate measurements of temperature are essential. There are several reasons to monitor temperatures in process control applications:

1. Precise temperature conditions are required when combining two chemicals to form a compound.

2. Over-temperature conditions that could cause excessive pressure must be avoided in an enclosed system to prevent ruptures or explosions.

3. Temperatures must be kept below freezing to prevent stored food from spoilage.

4. By ensuring that the heating system is consuming energy efficiently, fuel costs can be minimized and environmental conservation concerns can be met.

Temperature Scales

Scientists have developed scales to indicate temperature, such as Fahrenheit, Celsius, Kelvin, and Rankine. Each type of temperature scale has fixed reference points at which water boils or freezes, and numerical values that fall in between those points. Most industrial applications use the Fahrenheit and Celsius scales for temperature measurements. The other two scales are most frequently found in research and engineering applications.

Fahrenheit Scale   The first temperature scale was developed in the early 1700s by Gabriel Fahrenheit, a Dutch instrument maker. Though modified from its original form, it is widely used, especially in the United States. At sea level, the freezing point of water is 32 degrees, and the boiling point is 212 degrees.

Celsius Scale   The next temperature scale developed was designed by Anders Celsius. A similar scale was designed by Christin of Lyons who named it the *Centigrade* scale. Both designs use a numerical value of 100 degrees for the boiling point of water and 0 degrees for the freezing point of water at sea level. This scale is referred to as either the Centigrade or Celsius scale.

Each of these temperature scales can be converted to the other by the following equations:

$$C = 5/9(F° - 32)$$

$$F = (9/5 \times C°) + 32$$

Heat is thermal energy that has the ability to perform work. Thermal energy is rated in work units called calories and BTUs (British Thermal Units). A *calorie* is the heat required to raise the temperature of 1 gram of water 1° C. A BTU is the amount of heat required to raise the temperature of 1 pound of water 1° F.

10-6   TEMPERATURE INDICATING DEVICES

A number of industrial situations require an indication that a predetermined temperature has or has not been reached. For such situations, several different types of heat-sensitive materials have been developed solely for indication and monitoring purposes. These temperature sensing materials are made of crystalline solids.

The materials operate on the principle that when heated, a temperature will be reached at which the solids change color or melt to a liquid. The change provides a visual indication that the necessary temperature has been reached. Temperature indicators are available in the form of crayons, paints, pellets, and labels. They are applied directly to or placed near the object being monitored.

These heat-sensitive indicators are accurate to within 1 percent and respond within a few tenths of a second. Because they are inexpensive, they are preferred in situations where they will burn, such as a heat zone of an industrial oven, ceramic kiln, or for products that travel on conveyers through a furnace.

CRAYONS

The crayon is available in stick form. They are manufactured in 100 different temperature ratings, ranging from 100° F to 2500° F. The workpiece is marked with the crayon. When the predetermined temperature is reached, the crayon liquifies, notifying the observer that the workpiece has reached that temperature.

PAINTS

A paint indicator is a lacquer that dries to a dull finish. When the predetermined temperature has been reached, its finish turns glossy and transparent. Paints are often used on very smooth surfaces to which crayons cannot stick.

PELLETS

The pellet works on the same principle as crayons and paints. Pellets are used in applications where extended heating periods are involved or when oxidation of a workpiece might obscure a crayon or paint marking. Because pellets are bulkier than crayons or paint, they are used when visual indication must be observed at a distance.

LABELS

The label shown in Figure 10-6(a) has one or more heat sensitive indicators sealed under transparent heat resistant windows. Each indicator changes color at a specific temperature. Labels are available in non-reversible styles to show peak temperature and reversible styles to indicate changing temperatures.

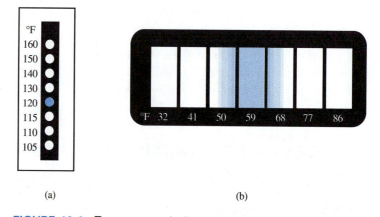

(a)   (b)

FIGURE 10-6   Temperature indicators made of crystalline solids

LIQUID CRYSTAL INDICATOR

The liquid crystal indicator shown in Figure 10-6(b) uses crystal material sandwiched between an adhesive backing and transparent film. The crystals change to different colors indicating different temperatures. Temperature is read by observing which patch has changed to a specific color.

LIQUID FILLED THERMOMETERS

Glass Thermometer

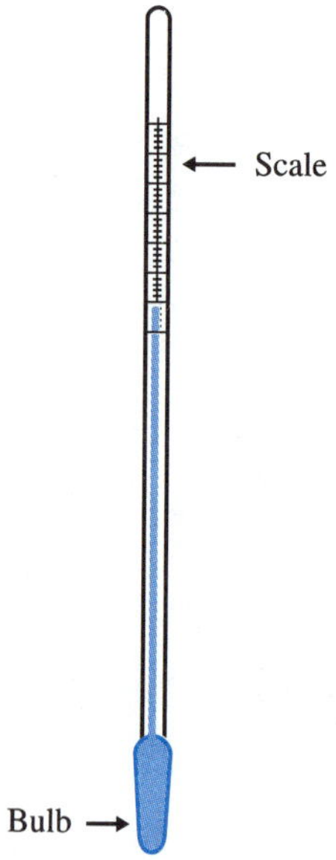

The thermometer shown in Figure 10-7 is a closed tube with a reservoir at the bottom and is partially filled with a liquid. It operates on the principle that materials expand when exposed to heat. For example, when the temperature increases, the liquid expands and rises in the glass tube. The temperature reading is taken by comparing the top level of liquid to the corresponding number on an adjacent temperature scale. Glass thermometers are very accurate and reliable. However, for industrial applications, they are too fragile and are not adaptable to recording or automatic control situations.

Filled-Bulb Thermometer

By modifying the thermometer principle, filled-bulb thermometers have been developed that are more durable than glass thermometers. They are also capable of

← Scale

Bulb →

FIGURE 10-7
Liquid glass
thermometer

providing feedback action for control purposes and recording temperature variations over a period of time. The filled-bulb thermometer is shown in Figure 10-8. The bulb is the primary sensor that detects changes in thermal energy. It also serves as the reservoir for the liquid or gas. For greater durability, the bulb is made of metal. The capillary is a tube that connects the bulb to the pressure-volume element. The element is a spiral tube that bends due to pressure changes in the filled-bulb thermometer system. The linkage is physically connected to the pressure-volume element. As the coil expands or contracts, it causes the linkage to move a needle over a temperature scale. The linkage may be attached to a pen which draws a line on a circulating chart with a temperature scale. The linkage can also be connected to an electrical component such as a potentiometer to provide feedback signals for a closed-loop system.

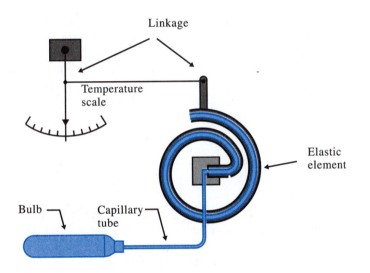

FIGURE 10-8   Filled-bulb thermometer

Bimetallic Thermometer

The bimetallic thermometer, as shown in Figure 10-9, is made of two dissimilar metal strips that are physically bonded together. Each metal has a different expansion ratio. As temperature changes, the strip will bend in the direction of the metal with the lower expansion rate. The deflection of the strip settles at a position that represents the temperature value. The strip can be attached to an indicator scale, recording chart, or linkage used to provide a feedback signal for a closed-loop system.

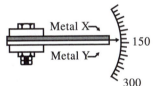

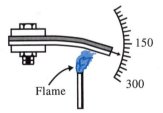

FIGURE 10-9   Bimetallic thermometer

10-7   ELECTRONIC SENSORS

There are two general types of electrical sensors used to measure temperatures: thermoresistive and thermoelectric. Thermoresistive sensors—thermistors and RTDs—change resistance as the ambient temperature varies. Thermoelectric sensors—thermocouples—produce a voltage proportional to the change in temperature. Table 10-1 compares these temperature sensors.

THERMISTOR

One type of temperature sensor is the **thermistor**. Its name is a derivation of the term *thermal resistor.* The thermistor exhibits a large change in electrical resistance when subjected to a relatively small change in temperature. Temperature variations can be caused either by a change in the ambient temperature external to the thermistor or, internally, by a change in the current through the thermistor.

Thermistors are constructed by using a paste-like metal oxide mixture to form certain shapes such as discs, beads, or rods. A set of two conducting wires is inserted into the paste. The mixture is hardened when exposed to heat by a process called *sintering.* The type of oxides, the proportions used, and the physical size of the thermistor determine the desired temperature and resistance ranges for the device.

Oxidized metals have characteristics similar to those of semiconductor materials. At lower temperatures, the valence electrons in the outer shell are strongly bound to each atom and function as good insulators. As the temperature rises, the thermal activity of the atom increases. Valence electrons gain sufficient energy to break away from the atoms. The electrons become free to take part in current that flows through the material. As temperature increases, more electrons become available and the resistance of the material decreases. This characteristic of the thermistor is called a **negative temperature coefficient**. The letters *NTC* are placed inside the thermistor symbol to indicate this characteristic.

How Thermistors Are Used

There are many circuits in which thermistors are used. A few of the more common applications are discussed below.

Temperature Measurement   The primary function of the thermistor is to exhibit a change in resistance as a function of temperature. In measurement instruments, this resistance is often converted into a voltage reading by using a voltage divider as part of a voltage divider network. The diagram in Figure 10-10 shows that the output is taken across the fixed resistor. As temperature increases, the thermistor resistance

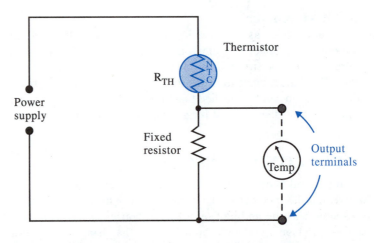

FIGURE 10-10 Temperature measuring voltage divider

decreases. Therefore the voltage across the resistor increases. If a meter with a scale that reads temperature is placed across the output terminals, its reading will increase as temperature increases. The output voltage as a function of temperature can be expressed by the following formula:

$$V_o = V_s \times \frac{R}{R+R_{TH}}$$

The resistance of variable R is the parallel equivalent resistance of the load connected to the output terminals and the fixed resistor.

When measuring ambient temperature, the circuit should be designed to keep the current through the thermistor low. If the current is too high, the thermistor will become much warmer than the surrounding temperature due to self-heating.

Another temperature measuring circuit is shown in Figure 10-11. R_1 and R_4 are precision resistors. Their values are selected to match the particular thermistor used. The variable null resistor is adjusted to balance the bridge so that the output is zero volts. Based on bridge theory, a voltage at the bridge output will develop as the thermistor resistance changes.

Temperature Compensation   The resistance of metals such as copper changes when subjected to temperature variations. These metals have a positive temperature coefficient. Such changes in resistance can affect the accuracy of sensitive measuring instruments, such as meters. To offset the temperature-resistance changes, the negative coefficient properties of a thermistor can be used.

Figure 10-12 shows a thermistor used to compensate for the resistance change of the coil. The resistance of the thermistor changes significantly more than that of the coil over the same temperature range. By placing a shunt resistor of the proper value

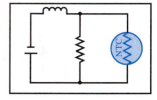

FIGURE 10-12 A thermistor used to compensate for temperature variations

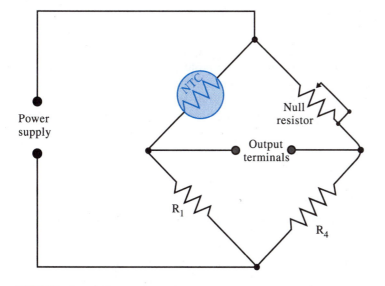

FIGURE 10-11 Temperature measuring bridge network

across the thermistor, the equivalent resistance of the parallel network provides a negative coefficient nearly equal to the positive coefficient of the copper. The network adds less than 15 percent to the total impedance of the circuit.

Surge Suppression   Cathode ray tubes in televisions and oscilloscopes use heater coils called *filaments*. The filaments emit the electrons for the beam that scans the display screen. When power is initially applied to the cold heater, its resistance is low. To prevent the filament from being damaged by a high surge of current, a thermistor is placed in series. The high starting resistance of the thermistor limits the current to a safe value. As current begins to flow, the thermistor self-heats and its resistance is reduced. At the same time, the temperature and resistance of the filament increase to normal operating values.

RESISTANCE TEMPERATURE DETECTORS

The resistance of electrical conductivity metals varies directly with temperature. Therefore, metals have a **positive temperature coefficient** (PTC). This means that as their temperature increases, their resistance increases.

Some types of metals are used in a temperature sensing device called a *resistance temperature detector* (RTD). Two metals commonly used in RTDs are nickel and platinum. Nickel is the most sensitive metal because it provides the greatest change of resistance for a given unit of temperature change. Platinum is used in applications that require a resistance change over wider temperature ranges.

RTDs are constructed by placing a coil of fine wire inside a housing to protect it from outside contamination or by placing a thin film of metal on a ceramic substrate. A laser beam is used to burn away the film until its resistance is at a prescribed value. The completed assembly is then sealed in a protective enclosure.

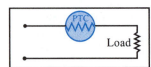

FIGURE 10-13 An RTD used as an overcurrent protection device

RTD Applications

Overcurrent Protection   The positive temperature coefficient characteristics of an RTD makes it an ideal overcurrent protection device. Figure 10-13 shows an RTD connected in series with a load. During normal operating conditions, the RTD resistance is low. Therefore its effect on current flow is minimal. When a short circuit or an overcurrent condition occurs, the RTD resistance goes high and limits the current to a low level.

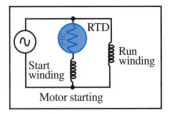

FIGURE 10-14 An RTD used to eliminate the start winding after the motor reaches full speed

Motor Starting   A single-phase AC motor has a start winding and a run winding that are connected in parallel branches. After the motor is running, the start winding should not be used. At full speed, a centrifugal switch opens the branch with the start coil.

Figure 10-14 shows how an RTD replaces the centrifugal switch. At ambient temperature, the initial resistance of the RTD is about 100 ohms. It allows sufficient current to flow through the start winding when the motor starts. By the time the motor is at full speed, the RTD is heated and its resistance is high. This reduces the current flow through the start winding to near zero.

THERMOCOUPLE

Figure 10-15(a) shows dissimilar metal wires joined at both ends. At each junction where the wires are in contact, they are exposed to heat from the surrounding ambient temperature. This causes a small number of electrons to drift from metal B and accumulate in metal A. The slight accumulation of electrons causes a small EMF to develop between the metals. Because the junctions are subjected to the same temperature, the same amount of voltage develops across them.

Suppose that heat is applied to the junction on the left. A larger voltage potential will develop across it than the one on the right. An equivalent circuit in Figure 10-15(b) is used to illustrate the result. Each battery represents the two junctions. The difference in voltage between the junctions forms a net voltage of .05 volts. This causes electron current to flow in the closed circuit formed by the wires. This phenomenon is called the Seebeck effect, named after its inventor, the German physicist Thomas Seebeck.

The device shown in Figure 10-15 is called a thermocouple, which is a transducer that converts heat into voltage. The amount of voltage developed by a thermocouple junction is directly proportional to the amount of heat applied to it. The higher the temperature, the greater the voltage produced. The voltage-temperature characteristics of a thermocouple are also linear. Equal changes of temperature produce equal changes of voltage. The metals used determine the polarity and voltage range of the junction.

Figure 10-16 shows the loop opened at the top wire. The junction exposed to the temperature to be measured is called the *hot junction*. The other junction in the loop is called the *cold junction*. The output voltage that appears across the opening will be proportional to the temperature difference between the junctions. If the temperature of one of the junctions is known, the voltage across the opening can be used to calculate the temperature of the other junction. The cold junction is considered the reference junction in a thermocouple because the temperature applied to it is a known value. A temperature display instrument is usually connected across the open leads. The value it displays is determined by the voltage developed across the terminals.

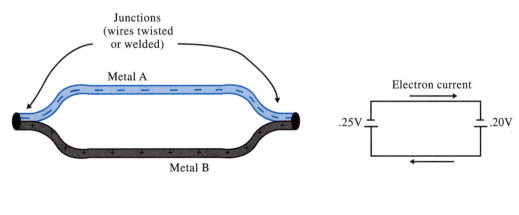

(a)   (b)

FIGURE 10-15 Thermocouple

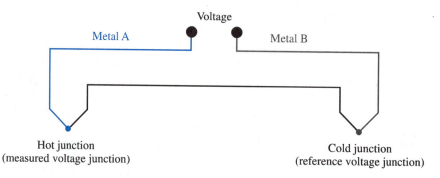

FIGURE 10-16   A thermocouple with hot and cold junctions

For a thermocouple to be used in a commercial application, two requirements must be met. First, the thermocouple must be connected to a temperature display instrument without changing the voltage between the measured and reference junction. For example, if wires made of a different type of metal from metals A and B connect the instrument to the termocouple's output terminals, then two new junctions form that alter the net voltage between the hot and cold junctions.

This problem is overcome by connecting the instrument as shown in Figure 10-17. Iron and constantan are the dissimilar metals used. Notice that the wires do not actually touch each other at the cold junction. Instead, both wires are connected to copper leads, forming two cold junctions. If both cold junctions are exposed to one temperature, the sum of their voltages will equal the voltage to that temperature.

The second requirement is that the cold junction must be constantly maintained at a certain temperature. This is impractical because the ambient temperature in an industrial location can vary as much as 50 degrees from winter to summer. Therefore, industrial thermocouple loops must be compensated to account for the temperature variations to which cold junctions are subjected.

An automatic compensation method is illustrated in Figure 10-18. It lists the voltages throughout the circuit. Both of the reference junctions and a resistor bridge network are integrated on a substrate. Therefore they are subjected to the same ambient temperature. One leg of the bridge, R_3, is a thermistor with a negative coefficient.

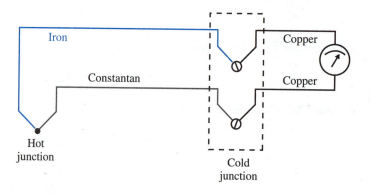

FIGURE 10-17   A thermocouple connected to a meter

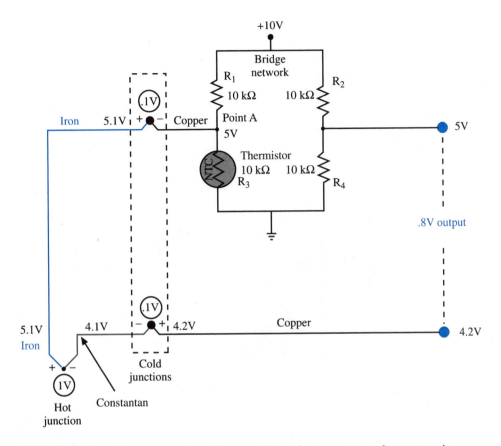

FIGURE 10-18 A thermocouple with a cold junction compensating network

Resistors R_1, R_2, and R_4 are not temperature-sensitive. As the ambient temperature changes, R_3 varies proportionately and unbalances the bridge. The voltage at point A changes the same amount as the two cold junctions, but at the opposite polarity; therefore, both voltages cancel. The only voltage change at the output should be the result of the temperature variance detected at the hot junction.

EXAMPLE

Refer to Figure 10-18. Make five assumptions:

1. The voltage produced by the measured hot junction is 1 volt as the temperature remains constant.
2. Every 100° F change of ambient temperature causes the thermistor to vary 750 ohms. At +100° F the termistor resistance is 10 kΩ.
3. Every 100° F change of ambient temperature causes each cold junction to change 0.1 volt.
4. The output of the thermocouple network is 0.8 volts.
5. At +100° F the thermistor resistance is 10 kΩ and each cold junction produces 0.1 volt.

PROBLEM

Suppose that the circuit in Figure 10-18 is exposed to an ambient temperature of 100° F. If the ambient temperature increases to 200° F, determine whether the output voltage remains at 0.8 volts.

SOLUTION

Figure 10-19 shows the voltage drop changes, which cause the output voltage drop to remain at 0.8 volts.

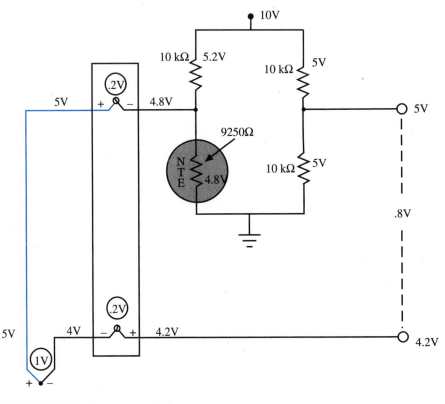

FIGURE 10-19 Example solution

Devices that perform automatic compensation for thermocouples are commercially available. They are called *cold junction compensators*. In recent years, computers have also been used to compensate thermocouples.

Some thermocouple metal combinations are in common use throughout industry. They have become standardized by the American National Standards Institute (ANSI), and are identified by letter designations. The more common types include:

1. Iron-Constantan (Type J)
2. Copper-Constantan (Type T)

3. Chromel-Alumel (Type K)

4. Chromel-Constantan (Type E)

5. Platinum-Rhodium Platinum (Types R, S, and B. The difference depends on the ratios of the two metals.)

Each type of thermocouple has different characteristics and its accuracy is affected by various conditions. The selection of a thermocouple is based on the following considerations:

1. Temperature range

2. Susceptibility to oxidation

3. Reducing atmosphere

4. Sensitivity

5. Accuracy

6. Cost

Thermocouples are used in industry to measure temperatures of ovens and furnaces, molten plastic vats, and nuclear reactor cores. Thermocouples respond quickly to temperature changes, are rugged, and have a wide temperature range.

Table 10-1 provides a comparison of the three types of temperature sensors.

PROBE ASSEMBLIES

In most applications, temperature-sensing elements should be protected from the environment surrounding the point of measurement. For example, if the sensor is immersed in a liquid, the leads or body may become corroded. Conductive liquids may short the two leads connected to the sensor body, resulting in false readings. The measurement of air temperature can be misread if the humidity is high or if the probe is exposed to wind, which causes a cooling effect.

These false readings can be avoided by enclosing the sensor body within a housing. Protective housings are made of glass, ceramic, epoxy, stainless steel, and other metals.

RADIATION THERMOMETRY

Most temperature instruments are invasive devices that make physical contact with the solids, liquids, or gases being measured. They make direct temperature readings as thermal energy is transferred by conduction to the sensing element.

It is also possible to take temperature readings without making physical contact by using a non-invasive device. A method called *radiation thermometry* infers temperature by measuring the thermal energy radiated from the surface of the measured body. The instrument used to make these readings is usually referred to as a radiation pyrometer. The term *pyrometer* is derived from this instrument's ability to measure high temperatures. Instruments of this type are ideally suited for applications where conventional sensors cannot be employed, such as:

1. When objects are moving, such as rolling mills in steel production, paper manufacturing, glass making, and conveyer belts.

2. Where temperatures are extremely hot, such as in furnace atmospheres.

TABLE 10-1   TEMPERATURE SENSOR COMPARISONS

Type/Range	Advantages	Disadvantages
Thermistor Resistive negative Temperature coefficient −40°F to 300°F	High sensitivity Most accurate Fast response Low cost Vibration resistant	Narrow temperature span Non-linear output
RTD Resistive positive Temperature coefficient −150°F to 1400°F	Linear output Large temperature span Large resistance range Interchangeability	Low sensitivity High cost Vibration

Thermocouple
Produces voltage or
   current proportional
   to temperature
   −300°F to 4200°F
	Linear output
	High temperature
	Wide temperature range
		Least sensitive
		Requires reference

3. Where non-contact measurements are required because of contamination, such as in food and pharmaceutical production.

4. Where corrosive and hazardous conditions exist, such as around high voltage conductors.

5. Where measurements are taken from a distance.

The principle of operation of radiation thermometry is based on the basic law of physics which states that every object at a temperature above absolute zero radiates electromagnetic energy. The frequency range of the electromagnetic waves includes visible light and lower-frequency infrared light. As the temperature of the object changes, the frequency also changes. For example, if the temperature rises, the frequency increases and the wavelength becomes shorter. This principle is illustrated by observing metal being heated. As it gets hotter, the color changes from red, to yellow, to white. The color change is a result of the frequency increasing.

By focusing the electromagnetic energy waves emitted by the measured object (target) on a detector element, measurements are taken. The signal from the element is electronically processed and the frequency is converted into a proportional temperature readout for display. Radiation thermometry theory is based on the assumption that the total energy emitted by a body is the result of its temperature. An object with this capability is referred to as a *blackbody*. Most objects, however, do not radiate energy from temperature alone. Instead, they also reflect and transmit (as fiber optics do) radiant energy, as shown in Figure 10-20. Therefore, the total radiated energy is the sum of emitted energy (E), reflected energy (R), and transmitted energy (T).

$$E + R + T = \text{Radiated Energy}$$

For a perfect blackbody, E=1, R=0, and T=0. For a non-blackbody object, the value for E is still 1.0, but the values for R and T are greater than 0. To account for the reflective and transmitted energy of an object, a term referred to as *emissivity* must be considered when making measurements. Emissivity is defined as the ratio of total energy radiated by an object to the emitted energy of a blackbody made of the same material at the same temperature. The emissivity of a blackbody is 1.0. The emissivity of non-blackbodies falls between 0.0 and 1.0. For example, the emissivity of carbon at 76° F is approximately 0.8. This value means that the carbon is radiating more energy than a carbon blackbody at the same temperature. Therefore, a pyrometer detector measuring the temperature of this target will record a reading higher than the actual temperature because it detects three types of energy instead of one. Many radiation pyrometers have an emissivity adjustment knob which allows the operator to compensate for the emissivity ratio factor. By setting the knob to 0.8, for example, the instrument electronically calculates this value and the frequency it detects to indicate the correct temperature of carbon. The operator uses a specific emissivity reference table to determine the adjustment setting of the instrument required for various materials. This table is developed in a research lab where measurements of emitted energy from holes drilled into various materials, and measurements of total radiated energy from non-blackbodies made of the same materials are taken at various different temperatures required for various materials. The holes become blackbodies because they are not capable of emitting reflective or transmitted energy. Some targets have very low emissivity. If the values are below 0.2, accurate measurements are not always possible. Examples of these objects are polished metallic surfaces that are reflective, and thin film plastic that transmits a high amount of energy.

Based on the different techniques used to measure radiant energy, there are three categories of instruments: broadband, optical detector, and ratio pyrometers.

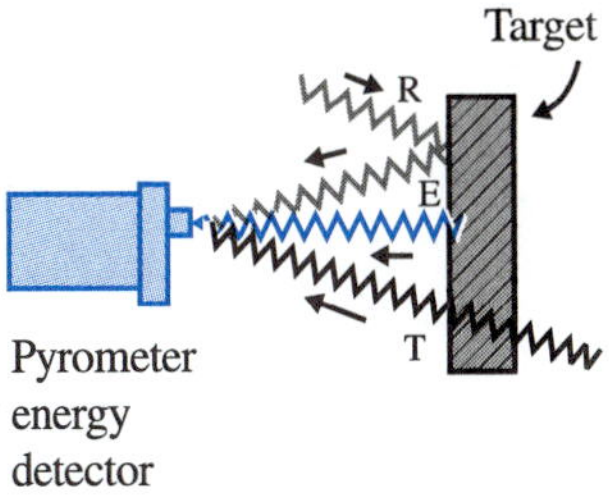

FIGURE 10-20   The target object radiating emitted (E), reflected (R), and transmitted (T) energy to the measurement detector

Broadband Pyrometers

A broadband pyrometer, shown in Figure 10-21(a) uses a lens system or sight tube which directs the radiation onto a blackened reference surface inside the instrument.

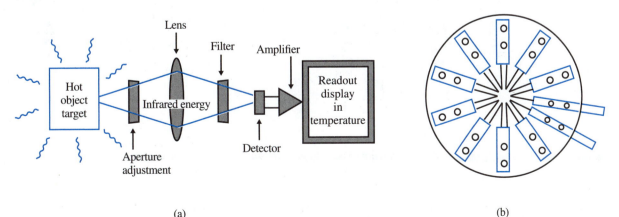

(a)   (b)

FIGURE 10-21   A simplified pyrometer temperature-measuring instrument and a thermopile

The filter is used to pass electromagnetic waves within a desired frequency range. The energy detector employs a device known as a *thermopile* to measure the temperature of the reference area. A thermopile, shown in Figure 10-21(b), consists of several thermocouples connected in series to provide greater sensitivity to small changes in temperature. The composite output of this detector is a DC voltage that is proportional to the amount of energy at its surface.

Optical Pyrometers

The optical pyrometer is shown in Figure 10-22(a). A viewfinder is positioned to allow observance of the target and the filament of a lightbulb simultaneously (Figure 10-22(b)). The object being measured is compared to the brightness of the filament. A current adjustment is made by the operator until they are both the same intensity, at which time the filament visually disappears into the background. When the brightness of both objects is equal, their temperatures are also the same. A scale on the current adjustment knob is calibrated to indicate the temperature of both the target and the reference filament.

Ratio Pyrometers

At all temperatures the target object radiates energy at different frequencies. Most pyrometers measure the dominant waves with the most energy.

The ratio pyrometer differs from other pyrometers by taking measurements of two different frequencies emitted by an object. First, the radiant energy of a blue wavelength is passed through a blue filter and its power strength is measured. Then the radiant energy of a red wavelength is passed through a red filter and measured. These measured values, along with the actual lengths of blue and red magnetic waves, are used in the following formula to determine a ratio quantity:

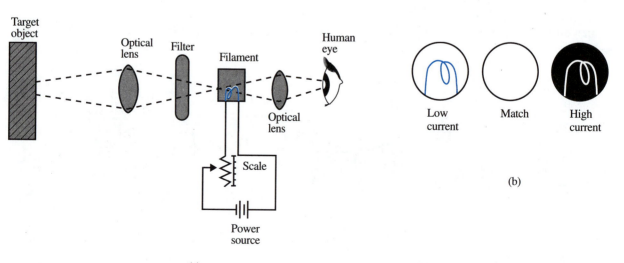

FIGURE 10-22 Optical pyrometer

Red Wavelength = 1.0

Blue Wavelength = 0.8

RP = Power Reading of Red Wavelength

BP = Power Reading of Blue Wavelength

$$\text{Ratio} = \frac{RP - BP}{1.0 - 0.8}$$

The target temperature is inferred from the ratio value electronically calculated by the instrument. Figure 10-23 graphically illustrates the power readings taken at the blue and red wavelengths for two different temperatures.

The advantage of the ratio technique is that it is less susceptible than other measurement methods to dust, steam, and other factors that distort readings.

Field of View

When using a pyrometer, the target being measured should completely fill the view of the instrument. Figure 10-24(a) shows a proper technique when the target is measured.

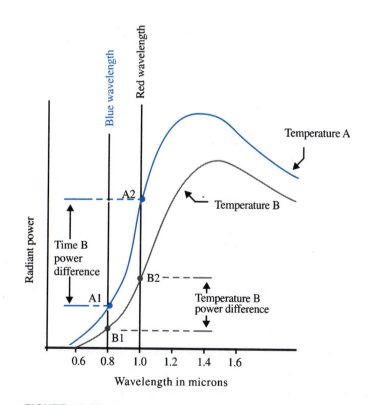

FIGURE 10-23 The power detection of two different wavelengths by a ratio pyrometer

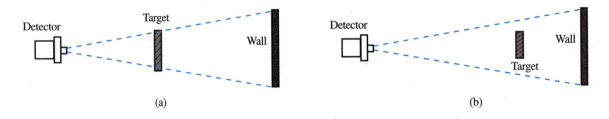

FIGURE 10-24 Field of view

The reading in Figure 10-24(b) will not be accurate because the pyrometer will also detect the temperature of the wall surface in the background.

CHAPTER PROBLEMS

1. Molecular movement creates heat known as _____.
2. Thermal energy moves from ____ objects to ____.
   a. warmer   b. cooler
3. List three types of thermodynamic movement.
4. Name a functional application for each of the following heating sources:

   Blast Furnace   Resistance Furnace

   Fossil Fuel Furnace   Induction Furnace

   Arc Furnace

5. An evaporating liquid that turns to gas ____.
   a. absorbs heat   b. gives off heat
6. The freezing point on a Celsius scale is _____ degrees.
7. How many calories are used to raise the temperature of 5 grams of water 10 degrees Celsius?
8. For what type of functional application is a temperature-sensing indicator made of crystalline solids used?
9. When heated, why does the liquid in a thermistor rise in the tube?
10. As a bimetallic thermometer element straightens, the ambient temperature is

    _____ (increased, decreased).
11. A thermocouple is a ____ device.
    a. thermoelectric   b. thermoresistive
12. The _____ (hot, cold) junction of a thermocouple is the reference point.
13. Give a functional application of a thermocouple.
14. The resistance of an RTD is _____ (directly, inversely) proportional to temperature.

15. RTDs are considered _____ (linear, non-linear) devices.

16. Give a functional application of an RTD.

17. The thermistor has a _____ (negative, positive) temperature coefficient.

18. Give a functional application of a thermistor.

19. List the three types of energy radiated from a non-blackbody.

20. How can emitted energy be radiated exclusively from a non-blackbody?

21. What does a thermopile consist of?

22. What type of pyrometer views a lightbulb and target simultaneously in a viewfinder?

23. Which type of pyrometer is the least susceptible to dust and smoke?

CHAPTER 11

Flow Control

CHAPTER OBJECTIVES

At the conclusion of this chapter, you will be able to:

○ Define flow.

○ Describe the importance of measuring and controlling flow in industrial processes.

○ List some types of materials measured for flow and how they are transferred.

○ Explain the difference between volumetric flow rate and mass flow rate.

○ List common measurement units of flow rate.

○ Describe the method used for measuring the volumetric flow rate and mass flow rate of solid materials.

○ List four factors that affect the flow rate of liquids.

○ Calculate the Reynolds number for a liquid.

○ Describe the operation of the following mechanical measurement instruments used to determine flow rate:

Differential Pressure	Lobed Impeller
Rotameters	Turbine Flowmeter
Rotary-Vane	

○ Describe the operation of the following electronic sensors used to measure level:

Coriolis Meter	Thermal Flowmeter
Rotor Flow Detector	Vortex Flowmeter
Electromagnetic Flow Detector	Ultrasonic Flowmeter

○ State a rule that describes the placement of flow sensors in a pipe system.

INTRODUCTION

Many types of industrial applications involve the flow of materials during the manufacturing process. *Flow* is the transfer of material from one location to another. The materials can be raw materials, products, or wastes in the form of solids, liquids, gases, or solids that float on liquids, called slurry. The flow, or movement of materials, is transferred through such components as pipes, hoses, channels, or conveyer belts. Flow can be continuous or sporadic, depending on the type of process being performed.

This chapter will first discuss the basic principles of flow, and then describe the operation of mechanical and electronic measuring instruments.

11-1 SYSTEMS CONCEPTS

Automated systems that control flow first determine flow rates or volume by various measurement techniques, and then use the data to regulate the movement. These systems employ a source, a path, a control function, an actuator, and a measuring instrument to operate.

An automated flow control system is illustrated in Figure 11-1. It shows a batch process machine that makes soft drinks. A computer-based controller turns the valves on and off to direct the flow. Flow sensors that measure flow rate and volume are located on each pipe. They send feedback data to the controller.

At the beginning of a batch, valve V_1 opens and water drains into the batch tank. When the flow sensor S_1 registers that a certain volume of water has passed through the pipe where it is located, the controller closes V_1. V_2 then opens until the required amount of water enters the carbonation tank. As V_2 closes, V_3 opens to allow CO_2 into the carbonation tank. When the sensor S_3 measures that a certain volume of gas has entered the tank, the controller closes V_3 and V_4 opens to fill the seltzer tank. Selected valves then open in a certain sequence to allow the necessary coloring and flavoring ingredients to drain into the batch tank. The volume of each coloring and flavoring fluid is monitored by sensor 4. When the required amount passes S_4, the valve allowing an ingredient to flow closes, and the next valve in the sequence is opened. After the sequence is completed and the final valve closes, pump 1 turns on to transfer the seltzer ingredients at a desired flow rate into the batch tank. Sensor S_5 monitors the rate of flow as the carbonized fluid passes through its pipe. The controller will vary the speed of the pump so that the flow rate is at the required value. When the desired portions of the formula are present in the tank, pump 1 stops and the motor starts to run an agitator that mixes the contents for the required period of time. When the motor stops, pump 2 transfers the finished product to the bottle-filling stage.

After the batch run is completed, V_1 opens and fills the tank with a supply of fresh water to rinse the tank. Then the motor is activated to aid the rinsing process by agitating the water. When the agitation cycle is complete, V_5 opens to allow the batch tank to empty. After V_5 is closed, the tank is clean and ready for the next production run.

REASONS FOR CONTROL

To provide good control in the process industries, accurate measurement of flow is essential. Three reasons to monitor the flow of materials are:

1. To ensure that the correct proportions of raw materials are combined during the manufacturing process.

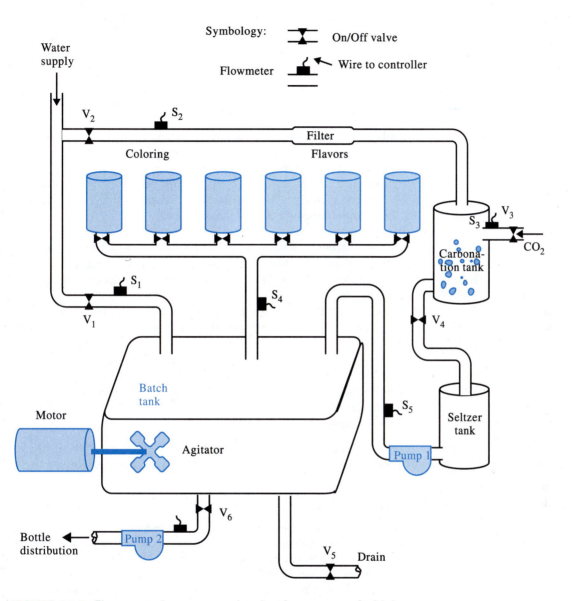

Symbology: ▷◁ On/Off valve

Flowmeter   ◀ Wire to controller

Water supply

V_2 S_2

Filter

Coloring   Flavors

S_3 V_3

CO_2

Carbona-
tion tank

S_1

S_4

V_1

V_4

Batch
tank

Motor

S_5

Seltzer
tank

Agitator

Pump 1

Bottle
distribution

Pump 2 V_6

V_5 Drain

FIGURE 11-1 Flow control system used to batch process soft drinks

2. To ensure that ingredients are supplied at the proper rate during the mixing and blending of the materials.

3. To prevent a high flow rate that might cause pressure or temperatures to become dangerous, overspills to occur, or machines to overspeed.

Flow measurements are also used to determine how much of a product is passed from the supplier to the customer. This application is known as custody transfer. Measuring flow accurately is essential in keeping records for accounting purposes.

The flow of materials is a response to an applied force. The force may be produced by a motor that drives a pump or a conveyer belt. Force may be supplied by pressure in a hydraulic system or an air compressor. Force is also produced by static head pressure.

11-2   FLOW UNITS OF MEASUREMENT

The measurement of flow is influenced by flow rate. **Flow rate**, by itself, is commonly measured. It refers to velocity, or how fast a material is moving. It does not provide information about the quantity of material that passes.

The quantity of flow can be determined by using flow rate along with volume or mass quantities. Two common types of classifications that are used to determine flow measurements are **volumetric flow rate** and **mass flow rate**.

Flow rate instrumentation is used to determine the amount of material that moves past a given point at a particular instant. Volumetric flow rate instruments are used to determine the volume of material that flows during a time period. The volume can be read as cubic feet, gallons, or liters. The time can be read per units of time, such as seconds, minutes, or hours.

For example, volumetric flow rate of a liquid can be determined from the velocity and the area of a pipe. The equation that uses the velocity and volumetric flow rate quantities is this:

$$Q = VA$$

where, Q = the volumetric flow rate in units of volume per units of time.
                 V = the velocity of the fluid.
                 A = the cross-sectional area of the pipe.

Mass flow rate instruments are used to determine the weight of materials that flows during a time period. The weight can be read in pounds, tons, grams, or kilograms.

11-3   SOLID FLOW MEASUREMENT

The solid materials that are measured for mass flow rate are typically in the form of small particles, such as powder, pellets, or crushed material. A conveyor belt is usually used to move these materials from one location to another.

Figure 11-2 shows one method of measuring the flow of powder as it is being transported by a conveyer system. The powder is released by the hopper and transferred to another location, such as a mixing vat, a storage tank, or the hold of a ship. A measurement is taken by a load cell device that determines the weight of a fixed length of the belt. Using inferred data, such as the weight measurement and speed of the belt, allows a calculation of the mass flow rate, as shown by the following formula:

$$F = \frac{WS}{L}$$

where, F = Mass flow rate in lb/min
                 W = Weight of a material on a section of length
                 S = Conveyer speed in ft/min
                 L = Length of the weighing platform

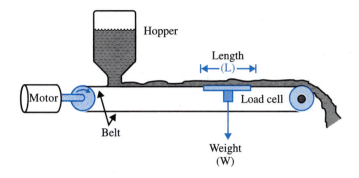

FIGURE 11-2 A conveyer system used to measure the flow of solid materials

PROBLEM

A tackinite ore conveyer system moves at fifty feet per minute, and the weighing platform is ten feet long. Determine the mass flow rate if the load cell measures 300 pounds of tackinite.

SOLUTION

$$F = \frac{WS}{L} = \frac{(300\text{lbs})\,(50\text{ft/min})}{10\text{ft}}$$

$$= 1500\text{lb/min}$$

The load cell that measures the weight is a strain gauge. Another instrument used to measure the ore weight is a linear voltage differential transformer (LVDT). Chapter 17 describes the operation of an LVDT in detail. Figure 11-3 shows a conveyer belt with a section that is allowed to droop due to the weight of the material it carries. The more

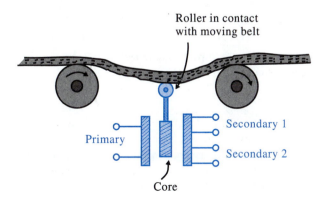

FIGURE 11-3 An LVDT used to measure the weight of materials flowing on the conveyer belt

the belt droops, the heavier the weight. The LVDT makes a weight measurement by reading the droop in the belt.

There are two ways to control the amount of solid material transferred by the conveyer system. The first is to control the speed of the belt. For example, if the belt's speed is increased while the powder from the hopper flows at a constant rate, a smaller amount of powder will flow per unit length of the belt. The second method is to keep the belt speed constant and to use a valve control at the outlet of the hopper to vary the flow.

11-4   FLUID FLOW MEASUREMENT

Liquids, gases, and vapors are classified as fluids. The accurate measurement and control of fluid flow is essential in industrial processing plants that use water, steam, gases, petroleum, acids, base solutions, and other types of fluid materials.

PIPE FLOW PRINCIPLES

The following terms must be understood when studying the principles of liquid flow control:

Velocity.   The velocity of a fluid is the speed at which it moves through the pipe. In the United States, the unit of measurement is feet per second.

Density.   The density of a fluid is its weight per unit of volume. Both temperature and pressure affect the density of fluids. High temperatures or lower pressure cause the fluid to expand so that the molecules move further apart, which causes the weight of a given volume to be less than it would be at a lower temperature or higher pressure.

Viscosity.   The viscosity of a fluid represents the ease with which it flows. A numerical unit of measure used to represent viscosity is called the *poise* or centipoise. A higher number indicates increased viscosity and more reluctance to flow.

The temperatures to which the fluids are exposed affect viscosity. With liquid, a lower temperature will cause the viscosity to increase, creating more reluctance, which slows the flow rate. With gases, a lower temperature decreases viscosity, and creates less reluctance to flow.

Pipe size.   The size of the pipe carrying a fluid affects the flow. The larger the diameter, the more easily the fluid will pass through.

Reynolds Number

In 1883, Sir Osborne Reynolds, an English scientist, submitted a paper to the Royal Society that described the effects of the preceding four factors on fluid flow. He also presented a numerical scheme that assigned values to express the fluidity of a moving liquid based on the influence of these four factors. The numerical value, known as *Reynolds number* or R number, is determined by the following formula:

$$R = \frac{VDp}{u}$$

where,   V = Velocity
          R = Reynolds number
          D = Pipe inside diameter
          p = Fluid density
          u = Liquid viscosity

The R number represents the ratio of the liquid's inertial forces to its drag (viscous) forces. The velocity, pipe diameter, and fluid density are inertial forces, and viscosity is the drag force. The R number is used to identify the type of flow currents that are likely to occur. These currents are illustrated in Figure 11-4: *laminar* flow (Figure 11-4(a)), *turbulent* flow (Figure 11-4(b)), and *transition* flow (Figure 11-4(c)). The information supplied by the R number is useful when determining the proper flowmeter for a specific application. For example, some flowmeters are designed to read laminar flow, and would give erroneous readings if they were measuring turbulent flow. By knowing a fluid's R number, a suitable flowmeter with a rating of the same value can be selected.

At very low velocities, R is low and laminar flow takes place. With this type of flow, liquid moves in layers. However, the fluids do not flow in uniform velocities across a given cross-section of the pipe. The layers in contact with the pipe wall move at velocities close to zero because of friction. As drag forces decrease further away from the pipe, the layers progressively travel at faster speeds as they near the center. The result is the parabolic shape of the velocity profile shown in Figure 11-4(a), which shows the laminar flow at an R value of 2000 or less. As the Reynolds number approaches 3000, the laminar flow becomes non-symmetrical, as shown in Figure 11-4(b). At very high velocities or low viscosities, the R value is high. When the R number reaches the range of 7000 to 8000, the uniform layers break up and develop into turbulent eddies that travel in all directions in the fluid stream, as shown in Figure 11-4(c). This mixing action tends to produce a flow rate velocity that is constant across the full stream profile. The R number range of 7000 to 8000 is the transition range between laminar and turbulent flow.

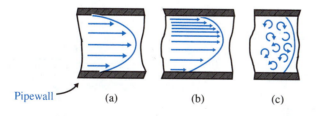

Pipewall   (a)   (b)   (c)

FIGURE 11-4 Flow currents of fluid

FLUID FLOWMETER CLASSIFICATION

One method of classifying flowmeters is to divide them into the following four categories:

1. Differential Pressure
2. Positive Displacement

3. Velocity

4. Direct Reading Mass

Differential Pressure Flowmeter

The differential flowmeter is the most common type of instrument used to measure the flow of fluids through a pipe. These instruments account for well over 50 percent of all flow measuring devices used in the process industry.

Differential Pressure Meters   Figure 11-5 illustrates the operation of this device. A restriction on the flow, called an orifice, is installed in the pipe between two flanges. An orifice is a metal plate with a hole of a specified size bored through it. The purpose of the orifice is to reduce the area that the fluid can flow through. According to the laws of conservation of energy, the mass (fluid) entering a pipe must equal the mass leaving the pipe during the same time period. Therefore, the velocity of the fluid that leaves the orifice is faster than the fluid that approaches it. According to Bernoulli's principle, as the velocity of a fluid increases, pressure decreases. The result is that there is more pressure on the incoming side of the orifice than the outgoing side. As the fluid flow rate increases, the differential pressure also increases. The orifice that converts fluid flow into differential pressure is also referred to as a primary element. There are several types of primary elements used, as shown in Figure 11-6, but the orifice style shown in Figure 11-6(a) is the most popular. A flowmeter also contains a secondary element. Its function is to convert the pressure difference into a measurement that is used to indicate fluid flow. Various types of detectors, such as piezoelectric sensors, are placed on either side of the plate to detect the pressure difference. The

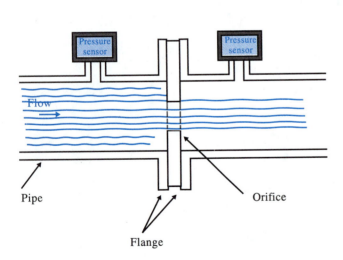

FIGURE 11-5 Differential pressure flowmeter

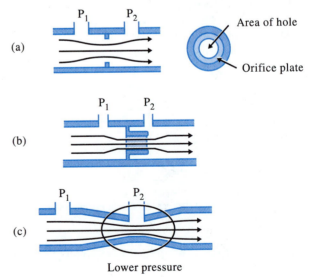

FIGURE 11-6 Three different types of flow restrictors used to convert pressure difference into flow measurements

outputs of the detectors are compared, and their differences are converted electrically into the actual flow value.

The disadvantage of the primary element's design in Figure 11-6(a) is that the plate has sharp corners on which solid materials can catch. Therefore, they are not used to measure slurries, dirty fluids, or corrosive liquids. To measure these types of fluids, alternative restriction devices have been developed.

The *flow nozzle* type, shown in Figure 11-6(b), has a constriction with an elliptical contour shape. Since there are no sharp edges at the inlet side, there is less friction to the flow. Therefore, because of its shape, it is used to measure corrosive liquids and can be used for fluids with small amounts of slurry.

The *Venturi tube* style in Figure 11-6(c) consists of a converging inlet section in which the cross section decreases in size. The high pressure reading is taken at the incoming portion of the pipe just before it converges. As the diameter becomes smaller, the velocity of the fluid increases, resulting in a decrease of pressure. The low pressure reading is taken at the location of the orifice with the smallest diameter. Since the Venturi tube has no sudden change in contour, solid particles tend to slide through its throat. Therefore, this style is recommended when measuring slurries and dirty fluids. The disadvantage of this style is that the resulting measurements are not as accurate as those based on plates with sharp-edged orifices.

Rotameters   The rotameter is a variation of the differential pressure flowmeter. It is also known as a variable area flowmeter. The rotameter is shown in Figure 11-7. It consists of a tapered metering tube that is vertically mounted and a float that is free to move up and down within the tube. The fluid to be measured enters the bottom of the tube, and exits at the top. Its operation is based on the variable area principle, where the flow raises the float to allow passage of the fluid.

When there is no fluid flowing through the meter, the float will settle at a location in the tube that has the same diameter as the float. As fluid begins to flow, it pushes the float upwards, allowing passage between the tube and float. The greater the flow, the higher the float is raised, and the area through which the fluid can pass is increased. The movement of the float is directly proportional to the flow rate. A marker on the float is used to identify a number on a measurement scale that indicates the flow rate.

When liquid flow is measured, the float is raised by a combination of the buoyancy of the liquid and the velocity force of the fluid. With gases, the float responds only to the velocity force of the gas. The float reaches a stable position when the upward force exerted by the fluid equals the downward gravitational force exerted by the weight of the float.

Differential pressure meters are suitable for use with most types of liquids and gases. They are simple in construction, have no moving parts, and are inexpensive. However, they are inefficient because the restrictions cause pressure losses in the system, and their pressure vs. flow rate is non-linear across the entire scale.

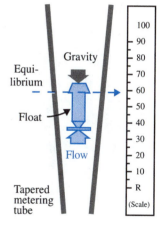

FIGURE 11-7 The rotameter used to measure the flow of liquids and gases

Positive Displacement Methods

Positive displacement (PD) devices are rotary instruments that mechanically make direct measurements to determine flow. They operate by separating the fluid into segments of known values, and passing them downstream through the pipe. Multiplying the count times the known volume of each segment provides a volumetric measure of flow.

Rotary-Vane Flowmeter   The most common type of PD meter is the rotary-vane flowmeter, illustrated in Figure 11-8. This meter operates by fluid entering each chamber section through the inlet port. As fluid fills the chamber, it forces the rotor to turn clockwise, as shown in Figure 11-8(a). The chamber is separated by spring-loaded vanes located in channels of the rotor body, as shown in Figure 11-8(b). As the rotor turns, the vanes slide in and out so that they make constant contact with the cylinder wall. The fluid is discharged when each chamber section reaches the outlet port, as shown in Figure 11-8(c). Since the volume of each revolution is known, the volumetric flow rate can be determined by multiplying the displacement times the RPM.

Lobed Impeller Meters   The lobed impeller flowmeter, illustrated in Figure 11-9, is a PD meter constructed with two carefully machined lobes that have very close clearances with each other and with the meter housing. The lobes are geared so that they rotate 90 degrees out of phase with each other. Since they are always in rolling contact with the housing and with each other, they form a seal. The force that turns the lobes is the flowing fluid. Since the volume of fluid required to turn the lobes one revolution is known, a volumetric measurement can be determined. By multiplying the displacement per revolution times the RPM, volumetric flow rate can be measured.

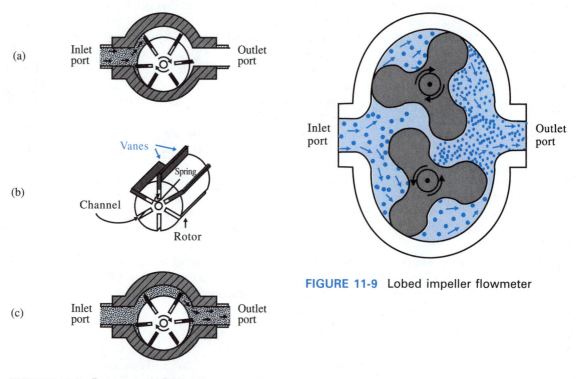

FIGURE 11-9 Lobed impeller flowmeter

FIGURE 11-8 Rotary-vane flowmeter

Limitations of PD Meters   The measurements taken from positive displacement meters are accurate. However, because they are self-powered, they extract some energy from the system. Also, since they consist of mechanical parts, they are prone to wear.

Velocity Meters

Velocity meters measure the velocity of fluid flow directly. A volumetric flow measurement is determined by the formula $Q=VA$, where Q is volumetric flow rate, V is volume, and A is area.

Turbine Flowmeter   The most common type of velocity meter is the turbine flowmeter, illustrated in Figure 11-10. Fluid flow causes a rotation of the turbine that is proportional to the flow rate. The output of the turbine flowmeter is a pickup coil. Stationary flux lines extend from a permanent magnet placed inside the coil to the area in which the turbine blades turn. Each time one of the ferrous blades passes through the magnetic field, the flux lines become distorted due to a change in reluctance. As the lines are being altered, they cut across the pickup coil which generates a pulse voltage by induction. The frequency of the pulses is proportional to the rotational speed of the turbine.

The output of the turbine flowmeter indicates volumetric flow rate. These flowmeters can also be used to record *totalization* measurements, that is, the total volume of flow. Each pulse generated by the meter is equivalent to a measured volume of liquid. By feeding each pulse into a counter, the total volume that flows can be recorded and displayed. Totalization measurements are used in applications such as batch processes.

Direct Reading Mass

Directly reading the mass, called mass flow measurement, provides the actual weight of the fluid during a given period of time. The readings are made directly, instead of

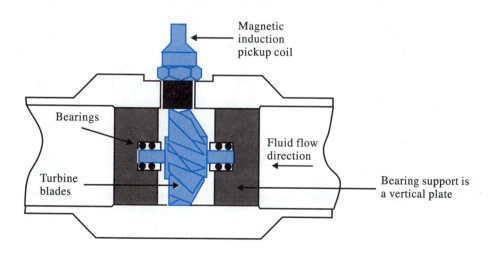

FIGURE 11-10 Turbine flowmeter that measures velocity

using inferred data from other variables. Since the measurement data is independent of solids, temperature, pressure, viscosity, and other factors that affect fluids, the readings tend to be very accurate. The conveyor system illustrated in Figure 11-2 is an example of a mass flowmeter.

11-5 ELECTRONIC SENSORS

Several electronic flowmeters have been developed: the Coriolis meter, the rotor flow detector, the electromagnetic flow detector, the thermal flowmeter, the vortex flowmeter, and the ultrasonic flowmeter.

CORIOLIS METERS

One type of device that measures mass of liquids is the Coriolis meter. It consists of a U-shaped tube for fluids to flow through, as shown in Figure 11-11(a). Fluctuating currents are sent through coils mounted near the tube. The magnetic forces they generate cause the tube to vibrate, similar to a tuning fork, as shown in Figure 11-11(b). As fluids flow through the tube, kinetic energy is produced by its speed and mass. The energy from the liquid tends to resist the vibrating motion of the tube, causing it to twist sideways, as shown in Figure 11-11(c). The degree of deflection is directly and linearly proportional to the mass of liquid passing through the U-tube. Magnetic position sensors are mounted on both ends of the tube to measure the amount of twist. The outputs from each sensor are conditioned into standard signals before they are sent to display units or to control equipment.

Coriolis meters are capable of measuring the mass of all types of fluids. However, their accuracy can be diminished if exposed to mechanical noise vibration.

ROTOR FLOW DETECTORS

Rotor flow detectors are inserted inside pipes by using tee or saddle fittings. They utilize a simple paddle wheel design to provide flow indication. Figure 11-12 illustrates the rotor flow detector. A permanent magnet is embedded in each of the four rotor blades. Fluid flow causes a rotor rotation that is proportional to the flow rate. Each pass by a magnetized blade excites a Hall effect device in the sensor body, producing a voltage pulse. The number of electrical pulses counted for a given period of time is directly proportional to flow volume.

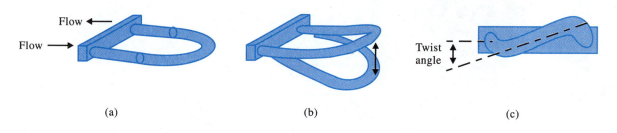

(a)   (b)   (c)

FIGURE 11-11 Coriolis mass flowmeter

This sensor can measure the flow rate of a wide variety of liquids including acids, solvents, and most corrosive fluids. They have a flow response of 0.3fps to 10fps in pipe sizes with diameters from 0.5 in. to 36 in.

ELECTROMAGNETIC FLOW DETECTORS

The electromagnetic detector is a transducer that converts the volumetric flow rate of a conductive substance into voltage. Figure 11-13 shows the electromagnetic flow detector. Major components are a flow tube, two electromagnetic coils mounted across from each other outside the flow tube, and two electrodes inside the pipe wall.

The electromagnetic flow detector's principle of operation is based on Faraday's Law of electromagnetic induction, which states that a voltage will be induced into a conductor when it moves through a magnetic field. The liquid serves as the moving conductor. The magnetic field is created by the energized coils, which produce flux lines perpendicular to the fluid flow. The induced voltage is measured by the two electrodes. This voltage is the summation of the voltage developed by each molecule in the flowing substance. As the fluid speed increases, the number of molecules a voltage is induced into also increases. Therefore, the amount of voltage produced is proportional to the flow rate.

Electromagnetic flow detectors are generally used to measure difficult and corrosive liquids and slurries such as acids, sewage, detergents, and liquid foods.

THERMAL FLOWMETERS

Flow detectors that use a paddle wheel or an orifice are susceptible to clogging. Also, their ability to detect flow at low velocities is limited due to the inertia of the wheel or the inability of the orifice sensors to detect small differential pressures.

Thermal flowmeters that use a thermistor have only a sensor tip that is inserted into the flow stream. They do not become clogged and can detect very low flow rates.

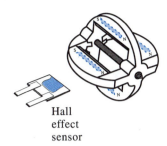

Hall
effect
sensor

FIGURE 11-12 Rotor flow detectors

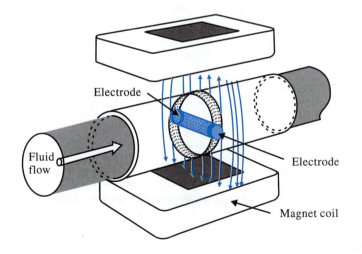

FIGURE 11-13 Simplification of the electromagnetic flowmeter principle

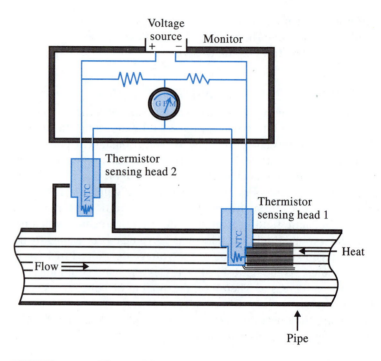

FIGURE 11-14 Thermal flowmeter

Figure 11-14 shows a thermal flow detector. It works on the principle of thermal conductivity. Thermistor sensing head 1 is mounted inside a pipe. As fluid passes, it carries away heat from the thermistor. The higher the rate, the cooler the thermistor becomes, increasing its resistance. The result is that the bridge becomes more unbalanced and the output voltage goes higher. A meter with a flow rate scale is connected across the output terminals and indicates the increase.

If the temperature of the fluid happens to change, so will the thermistor resistance. To prevent the flowmeter from giving a false reading, a second thermistor sensing head is used. Since both thermistors are in the pipe, the fluid temperature affects their resistances equally. Their placement in the bridge causes the resulting voltage changes to cancel each other. Therefore the only voltage at the output is the one caused by the flow rate. Thermistor 2 is shielded, so its resistance is not affected by the flow.

VORTEX FLOWMETERS

A liquid-measuring device called a vortex meter is illustrated in Figure 11-15. A blunt unstreamlined object such as a bar or strut is placed in the flow path of the fluid. As the liquid is forced around the obstacle, viscosity-related effects cause a series of vortices to develop downstream. The swirls are shed from one side of the obstacle and then the other in a predictable pattern. Within a wide range of Reynolds numbers, the number of vortices that appear downstream in a given period is directly proportional to the volumetric flow rate. A pressure detector placed downstream from the blunt object detects the vortices. The sensing element converts the pressure fluctuations into

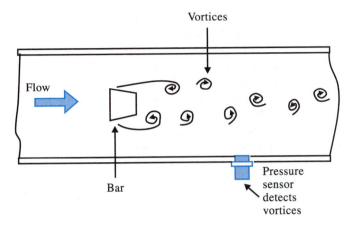

FIGURE 11-15 The vortex flowmeter

electrical pulse signals. The number of pulses within a time period is used to determine flow rate.

The vortex meter requires little or no maintenance because it is rugged, simple, and has no moving parts. However, since it introduces an obstruction in the pipe, it is limited to measuring only clean liquids to avoid clogging the pipe.

ULTRASONIC FLOWMETERS

A liquid-measuring device called an ultrasonic meter is shown in Figure 11-16. It operates on a principle of sound propagation in a liquid called the Doppler effect. As current pulses are sent by an oscillator through a piezoelectric transducer, it vibrates

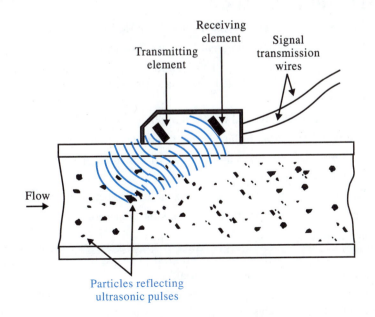

FIGURE 11-16 The Doppler ultrasonic flowmeter

and produces sound waves that are transmitted upstream into the flowing liquid. Each ultrasonic wave is reflected from particles or gas bubbles in the fluid back to a receiving element. The receiver is a piezoelectric device that detects pressure fluctuations created by the pulsating sound waves. This transducer converts the sound into electronic pulses that are processed by the measuring instrument circuitry.

Because the fluid is moving toward the receiver, the frequency of the reflected pulse received is higher than that of the transmitted pulse. The difference in frequency is proportional to the fluid velocity. As the velocity increases, the received frequency increases to create larger differences relative to the fixed frequency of the transmitter. Therefore, by measuring the frequency difference between the electrical pulses of the oscillator and the pulses at the receiver, flow rate can be recorded by the meter.

Because the ultrasonic flowmeter is placed outside the pipe, it is ideal for dirty liquids and slurries.

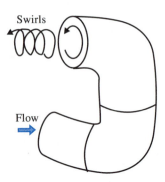

FIGURE 11-17 The swirling current produced by pipeline elbows

11-6 FLOWMETER PLACEMENT

As fluid encounters obstacles such as valves, or other geometric obstructions in the piping, the flow profile may become distorted and swirl, as shown in Figure 11-17. To minimize these conditions, the measuring device should be placed up to 20 pipeline diameters downstream from an obstruction.

CHAPTER PROBLEMS

1. List two units of measurement for volumetric flow rate.

2. List two units of measurement for mass flow rate.

3. A conveyer belt moves at 100 feet per minute and the weighing platform length is five feet. What is the mass flow rate if the cell measures 100 pounds?

4. The speed at which fluid moves through a pipe is called _____.

5. List two factors that influence the density of a fluid.

6. If liquid viscosity increases, the Reynolds number will _____ (increase, decrease).

7. Explain the importance of using Reynolds numbers for fluid flow.

8. The Reynolds number that represents the transition range between laminar

   and turbulent flow is _____ to _____.

9. A ____-styled orifice in a differential pressure flowmeter can be used to measure the flow of slurry material.
   a. orifice   b. flow nozzle   c. Venturi tube

10. Rotameters can measure ____.
    a. gases   b. liquids   c. both gases and liquids

11. A rotary-vane flowmeter _____ (is, is not) classified as a positive displacement device.

12. A velocity meter _____ (is, is not) capable of measuring mass flow.

13. The Coriolis meter _____ (is, is not) capable of measuring slurry material.

14. An electromagnetic flow detector is ____ sensor.
   a. an invasive   b. a non-invasive

15. To avoid distorted liquid flow, the sensor should not be placed within _____ pipeline diameters of an obstacle.

16. Thermal flowmeters use ____ to sense the temperature of the fluid flowing in the pipe.
   a. a thermocouple   b. an RTD   c. a thermistor

17. What kind of sensing element is used to detect the number of swirls in a vortex meter?

18. The pulses transmitted upstream in an ultrasonic flowmeter are ____ waves.
   a. electromagnetic   c. light
   b. sound   d. infrared

CHAPTER 12

Level Control Systems

CHAPTER OBJECTIVES

At the conclusion of this chapter, you will be able to:

○ Define level.

○ Describe the importance of measuring and controlling level in industrial processes.

○ Define *interface* and list three types of interfaces that may be measured for level indication.

○ List four level measurement units.

○ Define direct level measurement and list types and applications of this method.

○ Define indirect level measurement and list types and applications of this method.

○ Explain the difference between continuous and point level measurements.

○ Describe the operation of the following level indicator devices:

Rodgauge   Sight Glass

○ Describe the operation of the following mechanical measurement instruments used to determine level:

Float   Hydrostatic Pressure Detector

Displacement   Differential Pressure Detector

Bubbler   Weight Detector

Paddle Wheel Detector

○ Describe the operation of the following electronic sensors used to measure level:

Conductive Probes   Ultrasonic Sensors

Capacitive Probes

INTRODUCTION

In industrial process control, **level** refers to the height to which a material fills a container. The material can be either a liquid or a solid, such as granules or powder. The container can be a tank, a bin, a hopper, a silo, or a vessel.

Two types of techniques are used to control the level of materials in a container: On-Off and proportional. The On-Off method activates a device used to fill a container when the level is too low. When the desired level is reached, the filling operation stops. The proportional method maintains a desired level by filling the container at the same rate as the material it holds is removed.

Accurate measurements of level are essential to provide good control in the process industries. There are several reasons to monitor the level of materials in containers:

1. To ensure that enough material is available to complete a particular batch production process.

2. To prevent an industrial accident by overfilling an open container. Spilling caustic, hot, or flammable materials could be catastrophic.

3. To prevent the overfilling of a closed container or an enclosed system. This situation could cause an overpressure condition that may result in a rupture or explosion.

4. To determine an inventory of the materials in stock.

5. To prevent a heating element from overheating and being destroyed, by ensuring that a container holding heated liquid does not become empty.

12-1   A LEVEL CONTROL SYSTEM

An automated system is illustrated in Figure 12-1. It shows a liquid solvent distribution system located in a factory. Its purpose is to release the solvent into the system after the product in a batch process vessel has been emptied. As the solvent flows

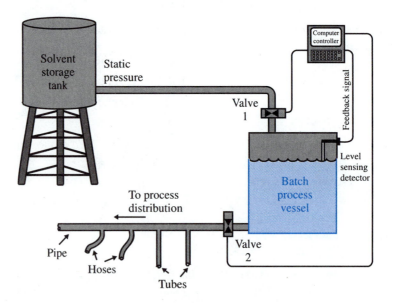

FIGURE 12-1   A level determination system

through connecting pipes, tubes, and vessels, it cleans the system in preparation for the next batch.

Before some type of automatic control can be implemented, it is necessary to determine the level with a mechanical or electronic sensing detector. The control function is initiated by a signal from the sensing device, which indicates the actual level of the batch process vessel. The solvent is transferred from a storage reservoir that functions as the source, through pipes that serve as the path, to the process vessel which is the load. Until the solvent level reaches the sensor, the controller keeps Valve 1 open to allow the fluid to pass through. When the level reaches the sensor, Valve 1 is shut off. The controller is programmed to keep the solvent in the vessel for a period of time while the chemical reaction by the liquid performs the necessary cleaning action. After the required time has elapsed, the controller opens Valve 2 to enable the solvent to drain into and clean the distribution system. Work is performed when the solvent is in the vessel, and again when it is released into the distribution system.

POWER SOURCES

Sources that provide the force required to transfer materials include pumps, static pressure tanks, and augers.

Pumps

A pump is primarily used to move liquids a required distance, or to an elevated level. The pump is powered by an electric motor, a combustion engine, or a hydraulic system.

Static Pressure Tanks

Static pressure tanks are employed to transfer liquids and solids a required distance, or to a lower level. The tank stores the material. Valves located near the output ports of the tank open and close the pathway, which controls the flow of material as needed. The pressure from the tank provides the force that moves the material. Gravity adds to the force if the flow is vertical.

Augers

The auger is used to move powders or granules in an upward vertical direction. The auger is driven by an electric motor, a combustion engine, or a hydraulic system.

TRANSFER SYSTEMS

There are various types of distribution systems that serve as the pathway for materials to be transferred from one location to another.

Pipes

A pipe is the primary vehicle used for transporting solids and liquids which need to be moved upward, downward, and horizontally.

Conveyer Systems

The conveyer belt is often used to transfer solid materials. The belt is powered by a motor. The movement is horizontal, or on an upward or downward slope.

12-2   METHODS OF MEASUREMENT

Level is measured by locating the boundary between two media, called the **interface**. The media can be liquid and gas, liquid and liquid, or solid and gas. An example of a liquid and gas medium is water making contact with air in an open vessel. A liquid and liquid medium is two liquids that do not mix, such as oil floating on top of water. An example of a solid and gas medium is powder in contact with air in an open vessel.

Level can be measured *directly* or *indirectly*. The direct method includes measuring with a float or a dipstick. Direct measurement devices are also referred to as *invasive* devices because the sensor is in direct contact with the material. The indirect method, also known as inferred measurement, means that some variable other than level is measured and used to *infer*, or indicate, a level measurement. For example, the level of a vessel can be determined by weight, pressure, volume, buoyancy, or electrical properties. Indirect measurement devices are also referred to as *non-invasive* devices, because no part of the sensor comes in contact with the material. Non-invasive devices are preferred when the material is corrosive, hazardous, or is at a high temperature or pressure.

Regardless of the method, level is measured either at a point value or continuously across a range.

POINT LEVEL MEASUREMENTS

Point level measurements detect if the interface is at a predetermined point. Generally, this type of detection is used to signal either a low-level limit when a vessel needs to be refilled, or a high-level alarm to warn of an overfill condition. The output of point level measurement devices typically produces On-Off, or 1 and 0 state digital signals.

CONTINUOUS LEVEL MEASUREMENTS

Continuous level measurement locates the interface point within a range of all possible levels at all times. The output of point level measurement devices typically produces an analog signal between 4 and 20 mA which is both proportional and linear to the level. The electrical signal can be converted into information that represents various quantitative values used to indicate levels. They include:

Height: In units of feet or meters.

Percentages: Percent full, or percent of measured span.

Volume: In gallons, cubic feet, or liters.

Weight: In pounds, tons, or kilograms.

A factor that must be considered when taking level measurements is the shape of the container. If the vertical walls of a tank are parallel, its volume can easily be determined. If the sides are irregular and not parallel, volume is more difficult to calculate.

12-3   LEVEL MEASUREMENT METHODS

Level measurements are made by a number of instruments that use different methods to make the readings. The instruments are classified either as visual observation systems, or as float and displacement systems, including buoyancy displacement, purge, hydrostatic head, differential pressure, weight, rotational suppression, electrical concepts, and ultrasonic radiation.

The selection of a specific method of measuring level is often based on the following considerations:

Material   Accessibility
Cost   Turbulence
Accuracy   Pressure
Level range

VISUAL METHODS

Visual method simply means that a direct measurement is taken by observing the location of the material's top surface in the container.

Rodgauge

A **rodgauge** is a dipstick that is inserted into the material being measured. It is the same type of device used to indicate oil level in a car. It has weighted line markings which indicate depth or volume. This device is very accurate and is often used during the calibration of other level measurement devices. It is not used to measure hazardous materials, to produce remote indication, or in vessels that are pressurized.

Sight Glass

The **sight glass** is a transparent tube connected to the side of a vessel. As the tank level changes, so does the level in the sight glass. This device provides a direct, local, and continuous measurement. Sight glasses are used for high pressure applications and for hot or corrosive fluids. They are not useful when foam or viscous liquids are used, because visual quality is obscured.

FLOAT AND DISPLACEMENT METHODS

Rodgauges and sight glasses give visual indication of level. However, they do not produce a feedback signal in automated control applications. The remainder of the measurement devices described in this chapter can provide feedback information through mechanical linkages or electrical signals.

Buoyancy Method

Float-type Level Indicator   The float-type level indicator is a spherical or cylindrical element that rides on the surface of the liquid as a means of detecting the level. A **float** is a direct, invasive method that provides either point or continuous measurement. By using different types of linkage connections between the float and the indicating device, such as the gauges in Figure 12-2, a continuous measurement reading can be taken. Figure 12-2(a) shows a high level; Figure 12-2(b) shows a lower one.

A float providing the point method of measurement is often connected to an electrical switch. The switch activates a light or buzzer for high and low limit indication. One type of float switch is shown in Figure 12-3. It consists of a magnetic toroid as part of a buoyant device with a guide tube, or stem, through its center. One or more

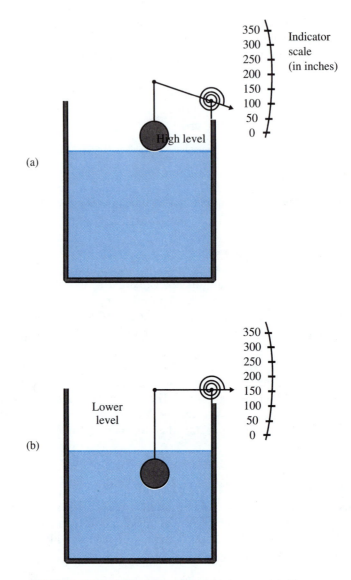

FIGURE 12-2 A float-type level instrument taking continuous measurements

Reed switches are placed inside the guide at desired levels. As the liquid level changes, the float moves. When it comes close to the Reed switch, the magnetic field closes the switch's contacts. The switch closure indicates a point level measurement.

Float devices can also be used for automatic control. By connecting the float element to the stem of a flow control valve with a linkage, it will allow water to replenish a vessel if the level becomes too low. An example is the float mechanism used in a household toilet tank. To provide more accurate control, a linkage can connect the float to the potentiometer of an electronic computing device. Small-diameter floats are

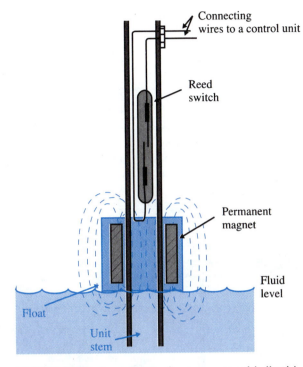

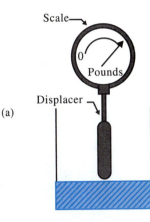

(a)

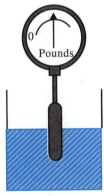

(b)

FIGURE 12-3 A magnetic float moves with liquid level to actuate the magnetic Reed switch within the unit system

used to measure higher density fluids. Larger floats are used for liquid-liquid interface detection or for reading lower density materials.

Displacement Method

Displacement Sensor The **displacement** level sensor is somewhat different than a float sensor in that its probe is weighted so it actually sinks in the measured fluid. Displacement level sensors operate on the Archimedes Principle: a body immersed in a liquid loses weight equal to the weight of the volume of liquid it displaces.

Figure 12-4 illustrates the operation of the displacement level sensor. The probe is suspended from a spring scale. In Figure 12-4(a), the level is below the displacer and the scale shows the full weight of the displacer. When the level rises as shown in Figures 12-4(b) and (c), there is an apparent loss of weight of the displacer. The weight change causes the displacer to move, thereby yielding a linear and proportional signal. Notice that the change in displacer movement is small in comparison to the change in liquid level. This device is limited to applications in open tanks and is capable of transmitting electronic signals for remote readings.

Displacement sensors are especially appropriate for measuring liquid-liquid interfaces, that is, two liquids that do not mix together, such as oil and water, different chemicals in the same container, and slurries. Two typical applications are monitoring

(c)

FIGURE 12-4
A displacement level sensor

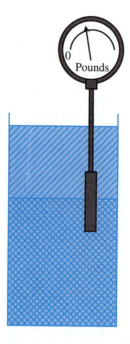

FIGURE 12-5 A displacement sensor that measures liquid-liquid interface

water condensation in fuel storage tanks and separating chemical emulsions in process systems. Figure 12-5 shows the displacement method used to measure a liquid-liquid interface in a tank.

Purge Method

Bubbler   The **bubbler**, or purge, is one of the oldest methods of level determination. This type of system can measure such materials as water, oil, corrosive liquids, molten metal, pulp, and fine powders. A bubbler system at both low and high levels is shown in Figure 12-6. It consists of a dip tube vertically immersed in the fluid of a tank with an open end placed close to the bottom. A tee connection joins the supply line, the bubbler pipe, and a pressure gauge.

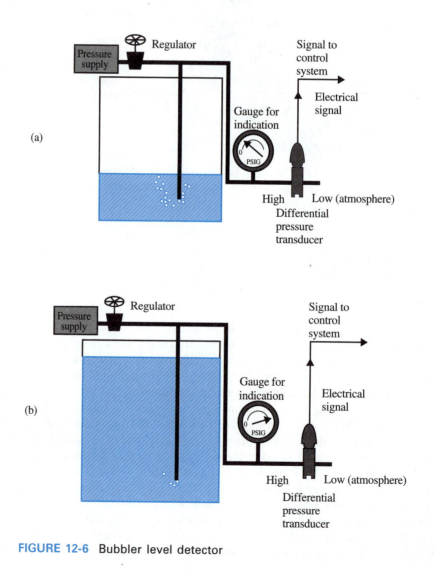

FIGURE 12-6 Bubbler level detector

The pressure at the bottom of the tank is hydrostatic head pressure. The more fluid in the tank, the higher the static head pressure. To make a level measurement, the air supply regulator must be adjusted so that its pressure is at least 10psi higher than the highest hydrostatic pressure to be measured. The result is a flow of air that passes through the dip-tube producing bubbles in the liquid. As the liquid rises, there is more static head pressure above the outlet of the tube. Therefore, backpressure at the bottom of the bubbler tube increases. The increase in backpressure allows less air flow through the bubbler tube and an increase in pressure at the gauge, causing a larger deflection. The increase in pressure is proportional to an increase in liquid level. Purge instruments are popular level detectors in the paper industry because they are self-cleaning and do not allow the pulp, measured in a vat, to clog the orifice.

Rotational Suppression Method

Paddle Wheel Detector   The paddle wheel uses motion to detect the level of either granular or powdered solid material. This invasive detector is a point-measuring device that spins freely when the level is below the paddle. As the level rises, the presence of the material is detected when it makes contact with the paddles and prevents the wheel from turning.

Hydrostatic Pressure Method

Hydrostatic Head Level Detector   The hydrostatic head level detector is a pressure detector that determines level in an open container using the indirect, or inferred, measurement technique. It operates on the principle that any column of material exerts a force at the bottom of the column due to its own weight. This force is called hydrostatic pressure, or head pressure. Hydrostatic pressure is determined by the following formula:

$$\text{Pressure} = \text{Height} \times \text{Density}$$

As the height of the material changes, there is a proportional change in pressure. By placing a pressure gauge at the bottom of the vessel, the level of material can be determined using the following formula:

$$\text{Height} = \frac{\text{Pressure}}{\text{Density}}$$

EXAMPLE _____

Water at 60 degrees F is stored in an unpressurized tank. A pressure gauge that displays readings in pounds per square inch indicates a value of 100psi. To determine the level of the water, divide 0.43, which is the weight of a 1 × 1 inch column of water 1 foot high (at 60° F), into the pressure value.

$$\text{Level (Height)} = \frac{100\,\text{psig (Pressure)}}{0.43\,\text{psi per ft. (Density)}}$$

$$= 143.2 \text{ ft.}$$

This type of pressure detector is capable of measuring the level of solids and liquids.

Differential Pressure Method

Differential Pressure Level Measurement

Hydrostatic pressure measurements determine the liquid level in an open container where the top is exposed to the atmosphere. When a liquid level inside a pressurized tank is determined using this method, the gauge will measure not only the fluid, but the pressure above the liquid as well.

The vessel pressure in the vapor space above the liquid can be compensated for by using a differential pressure transducer. This measurement device uses two pressure detectors. One is placed at the bottom of the vessel to make the high pressure measurement. The following formula indicates what factors influence its reading:

$$\text{High Pressure Measurement} = \text{Hydrostatic Head} + \text{Vessel Pressure}$$

The other sensor is placed in the vapor space to make the vessel pressure measurement. The differential pressure transducer subtracts the vessel pressure signal from the high pressure signal to produce a reading that represents the hydrostatic head proportional to the liquid level.

Weight Method

Level Measurements by Weight

The level of a material in an unpressurized container can be determined by obtaining an inferred measurement value of weight. This method is both non-invasive and very accurate. To measure the weight, devices such as mechanical springs are placed under the container and onto a supporting surface. As the springs compress, a linkage device connected to a pointer deflects to a calibrated position that indicates the weight.

Level in a container with parallel walls is found by following these five steps:

1. Weigh the container.
2. Weigh the container with the contents.
3. Determine the weight of the contents using the following formula:

$$\text{Contents Weight (lbs)} = \text{Measured Weight} - \text{Container Weight}$$

4. Determine the volume using the following formula:

$$\text{Volume (cubic feet)} = \frac{\text{Contents Weight (lbs)}}{\text{Density (lbs/cubic ft)}}$$

5. Determine level using the following formula:

$$\text{Level (feet)} = \frac{\text{Volume (cubic feet)}}{\text{Surface Area (square feet)}}$$

Weight is also detected by electronic sensors. The signals from these devices can easily be connected electronically to proportional weight, volume, and level readings for display. Weight level measurements can be used for solids and liquids.

12-4 ELECTRONIC SENSORS

There are three main types of electronic level sensors: conductive probes, capacitive probes, and ultrasonic sensors.

CONDUCTIVE PROBES

Conductive probe sensors are used in single or multiple point measurement systems to detect the presence of a conductive liquid. Figure 12-7 shows two conductive probe sensors, one with two probes and the other with five. Figure 12-7(a) shows the terminal housing, which contains the controller, and two projecting electrodes. A low AC voltage is applied to the electrodes as they are immersed in the liquid. The conductive liquid completes the electrical circuit of the control, which activates a semiconductor switch. When the level drops below the shortest electrode, the circuit opens, and the current flow stops. Figure 12-7(b) shows that by adding more probes, signals can be sent to a controller that automatically supplies liquid to the container. These signals will activate an alarm if the level becomes too high or too low.

CAPACITIVE PROBES

Capacitive probe sensors are shown in Figure 12-8. They are used for continuous level measurement. The principle of operation is based on the theory of capacitance.

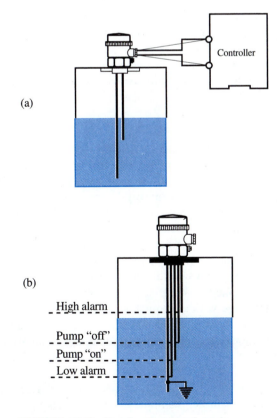

FIGURE 12-7 Conductive level probes

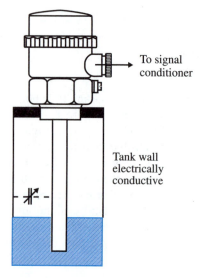

To signal
conditioner

Tank wall
electrically
conductive

FIGURE 12-8   Capacitive probe

The probe and the metal wall of the tank form the two plates of a capacitor, and the contents in the tank is the dielectric. When a non-metallic tank is used, a second electrode—referred to as a *counterelectrode*—is used.

When the level of the contents varies, it causes the capacitance to change. The change of capacitance is converted into a voltage signal which is proportional to the level. The medium must either be nonconductive or have a low dielectric constant.

ULTRASONIC SENSORS

The ultrasonic sensor, another type of continuous level detector, is shown in Figure 12-9(a). This type of sensor is similar to retroreflective optoelectronic sensors. Ultrasonic sound waves (above the frequency heard by humans) are developed by an oscillator. They are emitted by a transmitter toward the top surface of the medium and are reflected back to the ultrasonic signal receiver. The time it takes the waves to travel from the transmitter to the target surface and back to the receiver is measured. The time lapse between transmission and detection is proportional to distance. This data is calculated electronically and converted into a liquid-level measurement. The auto-focus mechanism of a camera works on the same principle.

The ultrasonic sending unit consists of a piezoelectric crystal sandwiched between two metal plates. An AC voltage with a frequency of 20K to 100KHz is applied to the plates. Because of its atomic structure, the side of the crystal connected to one polarity expands, and the other side contracts when the opposite polarity is applied. The high-frequency expansion and contraction of the crystal causes the surrounding air to emit ultrasonic waves.

By replacing the AC source with a voltage amplifier, a second assembly, made of the same components as the transmitter, can operate as an ultrasonic receiving unit. The incoming ultrasonic waves cause the diaphragm to vibrate. The result is that the

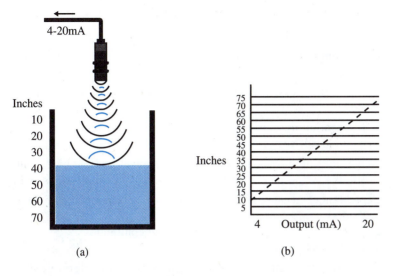

FIGURE 12-9 Ultrasonic sensors

piezoelectric crystal expands and contracts, which creates a high-frequency AC voltage between the plates. The graph in Figure 12-9(b) shows that the output current produced by the ultrasonic sensor is proportional to the distance it measures.

CHAPTER PROBLEMS

1. List two common units of measurement to describe height.

2. List two types of inferred values used to indicate level measurement.

3. List two reasons why it is essential to monitor the level of materials in a container.

4. Explain the meaning of the term interface.

5. A dipstick is _____ (a direct, an indirect) measurement device.

6. Point measurements are used to determine ____.
    a. high-level limits   c. both high- and low-level limits
    b. low-level limits

7. Using the weight of an object is ____ method of measuring level.
    a. an invasive   b. a non-invasive

8. List an indicator method used for level measurements that does not provide a feedback signal.

9. A float is ____ method of determining level.
    a. an invasive   b. a non-invasive

10. As fluid level rises, the weight of a displacement float indicator appears to ____.
    a. increase   b. decrease

11. As the fluid level increases, the rate of bubbles emitted from a purge device

    _____ (increases, decreases).

12. List a common mechanical instrument used to measure the level of solid materials.

13. Where is the pressure detector placed in a vessel to measure the level of a liquid?

14. A differential pressure to determine level is used in _____ (an open, a closed) vessel.

15. Weight measurements can be used to determine the level of ____ in a tank.
    a. solids   b. liquids   c. both solids and liquids

16. Capacitance measurements are made to determine the level by _____ method(s).
    a. point   c. both point and continuous
    b. continuous

17. The ultrasonic instrument uses ____ method(s) to determine the level of foods and pharmaceuticals in a container.
    a. invasive   c. both invasive and non-invasive
    b. non-invasive

CHAPTER 13

Industrial Detection Sensors

CHAPTER OBJECTIVES

At the conclusion of this chapter, you will be able to:

○ Define an industrial detection sensor and provide several examples of the types of applications for which it is used.

○ List the parts of a limit switch, provide examples of the types of functions it performs, and list precautions that should be followed when connecting it to machinery.

○ Explain the operation of an inductive proximity detector, describe the function of the sensor circuitry, and provide examples of its applications.

○ Explain the operation of a capacitive proximity detector, describe the function of the sensor circuitry, and provide examples of its applications.

○ Explain the operation of a Hall effect sensor and provide examples of its applications.

○ Describe the operational theory of the three components that make up a photoelectric sensor.

○ Describe the operational theory, characteristics, and application examples of the following photoelectric methods of detection:

Opposed Sensing	Convergent Sensing
Retroreflective Scanning	Specular Sensing
Diffuse Sensing	Color Mark Sensing

○ Properly interface electromechanical relays, solid-state relays, and analog sensor outputs to load devices.

INTRODUCTION

An industrial detection sensor is a specialized type of measurement device used in an automated system. Its function is to detect the absence, presence, or distance of an object from a reference point. The object to be detected is referred to as the **target**.

Applications of detection sensors are as follows:

1. Verifying when a machine part has reached a certain position.
2. Counting gear teeth or measuring the revolution of a shaft to determine rotational speed.
3. Verifying the proper placement of parts during an assembly line procedure.
4. Making edge guide measurements to detect the alignment of a manufacturing process.
5. Counting the number of products that are transferred on a conveyer belt.
6. Determining the size of a product passing an inspection point.

Industrial detection sensors are measurement devices that enable closed-loop equipment to perform process control and motion control operations. This chapter explains the operation of several types of industrial detection sensors, describes their characteristics, and provides practical examples of their use in industrial applications.

13-1   LIMIT SWITCHES

The most fundamental detection sensor is the **limit switch**. By using some type of lever, it converts mechanical motion into electrical signals. When the switch makes physical contact with a moving object, a set of electrical contacts are forced either open or closed. Limit switches are used in such applications as detecting the position of a machine shop carriage, counting parts on conveyer rollers, detecting size for sorting boxes on a conveyer belt, or limiting the travel of an elevator for safety reasons. In these applications, the limit switch causes a change in an operation such as starting, stopping, changing direction, recycling, slowing down, or speeding up movement.

Limit switches, as shown in Figure 13-1(a), have two main parts: the electrical contacts, usually sealed within an enclosed body, and a actuating mechanism. Three types of limit switches with different actuating mechanisms—a roller lever, a fork lever, and a push-roller—are shown in Figure 13-1(b). When the mechanism is moved, it causes the electrical contacts to actuate.

The contacts are either normally open (N.O.), normally closed (N.C.), or a combination of both N.O. and N.C. The contacts are designed to open or close quickly so that rebounding and arcing is minimized.

Certain rules should be observed when wiring the limit switch. It is important to make proper connections to multi-terminal limit switches. Also, the current rating of the switch should not be exceeded. Failure to observe these considerations can damage the switch or alter the operation of equipment, which can cause breakage or injury. Care should also be taken not to install the switch in a position where something may interfere with its operation. For example, it should not be placed where scrap material will move the lever or near liquid which flows between open contacts, because this can cause a false continuity condition.

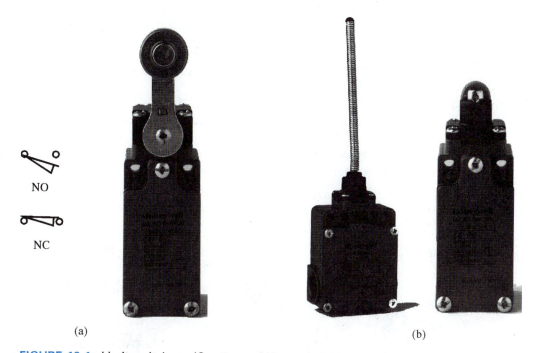

NO

NC

(a)

(b)

FIGURE 13-1 Limit switches (Courtesy of Honeywell Micro Switch)

13-2   PROXIMITY DETECTORS

Proximity detectors are electronic sensors that indicate the presence of an object without making physical contact. The detector normally does not respond by producing a linear output signal proportional to the distance of the object to the sensor. Instead, the output turns on or off. That is why these devices are commonly called *proximity switches.*

The two major types of proximity detectors are inductive and capacitive.

13-3   INDUCTIVE PROXIMITY SWITCHES

During World War II, limit switches were used in German tanks to detect the movements of parts. The switches worked adequately until the war expanded to the deserts of North Africa. Because sand granules became lodged between the contacts, the limit switches were unable to close. In their attempt to find a device impervious to sand, German engineers developed the inductive proximity detector. The inductive proximity switch detects the presence of metallic materials.

PARTS OF AN INDUCTIVE PROXIMITY SWITCH

Figure 13-2(a) shows a block diagram of the inductive proximity switch. It consists of an oscillator, a sensor head, a demodulator, a trigger, and an output stage.

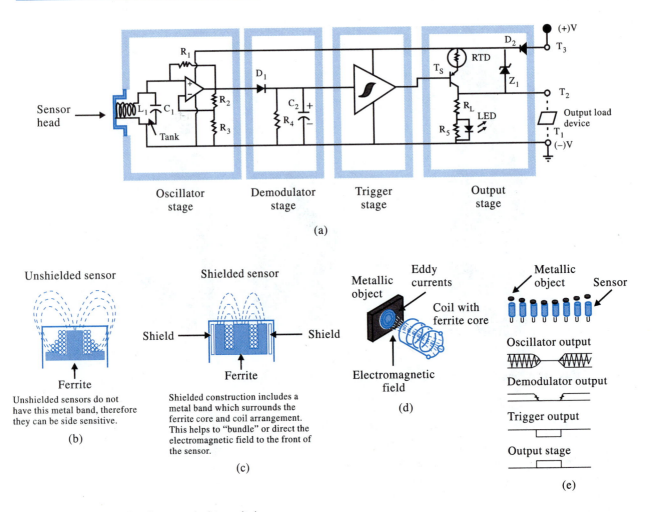

FIGURE 13-2 An inductive proximity switch

Oscillator

The oscillator consists of a closed-loop op amp configuration. It has a tank circuit connected to the non-inverting input lead. One branch of the tank circuit contains a capacitor and the other branch an inductor. Once power is applied, resonance develops in the tank, and the oscillator will produce a frequency between 100 to 1MHz.

Sensor Head

The coil of the tank inductor is wound around a ferrite core. As the oscillating current flows through the inductor, an electromagnetic field continually expands and contracts around the coil. The shape of the core causes the flux lines to bundle together. Figure 13-2(b) shows an *unshielded sensor,* which directs the flux lines to the front and side of the sensor head. This type of sensor head detects the *radial approach* of an object (an object from the side of the sensor head surface). Figure 13-2(c) shows a *shielded sensor* with a shielded metal band around the ferrite core. The shield directs the flux lines to the front of the sensor head. This type of sensor head detects the *axial approach* of an object (an object directly in front of the sensor head).

Demodulator

The oscillating signal is fed into the demodulator section. The demodulator is a filtered rectifier which converts the AC signal into a DC voltage level.

Trigger Stage

The output of the demodulator is connected to a Schmitt trigger input. The Schmitt trigger produces two voltages. If the demodulator decreases to a certain amplitude, it causes the trigger output to quickly change to a low voltage level. When the demodulator increases again to a certain amplitude, the trigger output quickly changes to its higher voltage level.

Output Stage

The function of the output stage is to drive a load *On* or *Off*. It consists of an NPN current sinking transistor or a PNP current sourcing device. The component functions depicted in Figure 13-2(a) are as follows:

Terminals

T_1	Ground terminal of sensor connected to the power supply
	Negative output terminal
T_2	Positive output terminal
T_3	Positive terminal of sensor connected to the power supply
T_S	Output switching transistor
RTD	Overcurrent protection
R_L	Load resistor
Z_1	Zener diode to limit voltage peaks across transistor
D_2	Protection diode to block current flow if the power supply polarity is reversed
LED	Indicator (light emitting diode), which turns on when the output terminals are energized

OPERATION OF AN INDUCTIVE PROXIMITY SWITCH

If metal is not detected, the following conditions will exist:

- The voltage amplitude of the oscillator and demodulator are at their highest level.
- The Schmitt trigger produces its high voltage level.
- The positive voltage applied to the base of the PNP transistor turns it off. The open transistor condition causes the load to be de-energized because there is no potential difference between the Output Terminals T_1 and T_2.

Eddy currents are induced into a metal object when it enters the oscillating field, as shown in Figure 13-2(d). The effect is like a shorted transformer secondary. If the inductor is considered the primary, current flow through it increases. The high current draw from the oscillator circuit results in a loss of its energy and, consequently, a smaller amplitude of oscillation.

- ○ The reduced voltage amplitude of the oscillator and demodulator causes the Schmitt trigger to switch to its lower voltage level.

- ○ The low voltage at the PNP transistor base turns it on. The voltage of terminal T_2 is raised to a potential close to the voltage of terminal T_3 because the transistor functions like a closed switch. The voltage difference between output terminals T_1 and T_2 energizes the load. The term *load* refers to the device to which the sensor is applied.

When the metal object leaves the sensing area, the oscillator regenerates, allowing the sensor to return to its normal state. The advantage of an inductive sensor is that it ignores non-conductive objects. Figure 13-2(e) pictorially shows the output signals of each stage when the object is at various distances from the sensor.

The proximity switch has a built-in characteristic called **hysteresis** or *differential travel*. This property means that the target must be closer to turn the sensor *on* than to turn it *off*. Figure 13-3 illustrates the advantage of hysteresis. Once an approaching object passes the On-point, the sensor remains on until it moves away to the Off-point. Without the differential gap, the output could potentially "chatter" on and off if a target was stationary at the fringe-sensing distance. The Schmitt trigger provides hysteresis. It switches high when the input increases to one threshold voltage level (Vth+), then switches low when the input reduces to a lower threshold (Vth−) than Vth+.

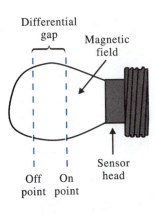

FIGURE 13-3
The hysteresis property of a proximity switch

13-4   CAPACITIVE PROXIMITY SWITCHES

The capacitive proximity switch detects the presence of conductive and non-conductive targets. This sensor is activated when an object enters the electrostatic field of a capacitor.

Figure 13-4(a) shows a block diagram of its construction. Its design is very similar to the inductive counterpart just described. The major difference is the oscillator used. The sensing element C_1 is placed in the feedback loop of the oscillator. Its electrodes are concentrically positioned, as shown in Figure 13-4(b). The other significant difference is that an NPN current transistor is used in the output stage.

Figure 13-4(c) shows a non-conductive object in the field of the capacitor. The capacitance increases because the object's dielectric constant is greater than air. The value to which it raises depends on the dielectric constant of the material that is sensed. Many solids and liquids cause the capacitance to change.

Figure 13-4(d) shows a conductive material within the field. The object becomes another electrode and forms two capacitors. Because the extra electrode causes the distance between the plates to reduce, the total capacitance increases.

OPERATION OF A CAPACITIVE PROXIMITY SWITCH

When an object is not present, the oscillator is inactive. The outputs of the demodulator and trigger are both zero volts. When a zero volt potential is present at the base of

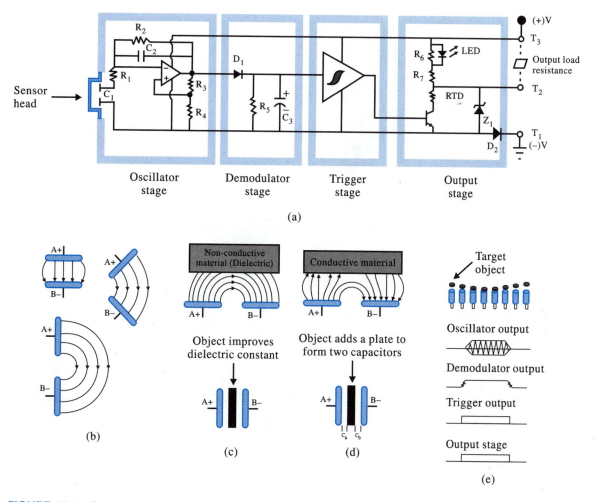

FIGURE 13-4 Capacitive proximity switch

the transistor, there is no bias voltage across the emitter and base. The transistor is in the cutoff condition and causes the supply potential to drop across terminals T_1 and T_2. Since no voltage is developed across terminals T_2 and T_3, the detector's output is de-energized.

As an object approaches, the oscillator begins to oscillate. The closer the object gets to the electrode plates, the faster the oscillations become. This action causes the Schmitt trigger to switch to its higher voltage level. The positive potential forward biases the base-emitter junction of the NPN transistor, causing it to turn on. The result is that the detector output energizes because most of the supply voltage drops across terminals T_2 and T_3. Figure 13-4(e) pictorially shows the output signals of each stage when the object is at various distances from the sensor.

Figure 13-5(a) shows a practical application of a capacitive proximity detector. Figure 13-5(b) shows a practical application of an inductive proximity detector. Table 13-1 compares capacitive and inductive proximity detectors.

Making sure the product is in the package   Detection of seal foil beneath cap

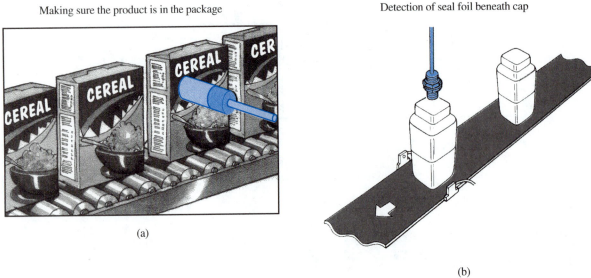

(a)

(b)

FIGURE 13-5 Proximity detector applications   (Courtesy of efector, inc.)

TABLE 13-1   **PROXIMITY DETECTORS**

Method of detection	Inductive shielded	Inductive unshielded	Capacitive
Configuration	Designed for flush mounting in metal. Available in cylindrical and limit switch sensor shapes.	Requires clearance around sensing end to prevent false signals from surrounding metal. Available in cylindrical, limit switch and small block, flat rectangular sensor shapes.	Requires clearance around sensing end to prevent false signals from surrounding mounting materials. Available in cylindrical and flat rectangular sensor shapes.
Advantages	Detect ferrous (iron, mild steel, stainless steel) and non-ferrous (brass, copper, aluminum) metals. Allows flush mounting to prevent impact damage to the sensor. Color and surface conditions of the target do not affect sensing. Most cost-effective option where appropriate.	Detect ferrous (iron, mild steel, stainless steel) and non-ferrous (brass, copper, aluminum) metals. Longer sensing distance than shielded sensors. Color and surface conditions of the target do not affect sensing.	Detect plastic, glass, liquids, leather, and wood as well as metals. Can be used to detect materials inside non-metallic containers.
Disadvantages	Reduced sensing distance. Usable only to 0.4 inch maximum.	Sensor is not protected from accidental impact damage. Usable only to 0.7 inch maximum.	Sensor is not protected from accidental impact damage. Usable to 0.9 inch maximum.

13-5   HALL EFFECT SENSOR

A Hall effect device is a sensor which detects the presence of a magnetic field. This sensor is a flat rectangular piece of P-type semiconductor (called a Hall generator) usually made of indium arsenide (In As). Hall effect sensors are four-terminal devices. The two end terminals are power supply connections. The terminals located on either side are the output leads.

When a power source supplies a constant current to the sensor, positively-charged carriers flow uniformly through the material. Figure 13-6 shows what happens when the semiconductor is subjected to a perpendicular magnetic field. The charged carriers are deflected to one side of the flat piece by an effect called the *Lorentz force*. The result is that an EMF (called *Hall voltage*) develops across the output terminals. The side to which the charged carriers move becomes positive, and the other side becomes negative. If the power supply or the magnetic field direction is reversed, the polarity of the Hall voltage will reverse.

Three variables determine the Hall voltage amplitude produced by the Hall generator:

1. The amount of current supplied by the power source.
2. The magnetic field strength to which the Hall device is subjected.
3. The physical size of the semiconductor material.

Since the current is held constant and the physical size does not change, the magnetic field strength determines the voltage amplitude. As the magnet is brought near, the flux density rises and the Hall voltage increases proportionally. This is known as the Hall effect, discovered by Edward H. Hall in 1879.

Hall effect devices come in two basic functional classifications: linear and digital.

LINEAR HALL EFFECT SENSORS

Figure 13-7(a) shows a diagram of a linear Hall effect device. It consists of a Hall generator, linear amplifier, and emitter-follower output transistor. All three sections are fabricated onto an integrated circuit chip.

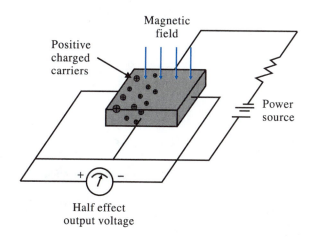

FIGURE 13-6 A Hall voltage produced by the Lorentz force

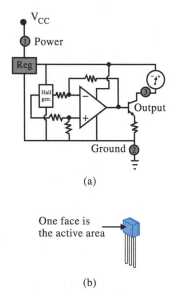

(a)

(b)

FIGURE 13-7 A linear Hall effect device

A power supply with good voltage regulation supplies a constant current flow. When flux lines intersect the semiconductor material, it produces an analog voltage that is directly proportional to the strength of the magnetic field. A differential operational amplifier is used to boost the Hall generator voltage above the millivolt level. Its output controls the input bias to the output transistor. Note that emitter followers have no voltage gain. They are used to produce a current gain sufficient to drive an output indicator.

Figure 13-7(b) shows the package style of a Hall effect sensor. Typically three leads are used because the negative leads of the power supply and output are connected together. One face of the package is the Hall sensor's active area. To operate, the magnetic flux lines for a prescribed pole must be perpendicular to the face of the package. If the polarity of the magnet is incorrect, the output of the sensor will be zero volts.

DIGITAL HALL EFFECT SENSOR

The linear Hall effect sensor can be modified to make it compatible with digital circuitry. Figure 13-8 shows the output of the differential op amp. It is connected to a Schmitt trigger, which feeds a signal into an open collector NPN output switching transistor.

As the magnetic south pole approaches, the Hall cell is exposed to increasing magnetic flux density. The Hall generator and the op amp outputs increase. At some point the threshold input of the Schmitt trigger is reached. Its output switches from low to high level voltage, which turns the switching transistor ON. The output (Terminal 3) goes to zero. The level at which the flux density's strength causes the output to turn on is called the *operating point*.

If the magnet is moved away from the Hall cell, the flux density decreases, and the Hall generator and op amp voltages decrease. Because of the Schmitt trigger hysteresis, the Hall output will not return to a low level unless the magnetic flux density falls to a value far lower than the operating point. When it does, the switching transistor turns OFF and the output voltage equals the power supply. The value at which the flux density's strength causes the output to turn off is called the *release point*.

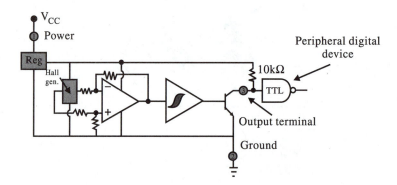

FIGURE 13-8 A schematic diagram of a digital Hall effect device

A frequent application of Hall effect sensors is to generate a digital output indicating the velocity, displacement, or position of a rotating shaft. The magnetic field that the Hall device detects for rotary applications can be supplied by either a magnetic rotor assembly or a ferrous vane rotor assembly.

Magnetic Rotor Assembly

Figure 13-9(a) shows a disc with magnets attached to a rotating shaft. A stationary Hall device is positioned so that it becomes activated by each pass of a magnetic south pole.

Ferrous Vane Rotor Assembly

Figure 13-9(b) shows a magnet and Hall sensor mounted at stationary positions with a small air gap between them. When the flux lines are uninterrupted, the Hall output is held *on* by the activating magnet. As the shaft spins, the vane will pass between the magnet and Hall effect device. When it does, the Hall generator turns off because the vane will form a magnetic shunt that distorts the flux away from the sensor.

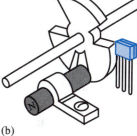

(a)

(b)

FIGURE 13-9 Magnetic rotor assemblies

13-6   PHOTOELECTRIC SENSORS

Photoelectric sensors use light to detect the absence or presence of an object. Detection occurs if a light beam is interrupted or reflected by the object being sensed.

Sensing with light became popular in the 1950s. The early photoelectric systems consisted of two elements: an incandescent lamp and a light-sensitive resistive device called a *photocell.* The lamp was placed so that its light could be projected across the sensing area to the photocell. These systems had three shortcomings. First, the bulb lost its intensity and became ineffective as it aged. Second, the filaments would break if exposed to temperature extremes or high vibration. Third, in order for the photocell to differentiate between the beam of the lamp and ambient light, both elements had to be carefully aligned and positioned within a limited distance. All of these problems were eliminated by the development of the semiconductor sensing devices presently used.

Photoelectronic sensors may be divided into three components: the light source, the light sensor, and the sensor circuitry.

LIGHT SOURCE

The **light source** supplies the light beam which is transmitted to the light sensor. The light source is also referred to as an *emitter* or *transmitter.* Light emitting diodes (*LEDs*) are used most frequently as light sources.

The LED is a PN junction semiconductor diode that emits light. When it is forward biased, electrons in the N-type material enter the P-type material, where they combine with excess holes. As a hole and an electron combine, a packet of energy called a photon is generated, which escapes the surface as light radiation. The type of semiconductor material used determines the wavelength or color of the light. For example, gallium phosphide produces green; gallium arsenide phosphide produces red. Yellow and blue LEDs are also available. Semiconductors made of gallium arsenide produce infrared

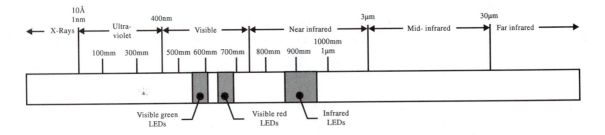

FIGURE 13-10   The light spectrum

light, which is invisible to the human eye. Figure 13-10 shows a light spectrum chart that indicates the wavelength of each type of light generated by LEDs.

Because LEDs are solid state, they will last nearly forever. Their light intensity does not deteriorate, and they are immune to vibration and shock. LEDs produce only one wavelength, which is a distinct advantage in some applications, and turn on immediately without a warm-up time. The one disadvantage of an LED is that the light intensity it produces is only 1 percent of the light generated by an incandescent bulb of the same physical size.

Visible and infrared LEDs are both used as the light source for photoelectric transmitters. Visible LEDs can be monitored visually during their operation, are easy to align, and can be used for sensing differences in color. Infrared LEDs produce a stronger light than visible LEDs. This characteristic gives them the capability of penetrating certain objects, such as cardboard cereal boxes, much the same way that light from a flashlight shines through a hand. Also, photodiodes and phototransistors used in the sensing component respond better to the infrared wavelength than to visible light waves. They are also popular for security applications because infrared light is invisible.

LIGHT SENSOR

The **light sensor** that detects the absence or presence of an object is called either the *detector* or the *receiver*. The most common types of light-sensitive components used for detection are photodiodes and phototransistors. A photodiode is a two-terminal junction device made of silicon that operates when it is reverse biased. A lens is placed on the main body to allow light to pass to the PN junction. Without light, the resistance is very high. When light falls on the junction, electron-hole pairs are generated and the resistance of the diode decreases. The brighter the light, the lower the resistance. Photodiodes are linear, operate at high frequencies, and detect reddish visible light and infrared light waves.

Phototransistors are two-terminal, three-junction devices made of silicon. A lens is built into the transistor case to allow light to strike the collector-base junction. When the reverse biased collector-base junction is exposed to light, the photons cause base current to flow. The stronger the light, the higher the collector current. Phototransistors respond well to infrared light. Because phototransistors amplify, they have greater light sensitivity than photodiodes but operate at much lower frequencies, are nonlinear, and have large temperature coefficients.

THE SENSOR CIRCUITRY

Figure 13-11 shows the block diagram of the emitter and detector for the photoelectric **sensor circuitry**.

Emitter Block

Compared to an incandescent lamp, an LED produces a very small amount of light. However, by modulating the LED, it is capable of producing more optical power than non-modulated incandescent sensors. Modulating the LED simply means turning it on and off at a very high frequency. By tuning the amplifier in the receiver to the frequency of the modulation, only that frequency is amplified. This principle of operation is similar to the way in which a radio receiver tunes to one station while rejecting all the other radio waves. In Figure 13-11, the oscillator circuit modulates the LED by turning it on and off at a high frequency.

Detector Block

The detector portion of the sensor in Figure 13-11 shows that a phototransistor is used to detect the light. As the lightwave strikes the transistor base, the signal produced is applied to the input of a high gain amplifier tuned to the modulated frequency. The output of the amplifier is demodulated by a filtered rectifier. The demodulation function reduces the problem of critical alignment, and it allows the sensor to be used in areas where ambient light levels are relatively high or when dirt, oil, or smoke obscure the lens. A Schmitt trigger is used to increase the switching speed for very high-speed applications, for hysteresis, for noise immunity, and for logic level outputs.

The receiving sensor unit is available with or without the demodulator. A sensor without the demodulator has a gain that is limited to the point at which the receiver recognizes ambient light. Therefore, it requires critical alignment because of the long focal point lens it uses. In contrast, a demodulated receiver ignores ambient light and responds only to the modulated light source. As a result, the gain of the amplifier may be turned up to a very high level. The high gain operation reduces the problem of critical alignment and enables the sensor to detect light effectively even when its lens is obscured by dirt, oil, or smoke. Some non-demodulated sensors are also referred to as ambient receivers. They are used to detect objects which emit their own light, such as red-hot metals, hot glass, or anything that emits infrared light energy that is many times stronger than ambient infrared light.

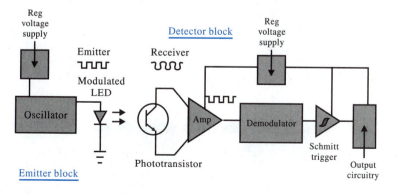

FIGURE 13-11   A block diagram of a light sensor

13-7   METHODS OF DETECTION

There are several ways in which the light source and receiving elements can be physically positioned to detect objects. These types of arrangements are referred to as **methods of detection**. The mode that is selected is based on sensing distance, the arrangement that yields the strongest signal, and mounting restrictions. They also depend on the characteristics of the object to be detected. For example, it is important to know whether the objects are opaque, translucent, or clear; whether they are highly or slightly reflective; and whether they are in the same position or randomly positioned as they pass the sensor.

There are two modes used to detect an object with photoelectronic sensors. The first mode, referred to as going from *light-to-dark,* uses a detector that sees energy coming from an emitter until it is obstructed by the object it is detecting. The second mode is referred to as going from *dark-to-light.* It uses a detector that looks for an energy source and sees it either through reflection or when an obstruction is removed.

Most sensing applications rely on one of six commonly used methods of detection: opposed sensing, retroreflective scanning, diffuse sensing, convergent sensing, specular sensing, and color mark sensing.

OPPOSED SENSING METHOD

In the opposed sensing method, the emitter and detector are positioned opposite each other. The light from the transmitter shines directly at the receiver, as shown in Figure 13-12. Because the target sensed by this method is usually opaque, the mode of detection is usually from light-to-dark. The object is detected when it breaks the light beam.

The opposed sensing provides the highest level of optical energy. Therefore this method resists lens contamination and misalignment, and it is capable of long scanning ranges. Opposed sensors are used in applications that require the detection of small parts, accurate positioning, and parts counting. The opposed scanning method is also known as *direct scanning, beam break,* and *through beam.*

It is possible to measure the size of an object when using an array of opposed beams, as shown in Figure 13-13(a). This series of beams is referred to as a *light curtain.* Light curtains are used in other applications, for example counting objects such as towels, as shown in Figure 13-13(b), or operating as a machine guard to provide personal safety, as shown in Figure 13-13(c).

The opposed mode should be used whenever possible because it is the most reliable optical sensing system. This method should not be selected for translucent or transparent targets that cannot block light. It should also be avoided when the objects are too small to interrupt 100 percent of the beam.

Emitter   Target   Receiver

FIGURE 13-12 Opposed sensing method

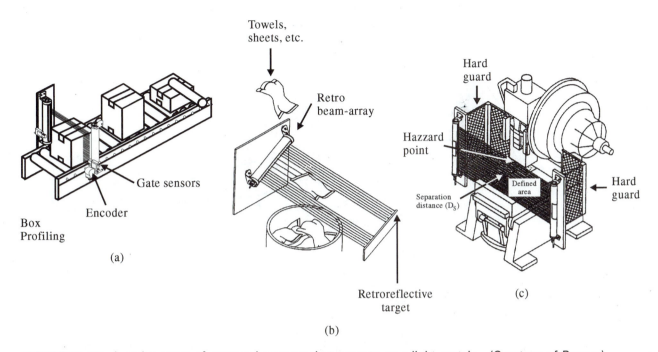

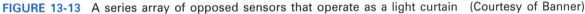

FIGURE 13-13 A series array of opposed sensors that operate as a light curtain   (Courtesy of Banner)

RETROREFLECTIVE SCANNING METHOD

The retroreflective sensing mode (also called *retrosensing*) is by far the most popular method of optical sensing. In this system, the emitter and detector are mounted next to each other in the same housing. The beam from the transmitter is reflected back to the receiver by a prismatic retroreflector that is mounted opposite the sensor unit, as shown in Figure 13-14. The retroreflector, which looks like a bicycle reflector, is made of 3M's Scotchlight® and Reflectolite®, materials originally designed for highway signs and markers. The mode of detection is from light-to-dark because the detection of opaque targets occurs when the light beam is broken.

Retroreflective sensors are sometimes used in spaces where scanning is only possible from one side. The increased flexibility of placement they offer is offset by a significant loss in signal strength. In addition to the inherent inefficiency of the reflector, the beam must pass through four lenses, twice at the sensor housing and twice at the reflector. Retrosensing is used in applications where the targets are large, the sensing environment is clean, and the scanning ranges are at a medium distance. Specific examples include conveyers, automated storage and retrieval systems, and bar code reading.

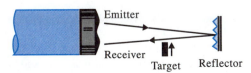

FIGURE 13-14 Retroreflective sensor

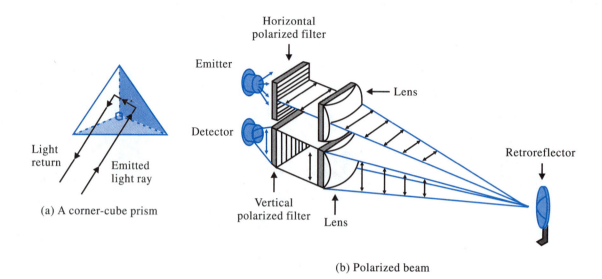

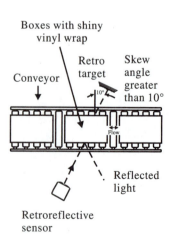

(a) A corner-cube prism

(b) Polarized beam

FIGURE 13-15 Retroflective sensors with polarization filters

Highly-reflective objects may go undetected if they return a similar amount of light as the reflector. Three techniques are used to prevent unwanted light from being reflected back to the receiver. The first technique affects the light emitted by the transmitter. A polarized filter inside the lens blocks the vertical light waves and passes only the horizontal waves. The second technique involves the retroreflector. The reflector is made up of many individual tiny specular inner surfaces called corner cubes, as shown in Figure 13-15(a). This configuration is similar to the inside of a box that has been cut diagonally in half from one corner to another. The three sides of the cube are located 90 degrees from each other. When a light beam enters a cube, it reflects off one surface to another surface, then reflects back to its origin. The result is that the beam that enters is rotated by 90 degrees when it returns to the receiver. The third technique is that a polarized filter is positioned behind the receiver lens so that it blocks horizontal light waves and passes only vertical waves. When the three configurations are used together, as shown in Figure 13-15(b), they block any reflected light from the target and any ambient light waves. Another method of solving reflection problems is to position the sensor housing at a skew angle to the object's surface, as shown in Figure 13-16.

The retroreflective scanning method is also known as *retro* or the *reflex* method. This method should not be used for detecting translucent or transparent objects that do not block light, or for sensing materials with shiny surfaces that may reflect as much light as the reflector.

FIGURE 13-16 Use of skew angle to control reflection

DIFFUSE SENSING METHOD

In the diffuse sensing method, the transmitter and receiver are contained inside the sensor housing, as with the retro sensor. The target is sensed when its position is in line with the light beam. As the object is struck by the transmitter beam, the light is scattered (or diffused), with enough light being returned to the receiver to indicate its presence. This dark-to-light mode of detection is opposite to that of the opposed and retro sensing schemes. Figure 13-17 shows a diffuse sensor.

From an installation standpoint, diffuse sensing is the most convenient method, because detection is accomplished from one side of the process without the need of a reflector on the other side. However, a few requirements must be met to make this method effective:

1. The object must be reflective to diffuse enough light back to the receiver.

2. The surface area of the object must be large enough to reflect sufficient light back to the receptor.

3. The sensor must be placed so that the nearest background surface is three times the distance from the sensor to the target. This avoids the problem that may arise if the background reflectivity is greater than that of the object to be sensed.

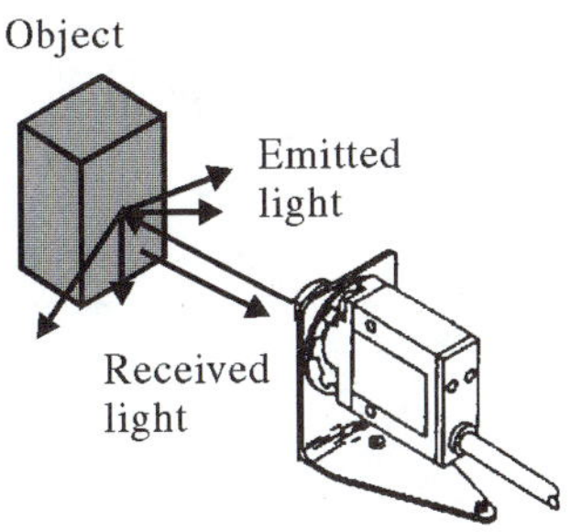

FIGURE 13-17 Diffuse reflection sensing method (Courtesy of Banner)

CONVERGENT SENSING METHOD

The convergent sensing method is similar to diffuse sensing. Both function on the principle of light being diffused from the target. Figure 13-18(a) shows the convergent sensor. The transmitter and receiver are set at the same angle from the vertical axis to capture the light. The sensor is designed to detect a target at one set distance. By using special lenses, the light source from the transmitter is focused to a narrow depth-of-field. (The **depth-of-field** is the distance on either side of the sensor's focus point.) Objects nearer or farther from the focal point will not activate the sensor.

By focusing the light beam onto a very small area, a convergent beam sensor directs light much more intensely than a diffuse sensor. This property enables the sensor to detect objects that are very small or that have very low reflectivity. Therefore, they are used for specific applications such as counting bottles, jars, or cans on a conveyer, where there is no space between adjacent products, as shown in Figure 13-18(b). They are also used to detect height differential, as with moving parts on a conveyer; to inspect for "parts-in-place" in an assembly operation; to sense accurate position of clear materials; or to detect the fill level of materials in a open container.

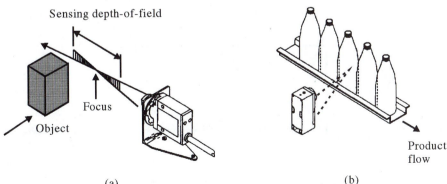

(a)   (b)

FIGURE 13-18 Convergent sensing method

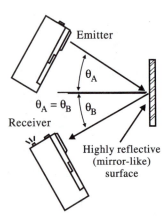

Emitter

θ_A

$\theta_A = \theta_B$ θ_B

Receiver

Highly reflective
(mirror-like)
surface

FIGURE 13-19 Specular sensing method, which detects the difference between shiny and dull surfaces

The convergent mode should not be used to detect objects that pass at an unpredictable distance from the sensor. To prevent an unwanted light reflection from a shiny background surface, the sensor should be rotated at an angle away from a perpendicular position.

SPECULAR SENSING METHOD

The specular sensing method is used to detect objects with mirror-like surfaces. Figure 13-19 shows that the transmitter and receiver are placed at equal angles from the object. To operate effectively, the distance between the sensor and target must remain constant.

The specular sensing method is used in applications in which it is necessary to differentiate between shiny and dull surfaces. A common example is sensing the alignment of a diffuse material such as cloth on a steel sewing machine table with a reflective surface.

COLOR MARK SENSING METHOD

Color mark detection is different than other types of photoelectric sensor methods. Instead of scanning an object as it passes an inspection point, color mark instruments detect the contrast between two colors. The two colors being sensed can be on the same surface or on separate objects.

Color marks are used extensively in packaging operations. For example, they are the reference point for indexing the cutoff location of packaging materials so that printed information always appears in the same location. Figure 13-20 shows a color mark printed on a toothpaste tube.

Convergent beam sensors with a visible light are commonly used to detect the color mark. The best source of light for color sensing is white, because it contains all colors. White is produced by clear LEDs or incandescent lamps. The greatest optical contrast is provided when a black color mark is printed on a white surface.

Visible LEDs with a red color are used most frequently to transmit the light in color mark applications because they are spectrally matched to a phototransistor. However,

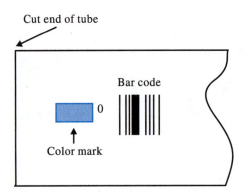

Cut end of tube

Bar code

0

Color mark

FIGURE 13-20 Color mark detection example

for the combination of red, orange, or pink against a light color, a visible red LED should not be used, because reflected red light from a red mark will be almost as great as the amount reflected by the background surface. A green LED should be used instead, even though it is considered a low-powered color.

13-8   PHOTOELECTRIC PACKAGE STYLES

Photoelectric sensors are available in two package styles: self-contained and remote. The lenses, LED, photodetector, and electronic circuitry are all a part of the self-contained package. Cables for providing power and data transfer are connected to the package. Remote photoelectronic sensors contain only the optical components of the sensing system. The circuitry for the sensor is at another location. For this reason, remote sensors can be placed in smaller and more hostile environments than self-contained sensors.

In some situations, the space is too confined or the environment too hostile even for remote sensors. For such applications, **fiber optic** conductors may be used. Fiber optics are transparent strands of glass or plastic that transfer light to and from such locations. Fiber optic "light pipes" are capable of conducting light around corners and operating when exposed to high temperatures or vibration. They are immune to magnetic noise. In Figure 13-21(a), the fiber optic transmitter and receiver are in the same housing, performing opposed sensing. The beam is sent to the site from the transmitter through one cable. The light spreads at a 60-degree angle from the emitter sensing head at the site. The light enters the detecting sensor head and returns to the receiver through the other optic cable if an object is not interrupting the beam. The distance between the emitter and detector sensing heads is up to 150mm for a diffused reflection sensor, and up to 500mm for the opposed sensing configuration. Figure 13-21(b) shows bifurcated fiber optic conductors that contain cables to transmit and receive signals simultaneously, performing diffuse reflective sensing.

Plastic fibers are used in applications where there is repeated flexing, such as on a reciprocating machine. However, their drawbacks are that they are only capable of conducting visible light, they cannot be exposed to extreme temperatures, and they can be damaged if exposed to chemicals. Glass fibers can withstand high temperatures and exposure to chemicals, but they break if bent too often or too sharply.

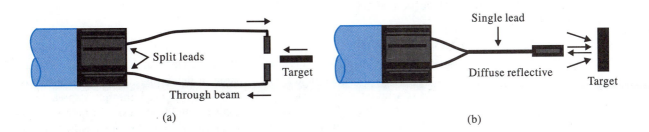

(a)   (b)

FIGURE 13-21 Fiber optic cables used to perform photoelectric sensing methods

13-9   OPERATING SPECIFICATIONS

Data specifications are available for each type of optical sensor. This information helps the user select the proper sensor by indicating how well it will operate under certain conditions. The data sheets provide information on sensitivity, excess gain, field-of-view, and sensor response.

SENSITIVITY

Sensitivity is a measure of the amount of change in light intensity that is required by the sensor to cause a switching action at its output. Sensitivity is the combined result of several design factors of the sensing device, such as:

1. The amount of amplification incorporated in the electronic circuitry.
2. The light power of the LED.
3. The size, shape, and quality of the lens.

Each of these variables affects the operating distance at which the sensor detects a target.

EXCESS GAIN

The excess gain specification is a measurement of the amount of light energy that falls on the receiver beyond the minimum amount of light required to operate the sensor amplifier. In equation form, it is expressed as:

$$\text{Excess Gain} = \frac{\text{Light Energy Falling on Receiver}}{\text{Amplifier Threshold}}$$

An excess gain of 1× (*one times*) indicates the minimum amount of light energy required to operate the amplifier. When the light is attenuated or decreased by environmental obstructions such as dirt, smoke, and other contaminants, the excess gain of the sensor is necessary to overcome the loss. For example, if 50 percent of the light is attenuated, an excess gain of 2× is required to overcome the loss of light. Table 13-2 provides some guidelines to help determine which sensor rating to use for certain conditions. In a modulated sensing system, voltage is used instead of light energy in the excess gain formula.

For applications that require the beam to penetrate paper-thin cardboard or similar materials, an excess gain of at least 50× is required. An example is the level sensing of the contents of a cereal box.

FIELD-OF-VIEW

Field-of-view is the dispersion angle in which the sensor can effectively sense light from the emitter. Some sensors are only capable of detecting a narrow beam within two degrees of where the light is aimed. Other types can sense light that is spread over a wide angle, exceeding 60 degrees. The broader the field-of-view of a photoelectric sensor, the easier it is to align. The narrower the field-of-view, the greater the light intensity becomes, enabling the sensor to increase its sensing range. Some narrow beam sensors detect light up to several hundred feet.

TABLE 13-2   **GUIDELINES FOR MINIMUM REQUIRED EXCESS GAIN**

Minimum Excess Gain Required	Operating Environment
1.5×	*Clean air:* No dirt buildup on lenses or reflectors.
5×	*Slightly dirty:* Slight buildup of dust, dirt, oil, moisture, etc. on lenses or reflectors; lenses are cleaned on a regular schedule.
10×	*Moderately dirty:* Obvious contamination of lenses or reflectors but not obscurred; lenses cleaned occasionally or when necessary.
50×	*Very dirty:* Heavy contamination of lenses; heavy fog, mist, dust, smoke, or oil film; minimal cleaning of lenses.

SENSOR RESPONSE

Whenever fast-moving objects are sensed, the response time of the sensor becomes important. *Sensor response time* is the maximum amount of time that elapses from an input transition mode (light-to-dark or dark-to-light) until the output switches. The required sensor response time is calculated by the following formula:

$$T \text{ required} = \frac{W - D}{V}$$

where,   T = Required sensor response time
               W = Size of the target
               D = Beam diameter at the sensing location
               V = Speed of the object that passes the sensor

EXAMPLE

Figure 13-22 shows an application of the sensor response calculation.

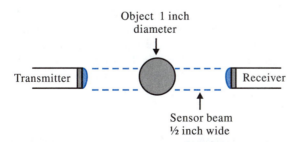

FIGURE 13-22 A sensor response example

○ The diameter of the target (W) is 1 inch.
○ The width of the light beam (D) is ½ inch.

○ The speed of the object (V) is 10 inches/second.

$$T \text{ response time} = \frac{W - D}{V}$$

$$= \frac{1 - .5}{10}$$

$$= 50 \text{ milliseconds}$$

The sensor selected for this application must have a switching speed of at least 50mSec to operate effectively. There is a correlation between sensor response time and excess gain. As the gain is increased, its switching speed slows down.

13-10 SENSOR INTERFACING

Before connecting the sensor output to a load, determine the current and voltage ratings of both devices. If the voltage and current requirements for the load are not matched to the voltage and current capacity of the sensor output, either the devices may be damaged or they will not operate properly.

When selecting a sensor to use for a particular application, first determine the type of load. The term *load* describes the device or equipment to which the sensor signal is applied. The first interface consideration is to determine whether the load requires a switched signal or an analog signal from the sensor's output.

SWITCHED SIGNAL

When the load requires a switched signal, the sensor must provide an *on* and *off* digital output. The switching device is either *electromechanical* or *solid state*.

Electromechanical Relays

Electromechanical relays are used in applications whenever the load requires more current than a solid-state switch can provide. This situation is most common when the sensor is directly connected to a device with a coil, such as the relay shown in Figure 13-23. The relay contacts on the left show the actual complementing output connections of the sensor. The switch on the right shows the N.O. and N.C. switching operation of the contacts.

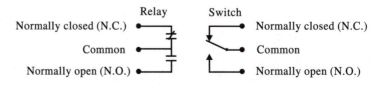

FIGURE 13-23 Electromechanical relay contacts

The following subsections make recommendations to prevent arcing and thereby prolong the life of the contacts.

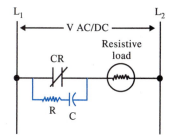

FIGURE 13-24
R-C "snubber"

Resistive Loads   When a large resistive load is controlled by a N.C. contact, a series R-C network is placed in parallel across the contacts, as shown in Figure 13-24. When the contacts open, the capacitor shunts the voltage away from the contacts. The resistor prevents the capacitor from discharging through the contacts when they close. The R-C network is called a *snubber.* The following formulas enable calculation of the component values:

Obtain the voltage value across the contacts when they are open: = E

Obtain the holding current value of the load: = I

To find the resistance value:

$$R = \frac{E}{10\,(Ix)}$$

$$\text{where } x = \left(1 + \frac{50}{E}\right)$$

The resistor should be ½ watt or larger.

To find the capacitor value:

$$C = \frac{I^2}{10} \text{ microfarads}$$

The working voltage should exceed E.

Snubber networks are available in single packages.

Inductive Loads   When an inductive load is controlled by a N.C. contact, a semiconductor clamping device is recommended to provide arc suppression.

When controlling a DC load, a diode is placed across the inductor with the anode connected to the common side, as shown in Figure 13-25(a). When the contact opens, the magnetic field around the inductor coil collapses as the coil attempts to keep the current flowing. Instead of arcing across the open contacts, current flows through the forward biased diode until the magnetic field is gone. A clamping diode with a high surge current rating and PRV of twice the supply voltage should be used.

When controlling an AC load, a metal oxide varistor (MOV) is connected in parallel with the load inductor, as shown in Figure 13-25(b). When the contacts open, the magnetic field collapses. As the induced current causes a voltage spike, the MOV clamps on as it changes from a high to low level. Induced current flows through the MOV instead of arcing across the open contacts. An MOV with the proper clamping voltage and sufficient surge current capacity should be used.

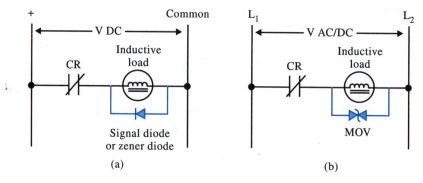

FIGURE 13-25 Arc suppression devices for inductive loads

Solid-State Relays

Most sensors use solid state circuits to perform their switching action. Specifically, transistors are used to perform the output switching action. Two types of transistor configurations are typically used:

1. Open collector NPN—current sinking
2. Open collector PNP—current sourcing

The type of input circuitry used by the load to which the sensor is connected dictates which type of configuration to use.

Current Sinking Output Sensors   A current sinking sensor is shown in Figure 13-26(a). The load is connected between the output of the sensor and the positive lead of the power supply. When the transistor turns on, its output terminal switches to ground and produces a voltage potential across the load. The sensor is called a sinking device because conventional current is drawn into the output in a manner similar to the way a heat sink absorbs thermal energy. Whenever a current sinking sensor is connected to a logic circuit, such as the TTL gate shown in Figure 13-26(b), a pull-up

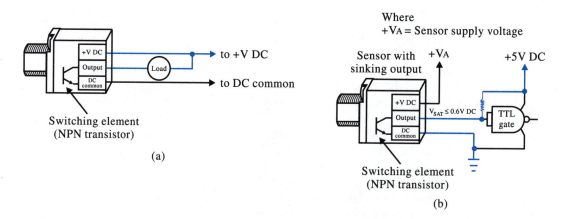

FIGURE 13-26 Interfacing a sinking sensor output with a load

resistor is required because of the open collector. Current sinking sensors are used whenever the load circuit interprets a *low* state as a logic 1. Chapter 16 describes this concept in more detail.

Current Sourcing Output Sensors   A current sourcing sensor is shown in Figure 13-27. The load is connected between the output of the sensor and the negative (common) lead of the power supply. When the transistor turns on, the output terminal switches to the +V DC potential and produces a voltage potential across the load. The sensor is called a sourcing device because conventional current is supplied from the output terminal (source) to the load. Whenever a current sourcing sensor is connected to a logic circuit, a pull-down resistor is connected from the output lead to ground.

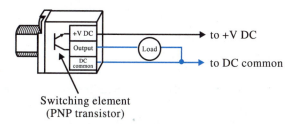

FIGURE 13-27   Current sourcing output

ANALOG SIGNAL

Sensors with analog signal outputs are usually used in process control applications. For example, thermal sensors, or *thermocouples*, produce analog signals to indicate temperature. The signal a thermocouple generates is directly proportional to the temperature it is sensing: The higher the temperature, the greater the voltage the sensor produces. In contrast, one type of photoelectric sensor produces an analog signal that is inversely proportional to the translucency of an object it is measuring.

Typical output signals produced by the sensors are as follows:

> 0 to +10 V DC Sourcing Voltage
> 0 to 20 mA Sinking Current

Electronic sensors that produce analog signals were described at the end of each chapter of Section IV, Process Control (Chapters 9 through 12).

CHAPTER PROBLEMS

1. The object to be detected by an industrial detection sensor is called a ____.
   a. retroreflector   c. target
   b. detector   d. emitter

2. Limit switches are used for which of the following applications? ____
   a. Size detection   c. Counting parts
   b. Limiting travel   d. All of the above

3. A sensor that is designed to detect an object exclusively in front of the sensor head uses the ____ approach.
   a. radial   b. axial

4. __ T __ F   The trigger stage performs the hysteresis function of the sensor.

5. As eddy currents are induced into a metal target by an inductive proximity detector, its oscillator circuit ____ energy.
   a. loses   b. gains

6. Another term used for hysteresis is _____ _____.

7. The capacitance of the capacitive proximity detector ____ when a non-conductive material is within its range, and ____ when a conductive material is within its range.
   a. increases   b. decreases

8. A voltage develops at the output terminals of a Hall effect sensor when a magnetic field is ____ its sensing range.
   a. within   b. outside

9. Which of the following variables affects the voltage amplitude of a Hall generator? ____
   a. Output capacity of the power supply.   c. Size of the generator.
   b. Strength of the target's magnetic field.   d. All of the above.

10. Infrared LEDs produce ____ light energy than visible LEDs.
   a. weaker   b. stronger

11. Phototransistors best respond to ____ light.
   a. visible   b. infrared

12. The process of rejecting ambient light and selecting an on/off frequency the same way a radio tuner operates, is referred to as _____.

13. __ T __ F   Photodetectors with a high gain are capable of detecting the level of cereal inside a sealed box.

14. Match the following methods of light detection with the mode of detection that they use:
   a. Light-to-Dark   ____ Opposed
   b. Dark-to-Light   ____ Retroreflective
                      ____ Diffuse
                      ____ Convergent
                      ____ Specular

15. Which method of light detection provides the highest level of optical energy?

16. Which method of light detection uses a disc that looks like a bicycle reflector?

17. Which method of light detection is the best photoelectric sensor to use for monitoring very small parts?

18. Which method of light detection uses a series of sensors to form a light curtain?

19. What is the most common light detection method used to differentiate between shiny and dull surfaces?

20. Fiber optic cables should be used to perform detection operations under which of the following conditions? ____
    a. The sensing space is confined.
    b. The environment is hostile.
    c. The sensor is exposed to magnetic noise.
    d. All of the above.

21. If emitted light is attenuated by 50 percent, an excess gain of _____× is required to overcome the loss of light.

22. Suppose the diameter of a target is two inches, the light beam width is one-half inch, and the speed of the object is five inches per second. What is the minimum response time required by the sensor?

23. Calculate the values of the resistor and the capacitor in a snubber circuit when the voltage across the contacts is 200 volts and the holding current is 100 mA.

24. Conventional current flows ____ the output of a current sinking sensor and ____ the output of a current sourcing sensor.
    a. into   b. out of

PROGRAMMABLE CONTROLLERS

Most automated equipment used in industry is controlled by programmable controllers. The programmable controller is a specialized computer designed for many types of manufacturing applications. These machines use a microprocessor that provides artificial intelligence enabling them to perform various functions formerly performed by people. The primary advantage of programmable controllers over hardwired circuits is their programmability. For example, an equipment function or manufacturing process can be altered by making a few entries on a keyboard instead of making costly rewiring changes within the controller section.

Because of the important role they play in industrial control, this three-chapter section describes the operation of programmable controllers. Chapter 14 describes the principles of relay ladder diagrams and the hardware circuitry of the equipment. Chapter 15 provides a generic approach to ladder diagram programming. Chapter 16 examines advanced programming instructions, programming languages, and applications of the programmable controller.

CHAPTER 14

Introduction to Programmable Controllers

CHAPTER OBJECTIVES

At the conclusion of this chapter, you will be able to:

○ Identify the symbols used in relay ladder circuits.

○ Draw a ladder diagram and label the parts of the configuration.

○ Design a ladder diagram that will perform a specified operation.

○ Draw a block diagram of the PLC and describe the function of the following sections:

Power Supply	Input/Output Modules
Processor Unit	Programming Unit

○ Describe the operation of the internal interface circuit inside the following I/O modules:

AC Input Module	DC Output Module

○ Draw the proper wiring connections of field devices to the terminals of input and output modules.

○ Describe the relationship between a terminal on an I/O module and a corresponding address location in memory.

INTRODUCTION TO PLC FUNCTIONS

Programmable controllers are computer-based devices capable of controlling many types of industrial equipment and entire automated systems. The programmable controller is also commonly referred to as a programmable logic controller (PLC) because its operation often requires many logic functions to be performed. Typical motion control applications for controllers are machine tools, materials handling equipment, conveyers, and robotics. Process control applications include food processing, wastewater treatment, chemical and petroleum production, and making paper.

The automotive industry is credited with the emergence of the PLC. Prior to the late 1960s, its automated equipment was primarily controlled by discrete inflexible circuits consisting of electromechanical relays and coils hardwired on panels. (*Hard-wiring* means that all of the components were manually connected by wires.) As annual model changes were made, modifications of the relay circuitry were necessary.

This effort was costly because it required shutting down an assembly system while the panels were rewired or replaced by new ones. The need for reduced downtime by a more rapid changeover scheme became a priority.

In the mid-1960s, Hydramatic, a division of General Motors Corporation, envisioned that a computer could be used to perform the logic functions then performed by relays. The company reasoned that circuit changes could be made using a keyboard, without performing extensive rewiring. The engineering team wrote a list of features of the proposed computing device. GM initiated the development of the computing device by specifying certain design criteria, including:

○ The device must be durable so that it can operate in the harsh environments (dirty air, humidity, vibration, electrical noise, etc.) encountered in a factory.

○ It must provide flexibility by implementing circuit modifications quickly and easily through software changes.

○ It must be designed to use a programming language in ladder diagram form already familiar to technicians and electricians.

○ It must allow field wiring to be terminated on input/output terminals of the controller.

GM used this list of specifications when it solicited interested companies to develop a device that met its design requirements. In 1968, the ruggedly constructed computer-based control, which became the programmable controller, was delivered to GM by Modicon.

The first PLCs were large and expensive. They were capable of On-Off control only, which limited their application to operations that required repetitive movements. Yet their initial design exceeded Hydramatic's requirements. For example, they could be installed easily, consumed less space and power than the wiring panels they replaced, and provided diagnostic indicators useful in troubleshooting.

Innovations and improvements in microprocessor technology and software programming techniques have added more features and capabilities to the PLC. These enhancements enable the PLC to perform more complex motion and process control applications, and with greater speed. Special purpose modules have expanded their capacity to include operations associated with PID control, bar-coding, vision systems, radio frequency communication, voice recognition, and voice generation.

Presently, more than a dozen manufacturers produce PLCs. Most of these companies make several models that vary in size, cost, and sophistication to meet the needs of specific applications. Regardless of size, cost, or complexity, all PLCs share the same basic components and functional characteristics.

14-1   RELAY LOGIC

Relay ladder logic is the language used by most programmable controllers. The term *ladder* is derived from the appearance of the diagram that resembles the rung of a ladder. The term *logic* is derived from the decision-making function that is performed by relays.

The first automated machines were controlled by mechanical devices. These mechanical control systems were replaced when electromechanical components were developed. These electromechanical components are still extensively used in industry today. They include *input* field devices such as push buttons, limit switches, and sensor detectors. These field devices are hardwired to a centrally located panel consisting of relays. The relay, shown in Figure 14-1, is an electromechanical component made of a coil and an armature, which operates a set of contacts.

These contacts can switch other circuits on and off, thereby performing the logic control function for the automated system. The output signals produced by the logic section are fed to a terminal strip. *Output* field devices such as indicator lamps or actuators such as valves are wired to screw connectors of the output terminals.

The wiring diagram in Figure 14-2 shows the circuit connection and its associated devices (switches, relays, and motors) in their relative physical location. Although this type of drawing provides information about how a circuit is actually wired, it does not show the diagram in its simplest form. A simplified method that illustrates electrical relationships between components and the circuit operation is the ladder diagram format. Figure 14-3 shows the same circuit as Figure 14-2 in ladder diagram form. This form is the most common method of documentation for electromechanical control systems. To simplify the circuit even further, the power portion is shown separate from the control portion, as shown in Figure 14-4. The two vertical lines are tapped off

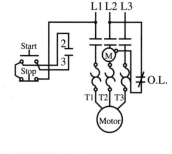

FIGURE 14-2 Wiring diagram

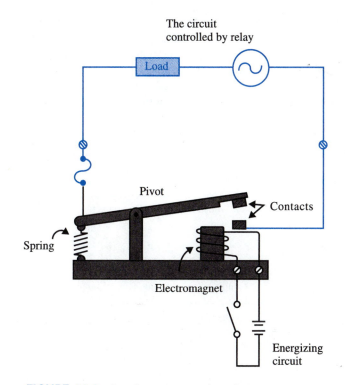

FIGURE 14-1 An electromagnetic relay

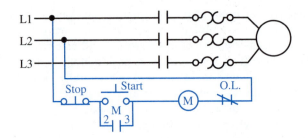

FIGURE 14-3 Ladder diagram

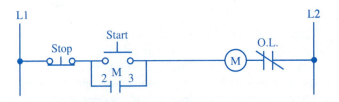

FIGURE 14-4 Simplified ladder diagram

lines L1 and L2 of three-phase power lines. They represent the potential difference between two voltages that supply power to the circuit. These vertical lines are referred to as *rails* because they resemble the vertical sides or rails of a ladder. The various field components and control relays are located horizontally between the two vertical lines. The horizontal lines are referred to as *rungs* because they resemble the rungs of a ladder. The input field devices are located on the left portion of the rung and are referred to as the *condition* section of the diagram, as shown in Figure 14-5. The output device is connected on the right portion of the rung and is referred to as the output section. Figure 14-5(a) shows a relay logic circuit; Figure 14-5(b) shows the equivalent ladder diagram used by programmable controllers.

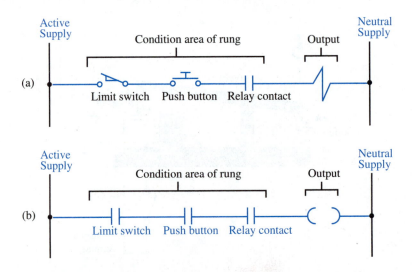

FIGURE 14-5 Ladder diagram rung

Numerous relay ladder symbols are used to identify components. The most common types are shown in Table 14-1.

Refer to Figure 14-6 as the following rules that apply to ladder diagrams are explained:

1. The vertical lines or rails represent the power lines. The voltage potential may be either AC or DC. The left rail is normally labeled L_1. L_1 is the hot lead for AC, and the positive line for DC. The right rail is normally labeled L_2. L_2 is the neutral lead for AC and the negative lead for DC voltages.

TABLE 14-1   RELAY LOGIC DIAGRAM SYMBOLS

Condition Devices			
Device	Normally Open Symbol	Device	Normally Closed Symbol
Push button		Push button	
Limit switch		Limit switch	
Relay contact		Relay contact	
Float		Float	
Thermal		Thermal	
Pressure		Pressure	
Flow		Flow	
Proximity		Proximity	
Toggle switch		Toggle switch	
Timer operated contact: Time closing Time opening		Timer operated contact: Time opening Time closing	

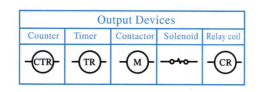

Output Devices				
Counter	Timer	Contactor	Solenoid	Relay coil
CTR	TR	M		CR

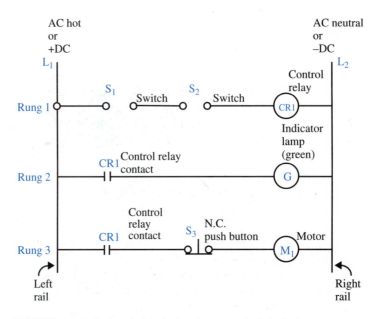

FIGURE 14-6 Relay ladder circuit properly labeled

2. The horizontal lines or rungs are labeled in numerical order from top to bottom.

3. The ladder diagram is read like a book, from left to right, and from top to bottom.

4. Whenever possible, the components are labeled in numerical order from left to right, and from top to bottom.

5. The components are shown in their normal condition, which means they are de-energized.

6. Contacts will always have the same letter and number designation as the device that controls them. These control devices include relay coils, timers, or motor starters.

7. A normally open contact closes when the device that controls it is energized. A normally closed contact opens when the device that controls it is energized.

14-2 RELAY LADDER LOGIC CIRCUITS

Ladder diagrams perform many types of control operations by making logic decisions. These logic functions include AND, OR, NOT, NAND, and NOR.

AND FUNCTION

In Figure 14-7, Switch 1 and Switch 2 are considered input devices and Lamp 1 is the output device. With S1 *and* S2 connected in series, they must remain closed for the lamp to remain lit. If either switch is opened, the lamp will turn off. This circuit performs the AND function.

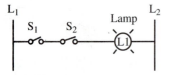

FIGURE 14-7 A relay ladder circuit that performs the logic AND function

OR FUNCTION

In Figure 14-8, Pressure Switch 1 and Flow Switch 2 are considered input devices and a solenoid is the output device. With S1 and S2 connected in parallel, the solenoid will become energized if S1 or S2 (or both S1 *and* S2) are closed. If both switches are opened, the solenoid will not be energized. This circuit performs the OR function.

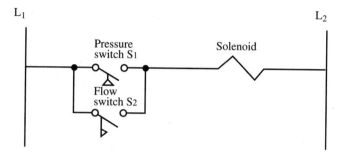

FIGURE 14-8   A relay ladder circuit that performs the logic OR function

NOT FUNCTION

In Figure 14-9, a normally closed foot switch is the input connected in series with a motor starter coil that operates as the output. When the input switch is *not* activated, the coil is energized. When the foot switch is pressed, the output is *not* energized. This circuit performs the NOT, or *inverter,* function because the output is always opposite the input.

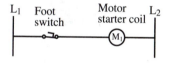

FIGURE 14-9   A relay ladder circuit that performs the logic NOT function

NAND FUNCTION

In Figure 14-10, a normally closed float switch and a normally closed temperature switch in parallel are the inputs, and a magnetic relay coil is the output. When one or both switches are not activated, the relay coil will be energized. If both switches are opened simultaneously, the output will be de-energized. The circuit performs the NAND function.

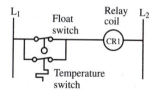

FIGURE 14-10   A relay ladder circuit that performs the logic NAND function

NOR FUNCTION

In Figure 14-11, three N.C. push buttons in series serve the inputs, and a timer is the output. When all of the switches are not activated, the timer will be energized.

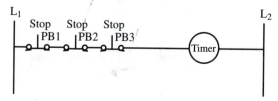

FIGURE 14-11   A relay ladder circuit that performs the logic NOR function

By pressing one or more of the stop buttons, the circuit will de-energize the load. The circuit performs the NOR function.

START/STOP INTERLOCKING CIRCUIT

The circuit in Figure 14-12 illustrates two very important functions of a ladder circuit: *stopping* and *latching*. It consists of a N.C. stop push button, a N.O. start button, a control relay coil, and a N.O. contact activated by the relay. The contact connected in parallel with the start button is called a *branch* connection.

When power is applied to the circuit, the relay coil cannot energize, because there is an incomplete path for current to flow due to the N.O. start button and normally open contact CR1. When the start button is pressed, a path for current develops from L_1 through the N.C stop button, through the start button, and through the relay coil to L_2. Because control relay 1 is energized, it closes the N.O. contact also labeled CR1. If the start button is released, the output coil remains energized, because there is an alternate path through contact CR1 for current to flow. By maintaining an energized circuit after the momentarily closed start button is released, a *latching* or *interlocking* function is performed. The circuit becomes de-energized when the series connected N.C. stop button is pressed.

Whenever possible, components in a ladder diagram are shown in order of importance. For safety reasons, the stop button has a higher importance than the start button. That is why it is placed in front of the start button.

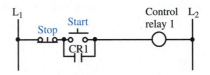

FIGURE 14-12 Ladder diagram for basic STOP/START circuit

14-3   BUILDING A LADDER DIAGRAM

Ladder circuits are primarily used to control a machine or some type of manufacturing process. To help the reader understand how to develop a ladder diagram, a drawing of a pumping station and the sample circuit that controls its operation are shown in Figure 14-13.

The function of the pumping station shown in Figure 14-13(a) is to pump water from a storage tank into a pressure tank. The ladder diagram that controls the pump system is shown in Figure 14-13(b). A N.O. push button switch is in series with the pump motor. Whenever the water level in the pressure tank is too low, the operator must keep the switch depressed to make the pump run until the tank is full. The operator then releases the push button to stop the flow of water into the pressure tank.

MODIFICATION 1

Figure 14-14 shows a modification to the original circuit, so that the operator is not required to manually keep the push button pressed while the pressure tank is filling

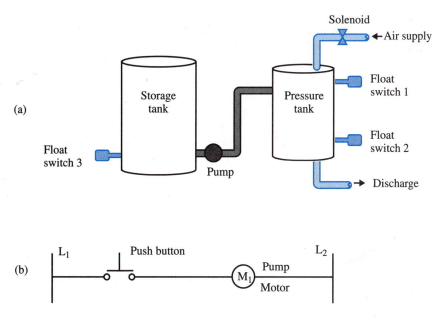

(a)

(b)

FIGURE 14-13 A water pumping system

up with water. Instead, the operator momentarily depresses the start button which completes the current path from L_1 through the pump motor to L_2. The pump motor M_1 performs two functions. First, it pumps the water into the pressure tank. Second, it also has a coil that, when energized, closes contact M_1, which is in parallel with the start button, thus maintaining the current to coil M_1. When the pressure tank is full, the operator then pushes the N.C. stop button which opens the circuit, stopping the pump.

MODIFICATION 2

By installing a float switch near the top of the pressure tank, the operator is not required to push the stop button. Instead, the operator needs only to push the start button, thus energizing the pump and starting water flowing into the tank. When the level of the water has reached N.C. float switch one (FS_1), its contacts will open, thus stopping the pump and the flow of water. The function performed by the float switch is that of an automatic stop. The ladder diagram in Figure 14-15 shows a N.C. contact added in series to fulfill the new operational requirements. The manual stop button remains connected in the circuit so that the operator can stop the pump at any time.

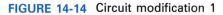

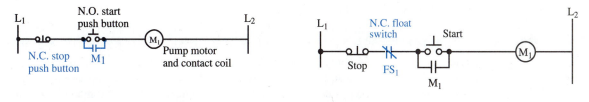

FIGURE 14-14 Circuit modification 1   **FIGURE 14-15** Circuit modification 2

MODIFICATION 3

By installing a N.O. float switch at the lower section of the tank, the pump system will also turn on automatically. This modification causes the pump to start whenever the water reaches a predetermined low level, thereby causing the N.O. contacts to close. Figure 14-16 shows the changes that are required on the control circuit to perform the operation. The contact (FS$_2$) is connected in parallel with the original start button, which enables it to start the motor. The parallel configuration provides a way for the start button to operate independently.

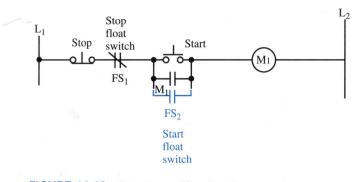

FIGURE 14-16   Circuit modification 3

MODIFICATION 4

Modification 3 enabled the pump to start automatically whenever the water level in the pressure tank reached a predetermined low level. However, suppose the water in the storage tank has dropped to a level too low to enter the pump orifice. A pump will be damaged if it is run without water flowing through it. As a fail-safe measure to protect the pump, a N.C. float switch (FS$_3$) is placed at an extreme low level of the storage tank. Its contacts will open whenever the water level drops to the set level of the float switch. Figure 14-17 shows that it is wired in series with the other start switches in the control circuit.

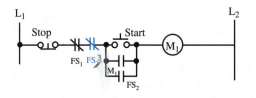

FIGURE 14-17   Circuit modification 4

MULTI-RUNG CONTROL

In many ladder circuits, an output relay coil in one rung will control the operation of circuits in one or more other rungs. An example is shown in Figure 14-18. A control relay is connected in Rung 1, and the contacts that it controls are located in Rungs 2 and 3. When SW1 is open, there is no current path between L$_1$ and L$_2$ to flow in Rung 1. Since the control relay is de-energized, the N.O. contact in Rung 2 remains open, and the N.C. contact in Rung 3 stays closed. Therefore, the red light is on and the green light is off. When switch SW1 is closed, the control relay is energized and causes the contact in Rung 2 to close and the contact in Rung 3 to open. The result is that the red light turns off and the green light turns on.

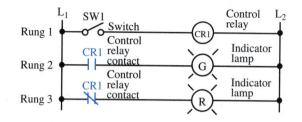

FIGURE 14-18 A multi-rung controlled circuit

INTRODUCTION TO PLC COMPONENTS

The modern version of the ladder logic system is the programmable controller. All PLC systems are comprised of the same basic building blocks that detect incoming data, process it, and control various outputs. The following sections discuss these blocks:

- ○ Rack Assembly
- ○ Power Supply
- ○ Programming Unit
- ○ Input/Output (I/O) Section
- ○ Processor Unit

With the exception of the rack assembly, the diagram in Figure 14-19 shows all of these sections and how they are connected.

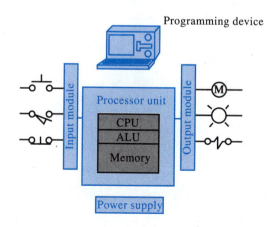

FIGURE 14-19 Major sections of a programmable controller

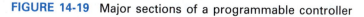

14-4 RACK ASSEMBLY

Most programmable controllers that have a large number of input and output terminals are constructed by using a variety of modules. These modules include the power supply, processor unit, and input/output modules. The modules are installed in a rack, such as the one shown in Figure 14-20. The PLC rack serves several functions. It physically holds the modules in place, and it also provides electrical connections between the modules by using a printed circuit board at the back of the rack assembly.

The modules are easily inserted into channels on the rack. They fit into sockets mounted on the motherboard to make electrical contact with the other circuitry. The ability to plug modules into the rack allows maintenance personnel to replace defective units quickly.

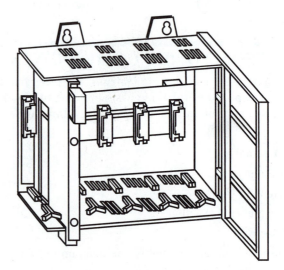

FIGURE 14-20 The rack assembly of a programmable controller

14-5 POWER SUPPLY

The **power supply** provides voltages that are necessary to operate the circuitry throughout the programmable controller. Some sections of the PLC require an AC voltage, such as AC input and output modules, or the field devices that are connected to them. Other sections require a low level DC voltage source. Figure 14-21 shows the section of the power supply that converts AC power to the DC voltage level required by the internal circuitry of the programmer.

1. The first section, on the far left, uses a step-down transformer that reduces the incoming line power of 120 or 240 volts to a lower level.

2. The second section uses two full-wave bridge rectifiers to convert the transformer secondary AC voltage to pulsating DC voltages. The top bridge develops a positive voltage, and the bottom bridge develops a negative voltage.

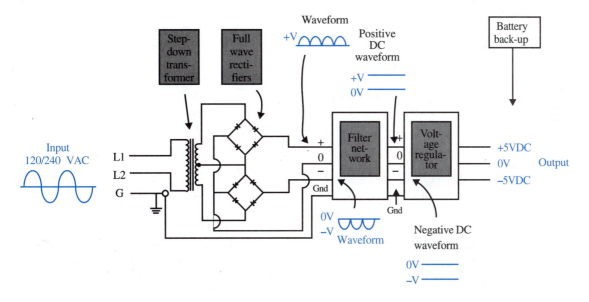

FIGURE 14-21 Power supply

3. The third section uses filter circuits to condition the pulsating DC output of the rectifier to become pure DC voltages.

4. The fourth section contains a voltage regulator which maintains a constant DC voltage if power line fluctuations or changing load demands occur.

The low DC voltage is required by the internal circuitry of the processing unit. It is also supplied to field devices and DC modules to which they are connected.

The power supply may also contain a back-up battery for the memory devices in the processing unit to retain data if an AC power failure occurs. Some PLCs include a battery indicator which indicates if the battery charge becomes low.

14-6 PLC PROGRAMMING UNITS

The **programming unit** of a PLC provides a way for the user to enter data, and to edit and monitor programs stored in the processor unit. The programming unit communicates with the processor unit by using a data communication link which transfers data in a serial or parallel fashion. Most programmers also perform troubleshooting procedures by simulating signals from input devices or by forcing output devices to energize through keyboard entries. This method is called *forcing* inputs and outputs.

Some programming units also have the capability of storing programs into their memory chips once they are written, and then loading them into the PLC at a later time. This method is called *uploading* and *downloading*.

Types of programming units used to perform these operations include hand-held programmers, dedicated terminals, and microcomputers.

FIGURE 14-22 Hand-held programming unit

HAND-HELD PROGRAMMERS

Hand-held programmers, such as the one shown in Figure 14-22, are very small and inexpensive devices. They typically have membrane keys for entering data and LCD displays that show one line of a ladder diagram. They are useful for programming simple ladder diagrams, or for troubleshooting on the factory floor because they are easy to carry. Their primary disadvantage is that the ladder diagrams are difficult to follow on the screen because only a small segment is shown at any given time. Many hand-held programmers are capable of forcing inputs and outputs, and of uploading and downloading.

DEDICATED TERMINALS

Dedicated terminals are designed to be used with only one brand of PLC. It has a keyboard for entering data and a display screen that can show several ladder rungs simultaneously. Dedicated programmers usually provide troubleshooting operations while the PLC is running. Their primary disadvantages are that they are too large to carry out to the factory floor, they are expensive, and they cannot be used by other PLC brands. Dedicated terminals can upload and download, or force inputs and outputs.

MICROCOMPUTERS

A microcomputer is a personal computer with a special circuit card. By using a special software package, data can be entered by the keyboard and the ladder circuits are displayed on the screen. This method is rapidly becoming the most popular way to program PLCs.

A primary advantage of using a computer for programming is that the time utilization capabilities of an existing PC are maximized. During the time the computer is not being used for PLC program development, it can be used for other purposes. Monetary savings are also possible because it is not necessary to purchase a hand-held or dedicated programming unit.

14-7   INPUT/OUTPUT SECTIONS

There are many types of external field devices and circuitry connected to the programmable controller. The purpose of the input/output (I/O) section of the PLC is to interface its internal circuitry to outside equipment.

The I/O section contains **input/output modules** that serve four basic functions:

1. *Termination.*   Each I/O module provides terminal connections to which field devices can be connected. Each terminal is assigned an identification number.

2. *Signal Conditioning.*   Most of the voltages used by field devices are not compatible with the low DC voltage data signals processed inside the programmable controller. The module converts the external signal to a voltage suitable for the PLC.

3. *Isolation.*   The factory floor is a very noisy environment. Noise is created by stray magnetic fields produced by devices such as large motors, welding

equipment, and contactors used to switch high currents on and off. If the flux lines created by these devices cut across any PLC conductors, unwanted voltage transients can be induced that may be falsely recognized as data pulses by the processor unit. A noise signal may be interpreted by the PLC as an instruction which could make it operate erratically, or cause an output device to move when it should not. Noise picked up by output lines could be fed back to the processor unit and cause damage. By using optical coupler circuitry, there is no physical connection between the input or output field devices and the circuitry inside the PLC. The module isolates the processor unit by blocking the noise and passing only those signals that are valid.

4. *Indication.*   Each terminal has an associate indicator. Its function is to illuminate when a voltage is applied to that terminal. I/O modules use either LEDs or neon light bulbs as indicators.

The most common type of signal interfaced to a PLC input, or from its output, is the discrete voltage. The discrete voltage is either present or absent. The signals applied to the input terminal are produced by On/Off, Open/Closed, or energized/de-energized conditions of field devices. Input field devices—which include push buttons, photoelectric switches, proximity detectors, and level sensors—are external instruments connected to the PLC. The controller interprets discrete field signals as logic states. For example, when a field device is on, it is interpreted as a logic 1 state, while an off condition is interpreted as a 0 state.

A logic 1 state produced by the controller turns an output field device on. A logic 0 turns it off. These signals from the controller are interfaced to discrete output devices external to the PLC. These field devices include relays, solenoids, valves, alarms, motors, and indicator lamps.

Input Modules

The PLC receives discrete incoming signals from switching devices. When a switching device is closed, the input interface module senses a voltage and converts it to a logic 1 for the internal circuitry of the processor unit. When the device is open, a voltage is not sensed at the input terminal. The module converts this input as a logic 0 for the processor unit.

There are three common forms of discrete input signals: 120VAC, high DC voltage, and low level DC voltage that are compatible with transistor-transistor logic (TTL) integrated circuits. An input interface module that converts 120VAC to logic signals used by the PLC is discussed below.

AC Input Module

The AC input module converts the presence of 120V applied to its input terminals to a logic 1 state. The absence of an AC voltage produces a logic 0. A typical circuit used to perform this function is shown in Figure 14-23(a) and discussed below.

1. A N.O. limit switch is connected between L_1 and Input Terminal. An AC potential difference between L_1 and L_2 is supplied to the circuit when the switch is closed.

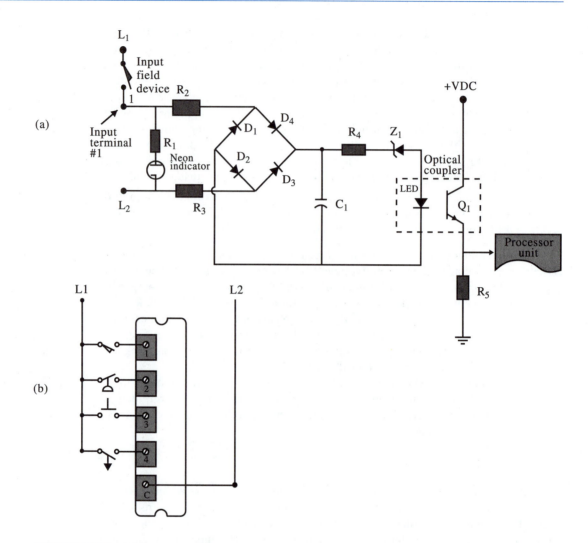

FIGURE 14-23 AC input module

2. A neon lamp and a series resistor are connected across the input terminal and L_2. The lamp is illuminated only when a voltage is present, thus performing the indication function. Since the lamp lights at approximately 50 volts, the remaining potential drops across R_1, which also is a current limiting resistor.

3. Resistors R_2 and R_3 are used to drop most of the incoming voltage. A remaining voltage of 5-12 volts is applied to a full-wave bridge rectifier that converts the AC voltage to a pulsating DC voltage.

4. The capacitor and resistor R_4 is used to debounce the input signal and to filter out electrical noise from the input line.

5. The zener diode Z_1 and resistor R_4 form a threshold circuit that breaks over and produces a sudden regulated stable DC voltage when the voltage of each pulsating alternation has reached a certain amplitude.

6. When a valid signal is produced by the threshold components, it forward biases the LED in the optical coupler. As the LED illuminates, the light it transmits turns on the phototransistor Q_1. The optical coupler performs the isolation function in the module.

7. As the LED turns on, the phototransistor turns on. The switching action of Q_1 provides continuity and allows current to flow through R_5. As a positive potential near 5 volts develops across R_5, a logic 1 state is supplied to the processor unit.

8. An AC potential is no longer present between L_2 and the numbered input terminal if the limit switch opens. Therefore, the LED in the coupler does not light and the phototransistor is off. Because Q_1 is in an open condition, there is no voltage dropped across R_5. A voltage that is near ground potential is supplied to the processor unit, and is recognized as a logic 0 state.

One input module can have 4, 6, 8, 10, 16, or 32 interface circuits, providing the same number of individual connections to receive input signals. Figure 14-23(b) shows how external connections are made to the AC input module. The connector (labeled "C") indicates that AC line L_2 is secured to a common terminal. Inside the module, the bottom input lead of each interface circuit shown in Figure 14-23(a) is connected to this terminal. The top lead of each input circuit is connected to one of the numbered input terminals. Several types of field switching devices are connected between AC line L_1 and the numbered terminals. A switch closure provides the continuity required to energize the input circuit.

OUTPUT MODULES

The outgoing discrete signals from the processor unit are transmitted to output field devices by the output module assembly. An example of an output interface module follows.

DC Output Module

The DC output module interfaces the logic signal from the processor with a DC output field device that operates at a potential greater than +5 volts. A typical DC module circuit is shown in Figure 14-24(a).

1. When the processor generates a logic 1, the LED in the optical coupler is forward biased. The light from the LED turns on the phototransistor.

2. As the phototransistor switches on, the potential at the base of the PNP transistor Q_1 becomes more negative, which causes it to turn on. As the transistor switches on, current flows through R_3 and a voltage near 5 volts develops at output terminal 4. The voltage drop that forms across R_3 activates the solenoid and also causes the LED indicator to light. The optical coupler performs the isolation function in the module.

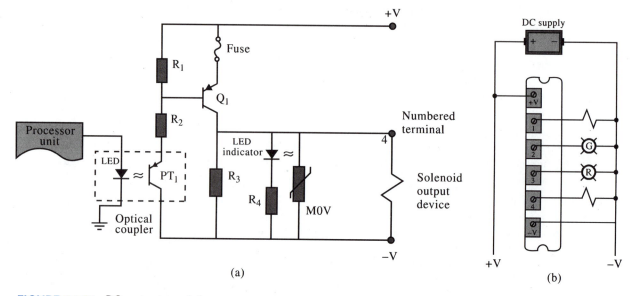

FIGURE 14-24 DC output module

3. If the processor's signal is a logic 0 the optical coupler's LED does not light and the phototransistor is off. Since the base current will not flow at Q_1, its off condition creates a voltage of 5 volts from its emitter to its collector. The load potential from the −V terminal to terminal 4 reduces to a 0 volt level, causing the solenoid to de-energize and the LED indicator to turn off.

Figure 14-24(b) shows how external field devices are connected to the DC output terminal. When one of the output circuits is energized by the processor unit, it enables current to flow from −V through the output field device, through Q_1 to +V.

MODULE TERMINATION

I/O modules are often inserted in pairs into the PLC rack assembly. This physical arrangement is referred to as a *module group*. Figure 14-25 shows an individual module. The bottom half of each module contains terminals to which field devices are connected. AC or DC power is supplied to each module at the top and bottom terminals. The uppermost terminal is labeled L_1 if it is an AC type module, or +V if it is a DC type module. The bottommost terminal is labeled L_2 if it is an AC type module, and C (common) or −V if it is a DC module.

There are eight I/O terminals located between the power supply connections. Each terminal has a two-digit octal identification number. The I/O terminals on the left module of each group are numbered 00-07, and 10-17 on the right module. Even though there are several module groups contained in a rack assembly, the terminals at each pair of I/O modules are labeled this way. Therefore, a five-digit address number is needed to distinguish a terminal in one module group from terminals with the same two-digit number in another module group. Module addresses are usually labeled with octal or decimal numbers. A five-digit octal address format is illustrated in Figure 14-26, and is described as follows:

First Digit: The first digit indicates if the module is an input or an output unit. Specifically,

Output Module = 0

Input Module = 1

Second Digit: The second digit indicates the rack number. Most PLCs have one rack, which is labeled 1. If needed, a second rack is labeled 2, a third rack 3, and so on.

Third Digit: The third digit identifies the module group number. This number is located above each module group on the rack. The number 0 is assigned to the group on the far left of the rack. The numbers increment for each pair of additional modules that are placed to the right.

Fourth and Fifth Digits: The fourth and fifth digits are octal numbers that identify the specific terminal address. There are 16 individual numbers for the terminals in each module group. The terminals on the left module in each group are numbered 00-07, and the terminals on the right module are numbered 10-17.

On the top half of each module shown in Figure 14-25, there are indicator lamps that are associated with the terminals below them. A lamp will turn on when a voltage is present at the terminal to which it is connected. Each lamp is labeled with the same number of the terminal that causes its illumination.

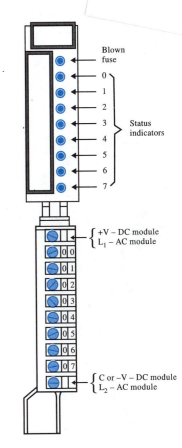

FIGURE 14-25 Field wiring terminals and indicator lamps on an I/O module

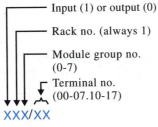

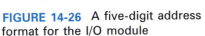

FIGURE 14-26 A five-digit address format for the I/O module

14-8 PROCESSOR UNIT

The **processor unit** coordinates and controls the operation of the entire programmable controller system. A processor module is usually located at one side of the rack assembly. It contains integrated circuit chips that include one or more microprocessors, memory chips, and circuits that enable data to be stored into and retrieved from memory. Some processor units have communication circuitry that provides interfacing between the processor and peripheral devices such as programmers, printers, and personal computers.

The processor is composed of three main sections: the central processing unit (CPU), the arithmetic logic unit (ALU), and the memory, as shown by the block diagram in Figure 14-27.

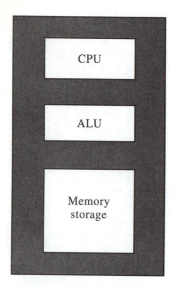

FIGURE 14-27 Block diagram of the processor unit

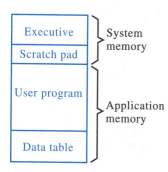

FIGURE 14-28 Block diagram illustrating memory sections

CENTRAL PROCESSING UNIT

The central processing unit is the brain of the PLC. The intelligence it provides is performed by one or more microprocessors, which are integrated circuits with tremendous computing and control capabilities.

The principle function of the CPU is to interpret and execute computer-based programs that are permanently stored in the processor's memory. These programs are written by the PLC manufacturer to enable the PLC to perform ladder logic instead of other programming languages. The CPU also coordinates the operation of the ALU and the memory. For example, based on the software program, the CPU determines what should be done in the ALU and the memory, and when it should be done. The CPU also performs other functions, such as self-diagnostic routines to determine whether the PLC is operating properly and communication with peripheral devices and other processors.

ARITHMETIC LOGIC UNIT

The function of the ALU is to perform mathematical calculations and make logic decisions.

MEMORY

The memory function of the processor stores programs and data that the CPU needs to perform various operations. The memory is organized into several sections according to the functions they perform. Figure 14-28 shows a block diagram of these sections. The function of each memory section is described in the following subsections.

Executive

The executive is a collection of system programs permanently stored in ROM memory devices. These programs enable the CPU to understand the commands it receives from the program instructions written by the operator. For those readers who are familiar with PCs, the ROM contains the same type of information as a DOS disk. In addition to supervising the operation of the PLC by executing control programs, the executive communicates with peripheral devices, and performs system housekeeping functions.

Scratch Pad

As the CPU performs various operations such as logic analysis, data manipulation, or mathematical functions, it is necessary to temporarily hold data as calculations are performed or decisions are made. The work area used to temporarily store the binary information used by the processor is the scratch pad. RAM-type memory chips are used to perform scratch pad operations. RAM memory is volatile, which means that if power supplied to these chips is removed, the contents will be lost.

User Program Storage Area

The programming instructions which are written by the PLC user are stored in this type of memory. The information contained in these programs is utilized by the CPU to perform the control functions of the PLC. The user program instructions create the ladder diagrams used to perform logic functions, mathematical calculations, data manipulations, and other operations.

Data Table

The instructions in the user program use data stored in this memory area to perform the desired operations. A block diagram in Figure 14-29 shows the four different elements of a data table. The functions they perform are as follows:

1. The *input image table* is used to store information about the On/Off status of the field devices connected to the input module.
2. The *output image table* is used to control the On/Off condition of the field devices connected to the output modules.
3. The *internal storage area* is used to control the On/Off condition of internal outputs. Internal outputs are used when there are a limited number of output terminals available, and the only function of the output is to control a contact, such as in a latching circuit.
4. Storage registers are used to store groups of binary numbers. For example, they store BCD values that are converted from binary numbers to decimal digits, such as accumulated values in a counter, or the number of seconds remaining in a timer.

Input image table
Output image table
Internal storage area
Storage registers

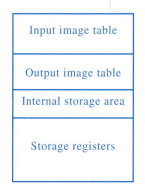

FIGURE 14-29 The four elements of the data table memory

Referring back to Figure 14-28, the executive and scratch pad sections are used by the CPU and are not accessible to the user. They are therefore categorized as *system memory.* The information in the user program storage area is entered by the PLC programmer. Most of the information loaded into the data table is entered two ways, from the programmer and from the On/Off status of the I/O terminals. These sections are categorized as *application memory*.

The processor memory consists of individual registers that store binary digits, or *bits* (*bi*nary digi*ts*) of data. The number of bits each register holds depends on the type of microprocessor used by the CPU. Groups of bits are referred to as *words.* Words can be made of 8, 16 or 32 bits. The 16-bit word shown in Figure 14-30 is the most common word length used by PLCs. The diagram also illustrates dividing 16-bit words into groups of 8 bits each, called *bytes.* In the octal format, the bits in the byte on the right, called the lower byte, are labeled 00-07, and the bits in the upper byte are

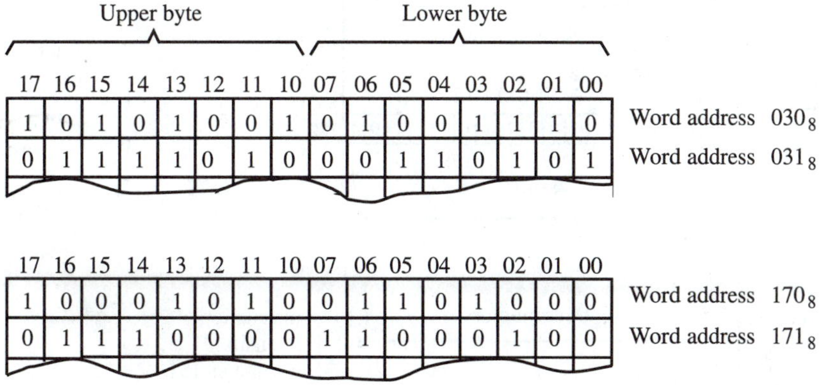

Upper byte
Lower byte
17 16 15 14 13 12 11 10 07 06 05 04 03 02 01 00
1 0 1 0 1 0 0 1 0 1 0 0 1 1 1 0
Word address 030₈
0 1 1 1 1 0 1 0 0 0 1 1 0 1 0 1
Word address 031₈
17 16 15 14 13 12 11 10 07 06 05 04 03 02 01 00
1 0 0 0 1 0 1 0 0 1 1 0 1 0 0 0
Word address 170₈
0 1 1 1 0 0 0 0 1 1 0 0 0 1 0 0
Word address 171₈

FIGURE 14-30 Octal word addresses and bit format in a PLC memory

labeled 10-17. Processor memories contain hundreds or thousands of registers, depending on the particular model or manufacturer. Each register is assigned a three-digit octal word address to identify its location, as shown to the right of the register. The register address, in conjunction with a two-digit octal address, identifies the location of any bit in memory.

14-9   RELATIONSHIP OF MEMORY WORD ADDRESS TO I/O MODULES

There is a relationship between the numbers assigned to the data table in memory and the number used by the I/O modules. Specifically, each memory register of the data table is organized to match the identification numbers of the I/O terminals in a module group. Figure 14-31 illustrates the addressing scheme, shows the internal connections of a PLC, and helps clarify the sequence of events that takes place when an input field switch closure causes an output field lamp to turn on.

The figure contains an input module, a partial memory map of the I/O image tables, an output module, and field devices. The ladder at the bottom shows an input switching contact connected in series with an output symbol. The ladder diagram is programmed into the user memory. A closure of the switch contact makes the rung true and causes the output to be energized.

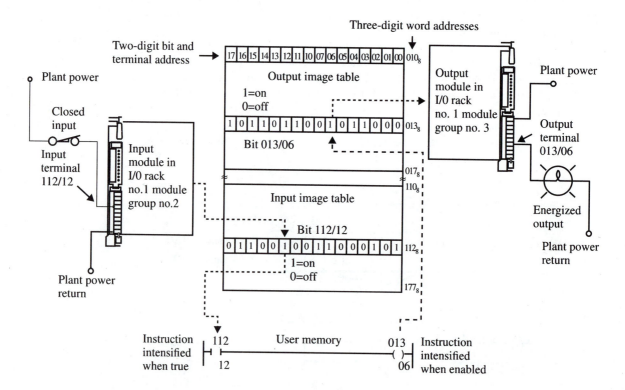

FIGURE 14-31 Relationship of bit addresses to input and output devices   (Courtesy of Allen-Bradley)

14-10   THE OPERATION SEQUENCE

The switch is connected to terminal 112/12 of the input module and the lamp is connected to terminal 013/06 of the output module.

1. When the switch closes, it causes a 1 to be stored into bit 112/12 of the input image table.

2. The input contact on the ladder rung is also assigned the number 112/12. This numerical indicator instructs the user program to close the contact if a 1 is present at memory bit 112/12.

3. The configuration of the logic circuit programmed into the user memory is designed to energize the output symbol when the rung is true.

4. Next, the user program is instructed to store a 1 at the bit in the output image table with the same address as the number assigned to the energized output device on the ladder rung.

5. The 1 at memory bit 013/06 causes the terminal on the Output Module with the same identification number to be actuated.

6. A voltage at output terminal 013/06 causes the lamp to turn on.

Figure 14-31 also helps illustrate several important concepts about the operation of the programmable controller:

1. The input field device is not wired directly to the output field device. Instead, it is connected to an input module.

2. Each terminal on an I/O module is assigned an identification number that corresponds to the address location of a bit in memory.

3. The open or closed condition of a switching device causes a 0 or a 1 to be placed in a bit of the input image table in memory. The energized or de-energized condition of an output coil in the ladder diagram places a 1 or 0 in the bit of an output image table. A 1 will activate an output field device; a 0 will cause it to be deactivated.

4. The configuration of the logic circuit is not developed by field wiring connections. Instead, the circuit design is created in the software program in the user memory. The rungs and components used are entered by the programming unit, and the circuit is displayed on the screen.

5. The On/Off input condition of the ladder circuit is determined by the open and closed positions of the field devices. Each contact symbol is programmed with the bit address of the input image table that is affected by the corresponding field switching device.

6. An output symbol on the ladder diagram is programmed with a number that corresponds to the address location of a bit in the output image table. A 1 at the bit will energize an output field device connected to the terminal with the same number.

If any circuit changes are required, a new configuration is created by simply modifying the program, not by rewiring the field devices.

CHAPTER PROBLEMS

1. The vertical lines on a ladder diagram are called _____ and the horizontal lines are called _____.

2. The vertical line on the left of a ladder diagram is connected to ____ of an AC source, and ____ of a DC source. The vertical line on the right is connected to ____ of an AC source, and ____ of a DC source.
   a. L_1 c. +DC
   b. L_2 d. −DC or Common

3. Whenever possible, the components are labeled from ____ to ____, and from ____ to ____ on a ladder diagram.
   a. top   c. left
   b. bottom   d. right

4. Input switches connected in ____ on a rung perform an AND operation.
   a. series   b. parallel

5. By maintaining an energized ladder relay circuit after a push button is released, ____ function(s) is/are performed.
   a. a latching   c. both latching and interlocking
   b. an interlocking

6. __ T __ F   More than one corresponding contact in the ladder diagram configuration can be activated by a single output coil.

7. List three types of programming devices that are used to enter a program into the PLC.

8. List the four functions of the I/O module.

9. How many I/O modules are in a module group?

10. __ T __ F   The input image table is an array of bits that stores the status of On/Off input conditions of switching devices connected to input module terminals.

11. __ T __ F   The input image table does not change the status of the bits while running a program.

12. Draw an input module terminal stip with one input field device, show all necessary connections, and identify potentials L_1 and L_2.

13. Which of the following functions are not performed by the CPU? _____
    a. Mathematical operations.
    b. Management of memory.
    c. Self-diagnostic routines.
    d. Communication with other processor units.
    e. None of the above.

14. The I/O sections provide the interface between which of the following? ____
    a. Field equipment and output modules.
    b. Field equipment and the CPU.
    c. Input modules and the CPU.
    d. Input modules and output modules.

15. Match the following sections of a typical PLC memory system with its use:

    ____ Executive   a. Interfaces with field devices.

    ____ Scratch pad   b. Where logic circuit is written.

    ____ User program   c. Supervises system.

    ____ Data table   d. Interim calculation area.

16.

List which of the following memory sections use RAM or ROM memory devices: _____

    a. Executive   c. User program

    b. Data table   d. Scratch pad

17. Groups of bits at each memory register are called _____.

18. The location of a memory register is identified by an _____ number.

19. Fill in the following blanks to identify what they represent in each I/O rack location.

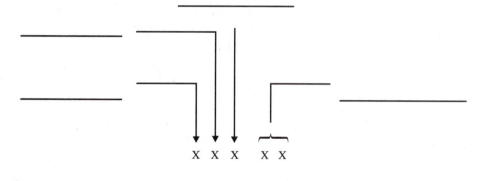

CHAPTER 15

Fundamental PLC Programming

CHAPTER OBJECTIVES

At the conclusion of this chapter, you will be able to:

○ Describe a typical processor scan cycle.

○ Describe the purpose of a software program used by PLCs.

○ List the four most common languages used by programmable controllers.

○ Identify a True and False logic condition at the rung of a ladder diagram.

○ Understand the Examine-On instruction.

○ Understand the Examine-Off instruction.

○ Program a PLC using the following types of instructions:

Relay Logic Inputs and Outputs	Data Manipulation
Timers	Arithmetic
Counters	

INTRODUCTION

The ability of a programmable controller to achieve machine control is the result of the program written into its memory and the power of its CPU. The **program** is simply a list of instructions that guides the CPU. The execution of the program is performed by the CPU through a three-step process called a **processor scan cycle**.

This chapter provides a more detailed explanation of the processor scan. The remainder of the chapter examines the different types of generic ladder logic instructions required to write a program.

15-1 PLC PROGRAM EXECUTION

Unlike the ladder relay circuit which monitors every input and output device throughout the circuit simultaneously, the PLC performs the same function in a sequence of steps called a *processor scan cycle*. Figure 15-1 shows one rung of a relay ladder diagram to help explain the following three steps that take place during the scanning process.

When the PLC is turned on, a program is fed into the CPU from the executive memory to run an internal self-diagnostic check on the system. If any part of the processor is not functioning properly, its fault indicator light illuminates. The diagnostic self-check also determines if there is a faulty memory or an improper connection with the I/O module.

If the self-diagnostic check determines that the system is operating properly, it starts its scanning operation.

Step 1: The first step of the scanning process is to *Update the Input Image Table* by sensing the voltage levels of the input terminals. The status of the input field devices is recorded in memory. A 0 or 1 is stored into the bit of the memory location designated for a particular input terminal. In Figure 15-1, a 0 is stored into memory bit 11311 when no voltage is sensed, and a 1 is loaded when a voltage is present.

Step 2: The second step of the scanning process is to *Scan Program Instructions* located in the user memory.

1. First, the CPU makes a reference to the input image table to find out the status of the inputs. In Figure 15-1, it finds out the condition of input 11311.

2. Second, the CPU reads the next instruction. The CPU makes decisions based on whether the input conditions are met and the type of logic function specified by the program. The program for Figure 15-1 would read:

    If a 1 is present at memory bit 11311, then write a 1 at bit location 01612 of the output image table.

3. Third, according to the decisions made, the CPU updates the output image table by recording a 0 or 1 at the relevant bit location in memory.

Step 3: The third step of the scanning process is to *Update the Output Terminals*. The CPU takes data from the output image table and sends it out to the real world through the terminals of the output modules.

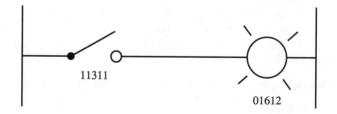

11311

01612

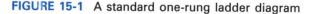

FIGURE 15-1 A standard one-rung ladder diagram

The diagram in Figure 15-2 graphically summarizes the three steps of the scan.

The three-step scanning process is continuous and is repeated many times each second. The time it takes to complete one scan depends on the size of the program and the clock speed of the microprocessor used by the PLC. If, for example, a program is

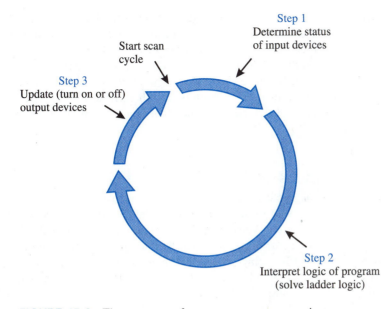

FIGURE 15-2   Three steps of a processor scan cycle

written that uses 1K words of memory, the time to scan the instructions once may take five milliseconds. Therefore, there will be 200 scans each second.

In some applications, the response time of 5msec by the output device to a change of an input condition may be too slow. To solve this high-speed requirement, many PLCs have special programs available that permit the scan to be interpreted so that the I/O devices can be updated immediately.

15-2   PROGRAMMING LANGUAGES

A software program written into the user memory provides a way for the user to communicate with the PLC. The program is a control plan that tells the processor what to do when certain conditions exist. Several different forms of programming languages have been developed by PLC manufacturers to enable instructions to be written. The four most common types of languages are:

Ladder Diagram
Boolean Mnemonic
Function Block
English Statement

Ladder Diagrams and Boolean Mnemonics are considered basic PLC languages. These basic languages are instruction sets that perform the most fundamental control operations, such as On-Off relay switching, timing, and counting.

The Functional Block and English Statements are considered high level languages. They are capable of executing more sophisticated control operations than the basic languages, such as analog control, diagnostics, and report generation. Most PLCs offer the basic relay ladder logic plus a combination of the other languages. The language used and the capabilities of the microprocessor and other hardware circuitry of the PLC determine the range of applications that can be performed.

Since the languages differ among PLC manufacturers, the instructions described in this book are generic. Because the ladder logic language is used to some extent by most brands of programmable controllers, it will be described in detail throughout the remaining portion of this chapter. The other languages will be covered briefly in the next chapter.

15-3   LADDER DIAGRAM PROGRAMMING

The ladder logic language closely resembles hardwired relay circuits. It is composed of symbols that are inserted onto ladder rungs. The symbols represent an instruction set that performs different types of On-Off operations. In general, the input conditions are represented by contact symbols, and the output instructions are represented by coil symbols. Figure 15-3 compares a relay diagram to a ladder diagram rung. Figure 15-3(a) shows a limit switch LS1 and relay contact CR2 that must be closed to energize relay coil CR3. Figure 15-3(b) shows an equivalent PLC ladder diagram with the input devices and the output device identified with their respective data table bit addresses. The address numbers correspond to the location of the I/O modules and the terminal to which each field device is wired.

Table 15-1 summarizes five major types of operations and shows the instruction symbols used within these categories. The format of the rung dictates which type of logic control is performed. For example, if the input contacts are placed in series, an AND function is performed. A parallel configuration performs OR functions. When the conditions of the input devices allow a true logic path in the rung, the rung condition is True. When there is is no true logic path, a False condition exists . A True rung condition energizes an output coil, which activates an output device that performs a desired operation.

The remainder of the chapter describes the instruction symbols listed in Table 15-1. These are commonly used to perform relay logic type operations. Example diagrams will be provided periodically to show how symbols are used in an actual ladder circuit. Each symbol in the diagram will be assigned a reference number which indicates the address of a corresponding bit located in memory.

The ladder logic instructions discussed in the following pages are generic examples. For detailed information on the instruction set of a specific controller, consult its manufacturer's manual. The manual will provide information on how to use each input and output instruction that is available in the PLC.

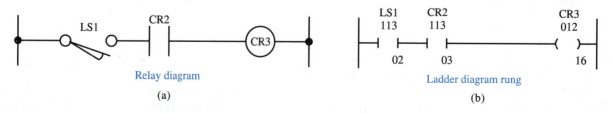

FIGURE 15-3 Comparing a relay diagram to a ladder diagram

TABLE 15-1   LADDER PROGRAMMING SYMBOLS

Operation Type	Basic Symbols	Description
Relay logic	——┤ ├——	Contacts (N.O.)
	——┤ / ├——	Contacts (N.C.)
	——( )——	Output coil
	——(L)——	Latched output
	——(U)——	Unlatched output
Timer and counter	——(TON)——	Timer On
	——(TOF)——	Timer Off
	——(CTU)——	Up-counter
	——(CTD)——	Down-counter
	——(CTR)——	Counter reset
Data transfer	——┤ GET ├——	Get data
	——(PUT)——	Store data
Data manipulation	——(=)——	Equal to
	——(<)——	Less than
Arithmetic	——(+)——	Addition
	——(–)——	Subtraction
	—(×)——(×)—	Multiplication
	—(÷)——(÷)—	Division

15-4   RELAY-TYPE INSTRUCTIONS

The relay-type instructions are Examine-On Contact, Energize Output Coil, Examine-Off Contact, Latch Output, and Unlatch Output.

Symbol: ┤ ├

Name: Examine-On Contact

This symbol represents an input condition instruction. It tells the CPU to examine a bit at the memory location specified by the address number listed with the symbol. The referenced address may represent the status of an external input or an internal program bit.

○ The Examine-On instruction is True when the addressed memory bit is a 1, meaning that a corresponding external input field device supplies a voltage at the input terminal.

○ The Examine-On instruction is False when the addressed memory bit is a 0, meaning that a corresponding external field device does not supply a voltage at the input terminal.

Examine-On instructions can be used alone as a single input device on a rung. Several of them can also be programmed in series to perform an AND operation, or in parallel to perform an OR operation. These instructions are located in the left portion of the rung. The field devices that provide input signals include switches, push buttons, and sensors.

Symbol: –( )–

Name: Energize Output Coil

This symbol represents an output action instruction. This instruction will be performed only if the condition (input) instructions preceding it provide a path of logic continuity. The Energize Output instruction tells the CPU to *set a bit to 1* or *reset a bit to 0* at the memory location specified by the address listed with the symbol. The referenced bit provides the action command signal for the output.

○ The output instruction sets the memory bit to a 1 when the rung condition is True. The result is that the corresponding field device energizes.

○ The output instruction resets the memory bit to a 0 when the rung condition is False. The result is that the corresponding field device de-energizes.

Output instructions are located on the right portion of the ladder-diagram rungs. Most PLCs allow only one output to be programmed on each rung. This instruction is considered as a non-retentive output, meaning that once the current passing through the coil stops flowing, it de-energizes. Field devices operated by this instruction includes lights, motors, and relay coils.

Output devices are not restricted to controlling external devices. They may also operate as internal coils that are not wired to any external device and instead are held in the computer's memory as an on or an off bit. The internal relay coil looks just like the coil symbol used for the external relay coil. An internal relay differs only in having a different address number than the real output. Internal relays are used when it is only necessary to control contacts in other rungs and not real external devices. The advantage of internal relay coils is that in large ladder programs, they reduce the number of output I/O modules required for an application.

○ When the output in the form of a coil is On, an Examine-On contact with the same address will be True, and a normally-closed (Examine-Off) contact will be False.

○ When the output in the form of a coil is Off, an Examine-On with the same address will be True, and a normally-closed (Examine-Off) contact will be False.

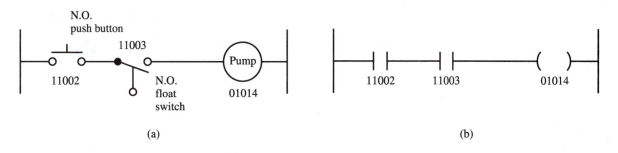

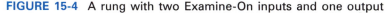

FIGURE 15-4 A rung with two Examine-On inputs and one output

There can be more than one corresponding contact in the ladder diagram configuration that is activated by only one output coil.

Figure 15-4 shows a one-rung relay diagram (Figure 15-4(a)) and its equivalent ladder diagram (Figure 15-4(b)) with two Examine-On inputs and one Energize Coil output. A normally open push button is connected to terminal 11002 of the input module, and a normally open float switch is connected to terminal 11003. A pump is connected to terminal 01014 of the output module. The float switch closes when the level of liquid in a storage tank reaches a given height, and the pump switch is manually closed by an operator. These actions energize the pump.

Branching
If one or more sets of parallel input conditions are required to energize an output device, branching instructions are used to develop the program. There are two branch instructions:

Branch Start: This instruction begins each parallel logic branch of a rung. The Branch Start is programmed immediately before the first instruction of each parallel logic path. The symbol on the keypad for this instruction is shown in Figure 15-5(a).

Branch End: This instruction completes a set of parallel branches. The Branch End is entered after the instruction of the last branch has been entered. The symbol on the keypad for this instruction is shown in Figure 15-5(b).

To write this diagram into the user memory, a programming unit is used to enter the required data.

Before writing the program, a mode switch on the processor unit module, shown in Figure 15-5(c), is placed in the program (PROG) position.

Branch Programming Steps

Step 1: Enter the Branch Start information by pressing the instruction key (Figure 15-5(d)). This is done because the CPU must be told that the primary rung has a parallel branch that will be connected to it.

Step 2: Press the Examine-On key to enter the first instruction in the primary rung (Figure 15-5(e)).

FIGURE 15-5 Branching instructions

Step 3: The Examine-On symbol will appear on the CRT screen with five zeros
above it. They will blink on and off to prompt the user to enter the input address. Enter the sequence shown in Figure 15-5(f).

Step 4: There are no more input examine instructions on the primary rung. Therefore to begin the next parallel branch, press the Branch Start key again (Figure 15-5(g)).

Step 5: Press the Examine-On key (Figure 15-5(h)). The symbol and five zeros will blink on and off to prompt the operator to assign an address to this device. Insert the appropriate address (Figure 15-5(i)).

Step 6: When the parallel branch is completed, enter a branch end instruction by pressing the Branch End key (Figure 15-5(j)).

Step 7: Since there are no further examine instructions after the parallel branch, proceed to the output. The primary rung is completed by entering an Energized Output Coil instruction (Figure 15-5(k)).

Step 8: Enter the address for the output coil by pressing the sequence shown in Figure 15-5(l).

There are limitations to the number of branches that can be entered for a single ladder rung. The exact limitations are dependent on the particular type of PLC being used.

After the circuit entries on the programming device have been completed, the mode switch must be changed to the RUN mode before the program can actually perform its operation. By placing the mode switch in the TEST mode position, the output device cannot be activated. Instead, the instruction symbols on the screen intensify as they become energized. By observing the symbols on the screen, the user can determine if the circuit is operating properly without the outputs being activated. When the TEST mode is used, the first two steps of the scan cycle are performed, and the third step is bypassed.

Symbol: –|/|–

Name: Examine-Off Contact

This symbol represents an input condition instruction. It tells the CPU to examine a bit at the memory location specified by the address listed with the symbol. The referenced address may represent the status of an external input, or an internal program bit. This instruction is used when the absence of the signal is needed to turn an output on.

○ The Examine-Off instruction is True when the addressed memory bit is a 0, meaning that a corresponding external input field device does not allow a voltage to be present at the input terminal. The result is the Examine-Off input will be True and allow logical continuity in its rung.

○ The Examine-Off instruction is False when the addressed memory bit is a 1, meaning that a corresponding external input device allows voltage to be present at the input terminal. The result is the Examine-Off input will be False. Therefore, logical continuity in its rung does not exist.

Symbol: –(L)–

Name: Latch Output

This symbol represents a retentive output action instruction. When its rung becomes True, it tells the CPU to set the bit to a 1 at the memory location specified by the address listed with the symbol. It is called a retentive output because once the rung condition goes False, the latch bit remains set. The bit becomes reset to a 0 by an Unlatch instruction programmed with the same reference address, or if system power is lost. Although most PLCs use Latch commands for internal and external outputs, some are restricted to latching internal outputs only.

Symbol: –(U)–

Name: Unlatch Output

This symbol is used to reset a latched output with the same address. It is the only automatic means of resetting a latched output. Therefore the Latch and Unlatch instructions should be used in pairs with each other. When its rung becomes True, the Unlatch instruction tells the CPU to reset the corresponding latch bit. In some situations the programmer may want the program to run through an entire scan cycle before the Latch instruction is reset. It is then recommended that the Unlatch instruction is placed before the Latch instruction.

APPLICATION EXAMPLE

An example of an application that uses the Latch and Unlatch instructions is a temperature alarm system for a greenhouse. If the temperature lowers to 32 degrees, an alarm sounds and stays on, even if the temperature rises above freezing. The alarm system notifies the greenhouse operator if any freezing condition occurred, regardless of the amount of time it lasted. The alarm is turned off when a push button is pressed.

Figure 15-6 shows two ladder rungs that perform the alarm function. The True/False condition of the Examine-On instruction in Rung 1 is connected to a

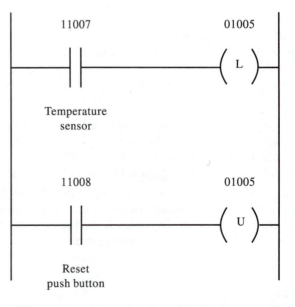

FIGURE 15-6 Using Latch and Unlatch instructions for a temperature alarm system

temperature sensor. Contact 11007 becomes True if a freezing condition occurs, causing Latch output 01005 to energize and activate the alarm. Regardless of what happens to the temperature, the Latch output remains on. When a push button makes the Examine-On contact 11008 in Rung 2 True, the Unlatch output 01006 energizes and unlatches the corresponding output in Rung 1.

15-5   TIMER INSTRUCTIONS

Timers are output instructions that are internal to the programmable controller. They are capable of providing timed control of devices that they activate or deactivate. Timing operations are used in many industrial applications. PLC timers perform various functions such as delaying an action, causing an operation to run a predetermined period of time, or recording the total accumulated time of continuous or intermittent events. They can also operate as astable or one-shot multivibrators.

A timer is activated by a change in the logic continuity of its rung. The rung condition is most often controlled by an Examine instruction. After the timed interval has expired, the timer output is energized, causing a normally open or normally closed contact it controls to change its logic state. These contacts can be used throughout the program as many times as necessary. The contacts associated with a specific timer are identified by using the same address number.

As each timing instruction is programmed into the PLC, as shown in Figure 15-7, the following information must be entered:

1. Symbol and Address
2. Time Base
3. Preset Value (PR)
4. Accumulated Value (AC)

Symbol and Address.   As a timer button on the programmer is pressed, an output symbol or rectangle appears on the screen, depending on the PLC. Inside it is an abbreviation indicating which type of timer has been selected. A three-digit number also appears to identify the address locations where the instruction is stored. A block of addresses is reserved in memory for timers.

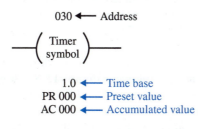

FIGURE 15-7 A timer instruction for the PLC

It is recommended that the first timer programmed be loaded into the lowest address of the block. Each additional timer should then be entered into succeeding memory locations.

Time Base.   After the symbol appears, a blinking cursor instructs the user to enter a time base value. A time base is a fixed time interval, usually 1.0 seconds, 0.1 seconds, or 0.01 seconds in length. One of the numbers is selected by the user.

Preset Value (PR).   After the time base data is entered, the blinking cursor instructs the user to enter a three-digit number to the right of the abbreviation *PR*, which represents *Preset Value.* The value entered indicates the amount of time that needs to expire before action is taken by the timer. This value denotes the number of time-base intervals to be counted.

Accumulated Value (AC).   Another display shows *AC=00,* which represents *Accumulated Value equals 00.* When the timer is activated, an internal clock causes the accumulated value to begin incrementing from 00 as it counts elapsed time-base intervals.

Two bits in the counter memory register are status bits. For explanation purposes, each bit will be assigned a number.

Enable bit (Bit 17).   The enable bit is set when the rung condition is True and resets when the rung condition is False. The enable bit is used to energize or de-energize one or more contacts. The contacts to be controlled are programmed by entering a five-digit number that includes the timer address, followed by 17.

Timer bit (Bit 15).   This bit is considered the primary output of the timer. When the accumulated value equals the preset value—AC=PR—the timer has timed out. The output bit then sets on or resets off. The on or off condition depends on the type of timer instruction selected. The output bit is used to energize or de-energize one or more N.O. or N.C. contacts located at other rungs. The contacts to be controlled are programmed by entering a five-digit number that includes the timer address, followed by 15.

Each timer uses two memory locations in the block of memory addresses reserved for timing devices. One address stores the accumulated value, the output bit, and the enable bit by the register shown in Figure 15-8(a). The three-digit value in the accumulator is stored at octal bits 00-07, and 10-13 in a Binary Coded Decimal (BCD) format. A decoder in the PLC converts each BCD value to an equivalent decimal digit to be displayed with the timer symbol. The output bit is located in octal bit 15 and the enable bit is located at octal bit 17.

The preset value is loaded into a register at an address 100 locations higher than its corresponding timer, as illustrated in Figure 15-8. This decimal value entered can range from 000 to 999. An encoder in the PLC converts the decimal digits into a BCD format. These bits are loaded into the first 12 bits of the register shown in Figure 15-8(b). Bits 15 and 17 are used to control contacts located in other rungs by producing a 0 or a 1.

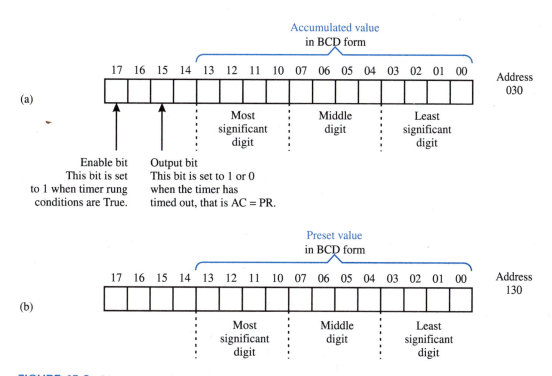

FIGURE 15-8 Memory registers that store data for timers

There are several types of PLC timers available. These include the Time-On Delay and the Time-Off Delay. The choice of which to use is dependent on the type of operation to be performed.

Symbol: –(TON)–

Name: Time-On Delay

When the rung condition becomes True, the timer begins causing the accumulator value to increment and bit 17 to set. When AC=PR, the timer stops timing and the output energizes as bit 15 sets. At the same time bit 17 resets. If logic continuity of the rung is lost before or after the timer has timed out, the accumulator value goes to zero, and bits 15 and 17 reset.

The Time-On Delay output is used to provide time-delayed action or to measure the duration for which some event is occurring.

APPLICATION EXAMPLE _____

An example of an application that uses a Time-On instruction is a high-speed transfer line of canned foods. An inductive proximity switch is used to detect the movement of the cans. If the cans become jammed upline, the movement stops. Since the cans will not be passing the detection point, the timer will time out. The output bit of the timer is programmed either to set off an alarm or to stop the machine until the problem is corrected.

Figure 15-9 shows the ladder diagram that controls the process. The inductive proximity switch causes the Examine-Off contact 11003 to become True each time there is a gap between the cans. The True condition of the rung starts Timer 030, and the accumulated value increments until the next can is picked up by the sensor. Each time a can is present, the Examine-Off contact goes False and causes the accumulated value to reset. If a new can does not pass within 40 milliseconds, the preset value 040 is reached by the accumulated value in the timer. The moment AC=PR, output bit 15 sets, creating a False condition at contact 03015 in Rung 2. A False condition of Rung 2 also develops, de-energizes output 01008, and stops the transfer line until the problem is corrected.

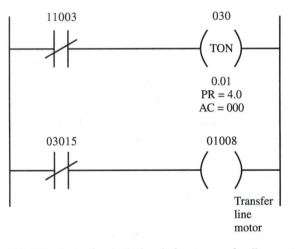

FIGURE 15-9　Control circuit for a transfer line

Symbol:　–(TOF)–

Name:　Time-Off Delay

While the condition of a rung is True, the accumulator value of a Timer-Off delay is reset to zero and bits 15 and 17 set. Once logic continuity of the rung is lost, the timer resets bit 17 to 0 and causes the accumulator value to begin incrementing. When AC=PR, the timer stops timing and the output energizes as bit 15 resets. If logic continuity of the rung occurs before or after the timer has timed out, it will not count and the accumulator value resets to zero.

The Time-Off Delay timer output is used to provide time delayed action.

APPLICATION EXAMPLE

A security light for a garage door opener is an example of an application for a Timer-Off instruction. Anytime the garage door is open the light is on. After the garage door is closed, the light remains on for 20 seconds while the driver walks to the house from the car.

Figure 15-10 shows an Examine-Off instruction addressed 11006 that is controlled by a N.O. limit switch, and a Timer-Off output addressed 035. When the

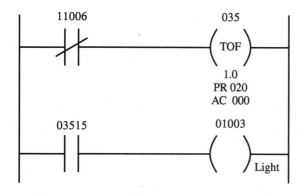

FIGURE 15-10 Security light control

garage door opens, the limit switch also opens and makes the logic condition of Rung 1 True, causing output bit 15 of the timer to set. The Examine-On contact 03015 becomes True and provides continuity to output 01003 which turns the light on. As soon as the garage door shuts, the limit switch closes, the condition of Rung 1 becomes False, and the Timer-Off instruction begins to time. While the timer is timing, bit 15 remains on, causing the light to stay lit. When the timer times out, output bit 15 is reset, causing Rung 2 to go False and the light to turn off.

Cascading Timers

If an application requires more time than the 999 seconds available from a single timer, two or more timers can be used to increase the maximum time value. The method of programming timers together to extend the time range is called *cascading*.

Figure 15-11 shows two timers that are cascaded to achieve a time of 1100 seconds. In this circuit, input device 11000 controls the timer in the top rung. The timer in the

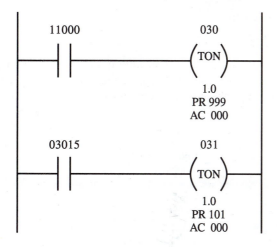

FIGURE 15-11 Cascading timers

bottom rung is controlled by an internal examine input that is activated by the timer on the top rung.

If the input device 11000 is True, the first timer starts incrementing. When the accumulated value reaches the preset time of 999, the timer times out and sets bit 15. Because the contact in the second rung has the same address as the timer in the first rung, it becomes True when bit 15 goes to a 1. At this moment the second timer begins to increment. When the count equals the PR value of 101, the second timer stops. The sum of accumulated values at each rung (999+101) equals the required time period of 1100 seconds. By cascading timers, virtually any desired time can be achieved.

15-6   COUNTER INSTRUCTIONS

Counters are output instructions internal to the programmable controller. A counter simply counts the number of events that occur, then stores and displays the accumulated value. There are two common types of counter functions performed by PLCs: up-counting and down-counting. (PLCs also perform the Counter-Reset instruction, which simply clears any accumulated values within the counter.) The choice of which one to use depends on the type of application to be performed. Programmable controller counters perform various applications such as counting the quantity of boxes passing a sensor on a conveyer belt, determining the number of parts left in a container, or keeping inventory of items in stock that are loaded into and removed from a storage facility.

To activate an up- or down-counter, either an external device or a software command must be used to control the logic continuity of the counter's rung. Each count occurs when a False-to-True transition is detected by the counter. The transition is sensed by the counter when its rung continuity changes from a False to a True logic condition.

As each counter instruction is programmed into the PLC, as shown in Figure 15-12, the following information must be entered:

1. Symbol and Address

2. Preset Value (PR)

3. Accumulated Value (AC)

Symbol and Address.   As a counter button on the programming unit is pressed, an output symbol or rectangle appears on the screen. Inside it is an abbreviation indicating which type of counter has been selected. A blinking cursor

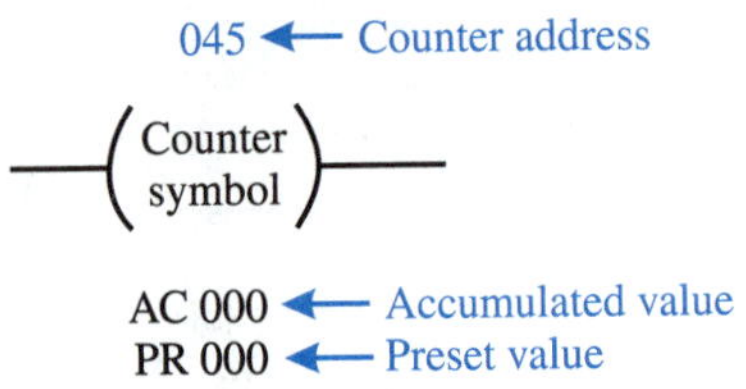

045 ◄— Counter address

—(Counter symbol)—

AC 000 ◄— Accumulated value
PR 000 ◄— Preset value

FIGURE 15-12   A Counter instruction for the PLC

appears, which prompts the user to enter a three-digit number to identify the address location where the instruction is stored. A block of addresses is reserved in memory for counters. It is recommended that the first counter programmed is loaded into the lowest address of the block. Each additional counter should then be entered into succeeding memory locations.

Preset Value (PR).   After the address has been programmed, a blinking cursor prompts the user to enter a three-digit number ranging from 000 to 999 at the right of the abbreviation *PR*. The value entered indicates the number of counts that occur before action is taken by the counter.

Accumulated Value (AC).   After the preset value has been entered, the cursor prompts the user to enter a three-digit number from 000 to 999 at the right of the abbreviation *AC*. Every time the counter is activated by a change in its rung condition, it changes by one count. An up-counter increments and a down-counter decrements.

Figure 15-13(a) shows a 16-bit register used for the up-counters and down-counters. Octal bits 00-07 and 10-13 store three digits of the accumulator value in the BCD

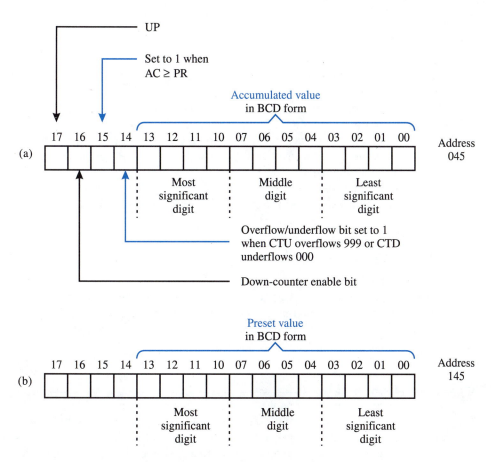

FIGURE 15-13 Memory registers that store data for counters

numbering format. The upper four bits in the counter's memory location are status bits. For explanation purposes, each bit will be assigned a number.

Enable bit.   The enable bit is set when the rung condition is True and resets when the rung condition is False. Bit 16 is the enable bit for a down-counter instruction. Bit 17 is the enable bit for an up-counter instruction.

Count Complete bit (Bit 15).   The count complete bit logic status is determined by comparing the accumulated value to the preset value. This bit is usually considered the primary output bit of the counter. Its status can cause a device that the counter controls to turn on or off after reaching a certain count.

Overflow/Underflow bit.   Unlike timers, which stop counting when the accumulated value equals the preset value, a counter will continue to count up (or down). Depending on which counter is used, the Overflow/Underflow bit sets when the accumulated value is greater than the maximum count of 999, or when it falls below the minimum count of 000.

Figure 15-13(b) shows the register that stores the preset value. Its address is 100 locations higher than its corresponding counter.

Symbol: –(CTU)–

Name: Up-Counter

The Up-Counter instruction will increment by one each time a counted event occurs. The number of events is recorded in the accumulator. The counts are activated by a False-to-True change of the counter's rung condition. When the accumulated value (AC) reaches the preset value (PR), the count complete bit (15) will set. Bit 16 will remain set if the counting continues beyond a value greater than the preset number. If the count goes beyond 999, the overflow bit (14) will set. A Reset instruction is required to clear the accumulated value.

Some PLC brands will reset to zero as soon as AC=PR. An application of an up-counter is counting the desired cereal boxes loaded into a case to indicate when it is filled to capacity.

Symbol: –(CTR)–

Name: Counter-Reset

The Counter-Reset instruction is used to clear the accumulated value of up-counters and down-counters to zero. When programmed, the reference address number of the counter to be reset must be entered with the CTR symbol. The referenced counter is cleared to zero when the CTR rung continuity becomes True.

APPLICATION EXAMPLE

Count-up instructions are often used when repeating a process. In Figure 15-14, for example, suppose that input device 11000 in Rung 1 drills five holes into a part. The number 5 is programmed as the preset value. Each time a hole is drilled, the count is incremented. When the accumulated value reaches five, the count complete bit will set. The Examine contact 03315 in Rung 2 becomes

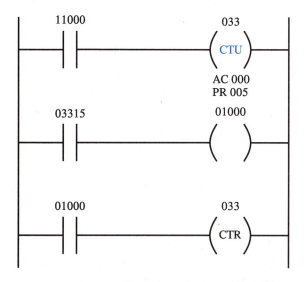

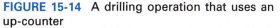

FIGURE 15-14 A drilling operation that uses an
up-counter

True and turns on output 01000. When energized, this output activates a robotic
arm, which removes the finished part and places a new part beneath the drill. It
also makes contact 01000 a logic True so that continuity is provided for the
Counter-Reset output of Rung 3. When activated, the CTR instruction resets
the accumulated value of the counter in Rung 1 before the first hole is drilled
into the next part.

Symbol: -(CTD)-
Name: Down-Counter

The Down-Counter instruction will decrement the count in the accumulator by one
each time a count event occurs. The number in the accumulator begins at any value
between 000 and 999 that is entered when the CTD instruction is programmed into
the PLC. Each count is activated by a False-to-True transition of the rung. Any preset
value can be entered between 000 and 999 during the programming of the CTD instruction. Bit 15 is set until the accumulated value goes below the PR number,
at which time it resets. If the count goes below 000, the underflow bit (14) will set.
A reset instruction is required to clear the accumulator value.

APPLICATION EXAMPLE _____

The Down-Counter instruction is often used to end a cycle. For example, suppose that only fifty parts are to be produced by an automation process. The ladder diagram in Figure 15-15 can be used to stop the process when the desired
number of parts is completed. The count-down preset value is set to 1 and the accumulated value to 050. When the PLC is put into the run mode, bit 03015 sets,
causes the Examine-On contact in Rung 2 to become True, and activates the

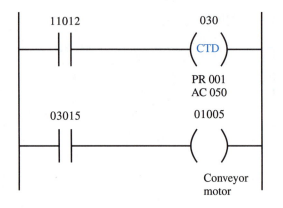

FIGURE 15-15   A conveyor operation that
uses a down-counter

automated production machine. A sensor that detects a finished part as it passes
an inspection point momentarily makes contact 11012 True. Each time a finished
part passes the sensor, the accumulated value decreases by 1. When the accumulated value equals 0, the count-down-done bit 03015 resets. As the Examine-On
condition of contact 03015 on Rung 2 becomes False, the automated production
machine is de-energized.

15-7   DATA MANIPULATION INSTRUCTIONS

Data manipulation is a category of instructions that enables words to be moved
within the memory. These instructions enable the PLC to perform more complex
operations than relay-type instructions, because they use multi-bit data to control outputs rather than only one bit. Data manipulation is divided into two categories: **data
transfer** and **data compare.**

DATA TRANSFER

Data Transfer instructions result in moving contents stored in one memory register to
another memory location. Only the memory addresses accessible to the user can be
used by these types of instructions.

To carry out the data transfer function, specific instruction commands are used to
read the data from one register and store the contents into another register. Examples
of data transfer commands are the GET and PUT instructions used in Figure 15-16.

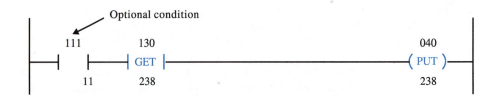

FIGURE 15-16   GET and PUT instructions

Symbol: ⊣GET⊢

Name: GET Word

The GET instruction tells the processor to fetch a three-digit number that is stored at the memory address programmed above the symbol shown in Figure 15-16. The number beneath the symbol displays the actual contents pulled from the register. The symbol can be located at the beginning of a rung, or with one or more examine instructions preceding it. The GET instruction will be activated only if the logic continuity of its rung is True.

Symbol: –(PUT)–

Name: PUT Word

The PUT instruction tells the processor to store a three-digit number at the memory location indicated by the number programmed above the symbol shown in Figure 15-16. This instruction must be programmed with a GET instruction to perform a data transfer operation. The three-digit number that it stores is listed beneath the symbol and is obtained from the GET instruction that precedes it. The PUT instruction is programmed in the output portion of the ladder diagram and only acts upon True rung conditions. It often has the same address as other instructions in the program.

In the ladder diagram in Figure 15-16, the GET instruction takes the contents (238) in memory location 130 and PUTs them into memory location 040. This operation only takes place if the Examine-On condition of contact 11111 is True.

The types of applications for which data transfer instructions can be used are presetting numbers into the registers of timers or counters, comparing data, displaying a number selected from a thumbwheel switch, or performing arithmetic operations.

APPLICATION EXAMPLE _____

An example of data transfer is to preset a counter. An up-counter is used in a factory that manufactures inductor coils. The counter controls the number of winds of wire around a core. When a 5 milli-henry (mH) inductor is produced, a lathe makes 400 revolutions to wind the coil. If the machine makes a 10mH coil, the lathe must make 800 turns before stopping. Figure 15-17 shows how to change the preset value for each different inductor by using push buttons.

Up-counter 035 is initially programmed with a preset value of 000. The GET instruction in Rung 3 is programmed to fetch 400 from memory location 20. The GET instruction in Rung 4 is programmed to fetch 800 from memory location 21. The PUT instructions in Rungs 3 and 4 are addressed to transfer the number from their preceding GET commands to preset the up-counter. To prepare the machine to make the 5mH coil, push button 11002 is pressed to make Rung 3 a logic True condition. The 400 is transferred from memory location 20 by the GET instruction to the preset value of counter 035 at address 135 by the PUT instruction. When start button 11000 is pressed, output 01100 starts the lathe turning. Every time the lathe makes a revolution, a sensor momentarily closes 11001, causing the counter to increment one count. When the accumulated value reaches 400 and AC=PR, bit 15 of counter 035 goes high. Contact 03515 of Rung 5 then closes and energizes output coil 01101. The result is contact 01101 of Rung 1 becomes False, de-energizes output 01100, and stops the lathe. It also

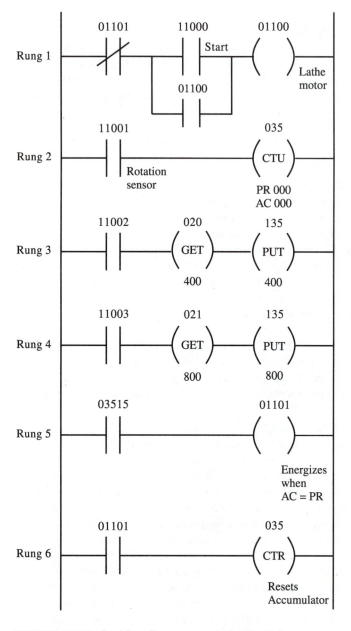

FIGURE 15-17 Ladder diagram control circuit for winding inductors

causes a True condition at contact 01101 in Rung 6 which activates the CTR output so that the accumulator value of the counter clears to 000.

To make a 10mH coil, the same procedure is repeated after the push button 11003 is pressed to load 800 into the counter, and then start button 11000 is pressed.

The GET and PUT instructions are also used to transfer a three-digit number to a down-counter. However, the number moved from the PUT command to the down-counter is placed in the accumulator register, whereas the value sent to the up-counter is placed in the preset register.

Thumbwheel Switch and BCD Display

A three-digit thumbwheel switch, shown in Figure 15-18(a), has three rotary knobs with ten different positions representing decimal digits 0-9. The switch has four terminals that produce a four-digit BCD output that corresponds to the equivalent decimal setting of each knob.

The BCD display has three seven-segment digits that are capable of producing a pattern to show any decimal number from 0 through 9. The digit that is displayed is the result of a corresponding BCD value applied to its four inputs.

A special input module is used to transfer the multi-bits of data from the three-digit thumbwheel switch assembly into a memory location of the PLC. An output module is used to transfer three BCD values from a different memory register to three BCD

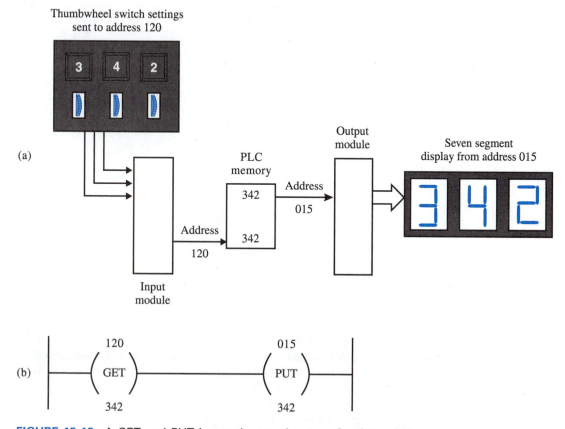

FIGURE 15-18 A GET and PUT instruction used to transfer data within memory

displays. By using the GET and PUT instructions in Figure 15-18(b), the three-digit number set by the thumbwheel switch can be read on the three-digit display. The numerical setting of 342 is fed through an input module to memory location 120. This address is displayed above the GET symbol and the contents are shown below. The 342 is also beneath the PUT symbol, and the value 015 above shows the address where it is stored. The BCD data from memory location 015 is fed through the output module to the three-digit display.

DATA COMPARE

The second category of data manipulation operations is Data Compare instructions. These commands instruct the processor to compare the numerical contents of two registers and to make decisions based on their values and the type of instructions used. Based on the result of the comparison, an output device can be activated or deactivated, or some other operation can be performed. The values compared are three BCD numbers that use twelve bits in a register. These numbers are displayed with the Data Compare symbol as three decimal digits. The Compare instructions must be programmed with a GET instruction. The GET instruction precedes the Compare instruction on the rung.

Two types of comparison instructions are Compare Equal and Compare Less Than.

Symbol: –(=)–

Name: Compare Equal

This instruction compares the number programmed below the Compare Equal symbol to the data fetched from the memory address above a GET instruction. If the values are equal, the rung is True and the output coil is energized, as shown in Figure 15-19.

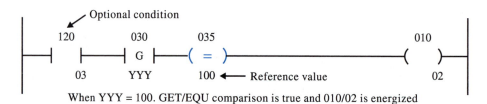

When YYY = 100. GET/EQU comparison is true and 010/02 is energized

FIGURE 15-19 An Equal To data comparison

Symbol: –(<)–

Name: Compare Less Than

This instruction compares the number programmed below the Compare Less Than symbol to the data fetched from the memory address above a GET instruction. If the contents at the Compare symbol are less than the number displayed below the GET symbol, the rung is True and the output coil is energized, as shown in Figure 15-20.

If an Examine instruction is programmed into the rung, it must precede the GET and Compare instructions and must be True to make the output coil energize.

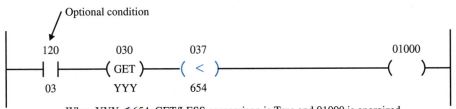

When YYY < 654, GET/LESS comparison is True and 01000 is energized

FIGURE 15-20 Less Than comparison

The order in which the Compare Equal To and Less Than instructions are programmed into the memory enables the following additional instructions to be developed:

○ Greater Than

○ Less Than or Equal To

○ Greater Than or Equal To

Compare Greater Than

To program a Greater Than data comparison, a Less Than symbol is used. By addressing the < instruction with the word that is to be compared, reverse logic is used to perform the > instruction, as shown in Figure 15-21. The value in word 031 (the Less value) is compared to the reference value of word 030 (the GET value), and the rung is only True when Examine-On contact 12002 is True and the Less value is greater than the GET value of 499.

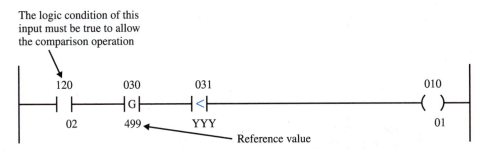

When YYY > 499, GET/LESS comparison is True and output 010/01 is energized

FIGURE 15-21 Greater Than comparison

Less Than or Equal To

To program a Less Than or Equal To data comparison, the rung is programmed with one GET instruction followed by Less Than (<) and Equal To (=) instructions in parallel, as shown in Figure 15-22. Output 01003 is energized if the number in address 030 is less than (<) 237, or equal to (=) 237. It is de-energized if the value goes to 236 or less.

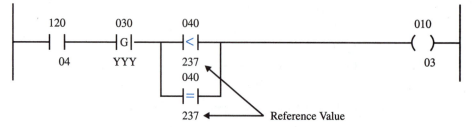

When YYY < 237 GET/LESS-EQUAL comparison is true and output 010/03 is energized

FIGURE 15-22 Less Than or Equal To comparison

Greater Than or Equal To

A Greater Than or Equal To data comparison circuit is shown in Figure 15-23. When the value in address 042 equals (=) or is greater than (>) the reference value 440 in address 030, the output is energized. The output is de-energized if the number is 439 or less.

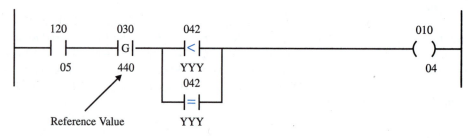

When YYY > 440 GET/LESS-EQUAL comparison is true and output 010/04 is energized

FIGURE 15-23 Greater Than or Equal To comparison

15-8   ARITHMETIC OPERATIONS

Most PLCs have the ability to carry out arithmetic operations. Instructions are used to tell the processor to perform the four basic mathematical functions: addition, subtraction, multiplication, and division. To make the calculations, the numbers in two registers are used. Some numbers are obtained by the GET instruction. Other arithmetic instructions simply reference the two registers using contact symbols. The contents are taken directly from specific memory addresses, counters, timers, or any other accessible word location.

APPLICATION EXAMPLE _____

A typical application of an arithmetic function is shown in Figure 15-24. A machine prints onto paper as it passes from a large supply roll to smaller rolls that are wound around cardboard cores. A pinch roller places a small pressure onto the paper to ensure that good contact is made with the print roller. All four rollers rotate at variable speeds. For example, when the winding is started, the

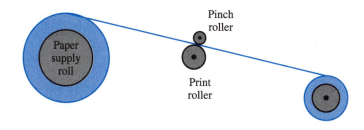

FIGURE 15-24 A printing machine that uses an arithmetic operation to control the speed

RPM increases until full speed is achieved. When the receiving roll is about full, the rollers decelerate until they come to a complete stop. When an empty core replaces the full roll on the receiver shaft, the printing machine starts up again.

As the printing roller turns, a sensor detects when it completes each revolution. The sensor signal is transformed into an RPM reading that is fed to a memory address in the PLC. Since the pinch roller is exactly half the circumference as the print roller, it must rotate at exactly twice the speed. To ensure that the proper RPM of the pinch roller is maintained, the PLC takes the RPM of the print roller and multiplies it times two during every scan cycle. The product of the calculation is fed to a converter that gives the motor of the pinch roller the proper current to rotate at twice the RPM of the print roller.

The following instructions describe a method to perform the arithmetic calculations. For each mathematical operation performed, calculations are made on two three-digit decimal numbers. Each digit in the memory registers is stored in a BCD format. If there is an Examine instruction in the rung, it must precede the arithmetic symbol, and it must be True before a calculation can be performed.

Symbol: –(+)–
Name: ADD

The ADD instruction performs the addition operation of three-digit numbers stored in two reference memory locations. These numbers are accessed by using data transfer GET instructions, as shown in Figure 15-25. When input device 11111 is True, the values stored in address 030 and 031 are added. The sum is displayed beneath the symbol

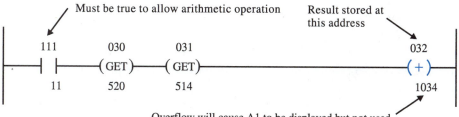

FIGURE 15-25 Add instruction

of the ADD coil and is stored in the register referenced by the address 032. If the sum exceeds 999, the overflow bit (14) of the ADD instruction will set to a 1, causing a 1 to appear in front of the sum beneath the symbol.

Symbol: –(–)–
Name: SUB

The SUB instruction performs the subtraction operation of three-digit numbers stored in two reference memory locations. These numbers are accessed by using GET instructions, as shown in Figure 15-26. When input device 11114 is True, the second GET word value from location 41 is subtracted from the first GET word value in address 40. The answer is displayed beneath the symbol of the SUB coil and is stored in address 042. If the subtraction result is a negative number, the underflow bit (16) of the memory register that stores the answer is set to 1, and the minus sign (–) will appear in front of the answer.

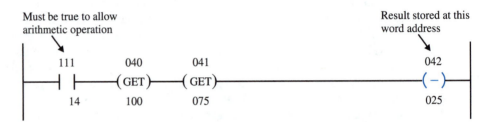

FIGURE 15-26 Subtract instruction

Symbol: –(×)–(×)–
Name: MUL

The MUL instruction performs the multiplication function of three-digit numbers stored in two reference memory locations. These numbers are accessed by using data transfer GET instructions, as shown in Figure 15-27. The number 123 in the first GET instruction is multiplied times the 061 in the second GET instruction. The product of the multiplication is held by the two MUL coils and is displayed as six digits beneath the symbols. The numbers above the symbols indicate the two addresses where the answer is stored.

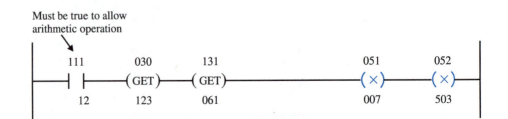

FIGURE 15-27 Multiply instruction

Symbol: –(÷)—(÷)–

Name: DIV

The DIV instruction performs the division operation of three-digit numbers stored in two reference memory locations. The numbers are accessed by two GET instructions, as shown in Figure 15-28. The number 050 in the first GET instruction is divided by 025 in the second GET instruction. The result of the division is held by the two DIV coils and is displayed as six digits beneath the symbols. The numbers above the symbols indicate the two addresses where the answer is stored.

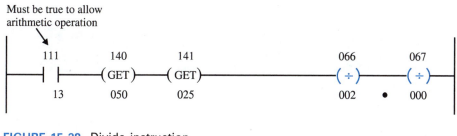

FIGURE 15-28 Divide instruction

15-9   WRITING A PROGRAM

When you write a program for a particular application, there are many different ways to achieve the same results. If other PLC users need to read the program you have written, it is important to document your work to help them understand the sequence of instructions. Use the following steps as a guideline to use your time efficiently and to ensure that the information you provide helps others follow your program.

Step 1: Choose the sequence in which you want the input and output devices to operate. Decide what the devices must do and what conditions must be True before these devices can begin operating.

Step 2: Write a description and make a drawing that shows the sequence and conditions for energizing each output device. The diagram of the mechanical and electrical components will help visualize the operation, making it easier to write the program.

Step 3: Use the description and drawing to write the ladder diagram program.

Step 4: Connect and label the input and output devices.

Step 5: Make a written record of each address used and what each address represents. The record should contain the addresses of all inputs, outputs, timers, counters, and other devices used in the program. This information is especially useful in troubleshooting.

Step 6: Enter the program into the programmable controller.

CHAPTER PROBLEMS

1. A _____ is a list of instructions that guides the CPU.

2. List the three steps of a processor scan.

3. If 10K words of memory are scanned in 10 milliseconds, how many scans will occur every second? _____

4. List four common types of programming languages.

5. Using Figure 15-29, indicate the logic condition of the program instruction when the corresponding input switching device in the left column is in the position shown.

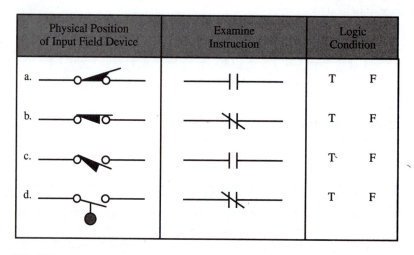

FIGURE 15-29

6. When an Examine-Off symbol is the only input on a ladder rung, the output will be energized when a corresponding N.O. push button is ____.
   a. open   b. closed

7. Branching instructions are used to ____.
   a. create a parallel circuit   c. create a series circuit
   b. develop a logic AND circuit   d. enable all of the above

FIGURE 15-30

8. Match the key position in Figure 15-30 with the appropriate description. Write the correct letter in the blank.
   PROG ____   a. Program is executed; outputs are disabled.
   TEST ____   b. Program is entered; outputs are disabled.
   RUN ____   c. Program is executed; outputs are enabled.

9. Choose all of the following statements that apply to a single coil: _____
   a. It could be used as an internal output.
   b. Its contacts could be used as an Examine-Off input in a rung.
   c. It could be used to drive several field output devices in more than one rung.
   d. All of the above.

10. __ T __ F   An address for a given input or output can only be used once in a ladder program unless additional wiring is done.

11. __ T __ F   A Latch Output instruction can be used to replace an interlocking branch circuit.

12. Indicate which of the following functions are performed by a timer circuit: _____

    a. Delay-On action.
    b. Cause an operation to run a specific period of time.
    c. Record accumulated time of a continuous event.
    d. Record accumulated time of intermittent events.
    e. Operates as an astable multivibrator.
    f. Operates as a one-shot multivibrator.
    g. All of the above.

13. List four types of information that must be programmed along with the timer symbol.

14. When more time is needed than the maximum amount provided by one timer, two or more timers can be programmed together. This programming technique is called ____.
    a. stacking   c. multiplying
    b. cascading   d. piggy-back

15. List three types of information that must be programmed along with the counter symbol.

16. An up-counter ____ and a down-counter ____.
    a. increments   b. decrements

17. List two categories of data manipulation instructions.

18. A thumbwheel switch produces a ____ output that corresponds to its decimal digit position.
    a. binary   c. hexadecimal
    b. BCD

19. A BCD display produces a pattern that will show which single digit numbers?

20. List four types of arithmetic functions performed by the PLC.

21. Numbers operated on by arithmetic instructions are taken from which of the following sources: _____
    a. Specific memory addresses   d. Thumbwheel switches
    b. Counters   e. All of the above
    c. Timers

CHAPTER 16

Advanced Programming and PLC Applications

CHAPTER OBJECTIVES

At the conclusion of this chapter, you will be able to:

○ Draw a ladder rung that represents a Boolean statement and an English statement language.

○ Describe how to interpret the data programmed into a functional block symbol.

○ List reasons for using jump instruction commands.

○ Describe how data bits are moved through a shift register to perform automated control functions.

○ Develop a function table that can be used to program a sequencer to perform data manipulation operations.

○ Draw wiring diagrams to show how sourcing and sinking I/O modules are wired.

○ Describe the operation of analog input and output modules.

○ List the different types of Special Purpose I/O modules and describe what types of functions they perform.

INTRODUCTION

Programmable controllers were initially designed to replace hardwired relay logic circuits. Presently, their operation is no longer restricted to relay equivalent on/off functions. As technological advancements have been made through software, hardware, and microprocessor developments, the capabilities of the PLC have expanded. Advanced programming languages make PLCs more powerful and easier to use. Many of the operations they must now perform require some of the same capabilities as a computer system. Advanced PLCs may contain specialized modules enabling them to perform very specific operations for a variety of motion and process control applications.

In this chapter, alternative types of programming languages will be explained. Also, various PLC modules and the operations they perform will be described.

16-1   ALTERNATIVE PROGRAMMING LANGUAGES

Alternative programming languages include Boolean algebra, functional block language, and English statement language.

BOOLEAN ALGEBRA

Boolean algebra is a mathematical logic system based on two states: True and False. It was developed in 1849 by an Englishman named George Boole. Its purpose was to help clarify the art of logical reasoning used in philosophy. Ladder circuits can also operate by making logic decisions based on True-False conditions. Because of this relationship between ladder logic and Boolean logic, Boolean equations can be used as a language to program PLC diagrams.

The Boolean language is based primarily on the Boolean operators *AND, OR,* and *NOT.* A capital letter represents the wire label of a signal at the input. A multiplication sign (·) represents the AND operation, an addition sign (+) represents the OR operation, and a bar placed over the input label ($\overline{A}$) represents the NOT operation.

The following abbreviations inside brackets provide a sample of symbols used when programming the Boolean language. These symbols are printed on the keys of a keypad. To the right of each abbreviation is a description of the instruction it represents. These commands are used to develop the circuit.

[LD]   Load or Store

[AND]   AND Function

[OR]   OR Function

[NOT]   NOT Function

[OUT]   Output

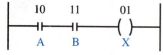

FIGURE 16-1 A ladder diagram that performs the AND function

A ladder diagram that performs the AND function is shown in Figure 16-1. Since the contacts are placed in series, the output is energized if input contacts A *and* B are closed. This AND function can also be expressed as A · B = X. The formula states that if A *and* B are True, then X must be True.

To enter the one-rung program for the diagram, type the following:

[LD] [1] [0] [AND] [1] [1] [OUT] [0] [1]

A ladder diagram that performs the OR function is shown in Figure 16-2. Since the contacts are placed in parallel, the output is energized if input contacts A *or* B are closed. The OR function can also be expressed as A + B = X. The formula states that if A *or* B are True, then X must be True.

FIGURE 16-2 A ladder diagram that performs the OR function

To enter the program for this diagram, type the following:

[LD] [1] [2] [OR] [1] [3] [OUT] [0] [2]

A ladder diagram that performs the NOT function is shown in Figure 16-3. The NOT function operates like a set of normally closed contacts. If input A is not activated, it remains closed and the output will remain energized. If the input is activated, it opens and the output de-energizes. The NOT function can also be expressed by the following formula: $\overline{A}$ = X, or A = X.

FIGURE 16-3 A ladder diagram that performs the NOT function

To enter the program for this diagram, type the following:

[LD] [NOT] [1] [4] [OUT] [0] [3]

A typical start/stop circuit is shown in Figure 16-4(a) and an equivalent digital logic circuit in Figure 16-4(b). The circuit uses both AND logic and OR logic. The Boolean operation is $A(B + C) = X$. It states that if contact A is closed, *and* either B *or* C is closed, output X is energized, as proven by the truth table in Table 16-1.

To enter the program for this diagram, type the following:

[LD] [1] [5] [AND] [1] [6] [OR] [1] [7] [OUT] [0] [4]

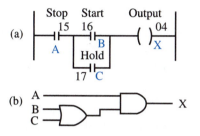

FIGURE 16-4 A start/stop circuit

TABLE 16-1 **TRUTH TABLE FOR START/STOP CIRCUIT**

Start	Stop		Hold		Output
A	(B	+	C)	=	X
0	0		0		0
0	0		1		0
0	1		0		0
0	1		1		0
1	0		0		0
1	0		1		1
1	1		0		1
1	1		1		1

FUNCTIONAL BLOCK LANGUAGE

The functional block is a high level language that uses the ladder diagram format to program complex functions. The blocks are a part of an instruction set used in conjunction with standard logic symbols to perform the desired operation.

When the block instruction is being programmed, a rectangular box appears on the screen. Inside the box are several parameters associated with the instruction that must be entered. These parameters include preset values for counters and timers, numerical data used by compare functions, and numbers to be calculated by arithmetic instructions. The parameters are loaded into registers within the memory section of the processor. Depending on the block type, there will be one or two input control lines to activate or deactivate the function, one or more parameters used to control the function, and one or more output lines.

An example of a functional block operation is the timer instruction shown in Figure 16-5. When the instruction is being programmed, the block will appear when the timer button is pressed. A cursor flashes to prompt the user to enter the preset and time-base (0.01, 0.10, or 1.0) parameter values. The accumulator (AC) stores the time while the timer is activated. The block instruction has two input lines labeled *Control* and *Enable/reset*. The timer will time as long as both inputs are True. If the control line goes False, the timer will stop counting. When the control line becomes True again, the timer resumes timing. If the enable line goes False, the timer stops and the accumulated value will reset to zero. There are two output lines. Output AC = PR will be 0 until the accumulated value equals the preset value. The operation of the other output, AC ≠ PR, is exactly the opposite. Either output line can be used to energize a coil or another output device.

Another example of a functional block operation is the counter instruction in Figure 16-6. When the instruction is being programmed, the block will appear as a rectangle on the screen. A blinking cursor inside the box prompts the user to enter a desired preset value. The accumulated value (AC) begins at a count of 000. The counter has three inputs labeled *Up, Down,* and *Reset*. Any Off-to-On transition applied to the Up input causes the counter to increment the accumulated value by one. An Off-to-On transition at the Down input will decrement the count. A 1 applied to the Reset input line causes the count in the accumulator to become 000. The count remains zero if an Off-to-On transition takes place at the Up or Down inputs when the Reset line is High. Switching contacts can be connected to the input lines to control the operation of the counter. The counter has two output lines. The line labeled AC = PR remains Low until the count in the accumulator reaches the same number as the Preset value, when it goes High. The operation of the other output line, AC ≠ PR, is exactly opposite. The output lines can be used to energize a coil or other output device.

Arithmetic functions, data manipulation, and sequencer instructions are also performed by functional block languages.

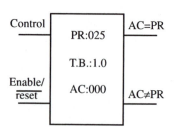

FIGURE 16-5 A functional block that performs a timer operation

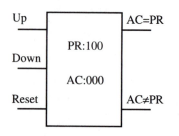

FIGURE 16-6 A functional block that performs an up/down counter operation

ENGLISH STATEMENT LANGUAGE

The English statement language used by PLCs is derived from instructions used to program computers. These statements are part of an instruction set that is used in conjunction with standard logic symbols to perform the desired operations.

English statements are instructions similar to those of the widely used BASIC computer language. Each statement represents a control command that tells the PLC what to do. The advantages of using this language are that it provides more computing power than ladder or Boolean instructions, and it is considered easier to understand by many users.

There are several English statement languages available. One example is a language that uses action statements to perform arithmetic and control functions. The action statements are:

LET   (Arithmetic)

SET   (Magnitude Comparisons)

GOTO   (Controls the Scan Sequence)

CALL   (Accesses Subroutines)

Each statement must begin with a four-digit number.

An element of the instruction set frequently used with an action statement is the IF-THEN qualifier. The IF-THEN qualifier allows the programmer to use conditional statements. The logic of a conditional qualifier is that IF condition *A* is true (contact closed), THEN perform action *B*. The IF-THEN is used when a LET, SET, GOTO, or CALL statement is to be executed only when the conditional contact is closed. When the conditional (IF) contact is open, the action statement will not be executed.

An example of a circuit programmed by using English statement commands is shown in Figure 16-7. The LET statement is used to perform addition. The CPU interprets the commands as follows: "If N.O. contact 1000 is closed, then add the number at address V8 to 5, and place the sum at ADDRESS V10."

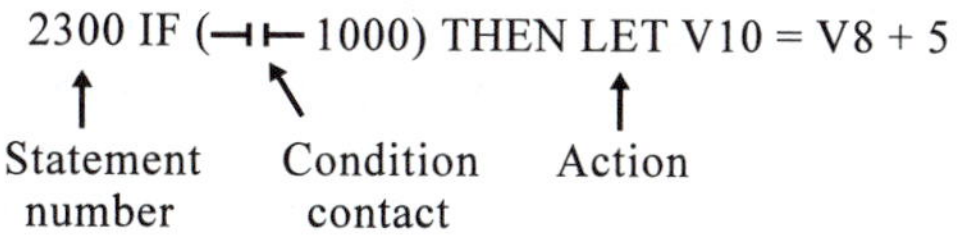

$$\text{2300 IF } (\dashv \vdash 1000) \text{ THEN LET V10} = \text{V8} + 5$$

| Statement number | Condition contact | Action |

FIGURE 16-7 English statement language format

16-2   JUMP COMMANDS

Most programmable controllers have instructions that allow the normal sequential program execution to be altered if certain conditions exist. Output instructions that perform this function are referred to as *override* commands. An example of an override command is a *jump* instruction. When this command is performed, a portion of the program is skipped over, as shown in Figure 16-8. The advantage of the jump

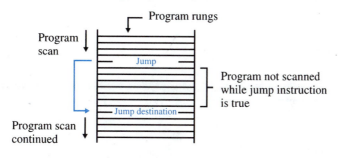

FIGURE 16-8 Jump operation

instruction is that it allows the PLC to hold more than one program and scan only the portion of the program needed to perform the desired action. This reduces scan time, allowing more scans to take place, so that program information is updated more frequently. A set of rung conditions always precedes the output symbol with a JMP label.

To activate a jump command, the JMP output is energized if the rung condition is True. A two-digit octal number above the JMP symbol instructs the PLC where it should resume the program execution after the jump has taken place. The label input instruction (LBL) with the same two-digit reference number as the JMP symbol identifies the destination of the jump operation, as shown in Figure 16-9. It is placed as the first condition instruction in the rung. The two-digit reference numbers are not the same as those used by examine instructions for addresses. Instead, numbers 1 to 16 are used. The number assigned to the first jump and label (JMP and LBL) programmed into the ladder diagram should be 1. Any additional numbers assigned are based on the number of preceding jump and label instructions. There can be more than one jump instruction pointing toward the same label instruction. Since the controller will not scan program rungs between a True jump instruction and its label instruction, the operation of timers may be thrown off if they are located on rungs that are jumped over.

An application example of a jump and label instruction is a multi-step operation where some steps are not always needed. Suppose several types of parts are moving down a conveyer and each must be assembled in a different way. An optical sensor is used to determine which kind of part passes a detection point. Each type of part inspected makes a different rung True. The rung that is activated causes the program to jump to the subroutine which contains the assembly operation for the part that is present.

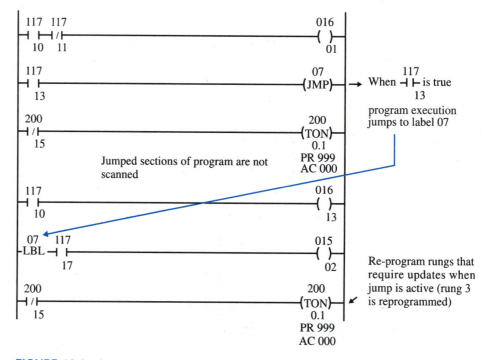

FIGURE 16-9 Jump and label operation

Another type of jump operation is performed by the subroutine (JSR) instruction. This command is used to direct the controller to a mini-program file within the program if certain conditions exist. A two-digit octal number above the JSR label tells the controller where to jump. The rung condition must be True to activate this command. An LBL instruction with the same reference number is located at the first rung of the subroutine.

An instruction labeled RET located in the last rung of the subfile tells the processor that the subprogram has been scanned and executed. When encountered, the controller returns to the main program and begins the ladder rung immediately following the JSR instruction that initiated the subroutine. Figure 16-10 shows a sketch of how the controller scans through memory when a subroutine instruction is encountered.

A subroutine instruction may be used in a program that controls stoplights at an intersection with a crosswalk button. The main program controls the light sequence to allow cars to pass first in one direction and then the other. This cycle continues until the main program is interrupted by the closing of the crosswalk button. The button is an Examine-On contact which makes the rung with the JSR instruction True. As the program jumps to a subroutine, it turns on the crosswalk light and the stoplights turn red for twenty seconds. When the time expires, the processor exits the miniprogram and returns to the main program to resume normal control of the lights at the intersection.

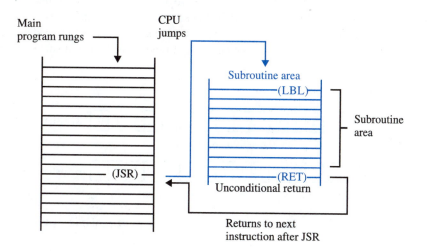

FIGURE 16-10 Jump to subroutine

16-3   DATA MANIPULATION

To energize the terminal of an output module using ladder instructions, a single bit in a memory register must be a 1. The binary state is usually determined by the conditions of input examine instructions and the circuit configuration to perform the desired logic function. Instead of using individual input contacts and logic circuits to control field devices, a set of instructions that maneuvers bits in a chosen register can perform the same function. These commands insert bits one at a time and set or clear individual bits. They can also move all the bits in one or more registers simultaneously. For example, suppose a sign that uses 256 light bulbs to display different patterns is controlled by a PLC . Sixteen registers that contain 16-bit words control the lights instead of 256 rungs.

Instructions that perform these multiple control operations are used extensively in all types of automated systems. Shift registers and drum/controller/sequencer instructions perform these types of functions. These are discussed in detail in the following text.

SHIFT REGISTERS

Instructions have been developed to move bits within a memory word or from one word to another. When this type of data transfer occurs, the memory devices are referred to as shift registers. The bits may be shifted forward (left) or reverse (right).

Figure 16-11(a) shows a 16-bit word used as a forward shift register. A clock pulse applied to the register causes its contents to shift. Data can be entered from the right, one bit at a time. The first diagram shows the stored data prior to a clock pulse, and Figure 16-11(b) shows how the register looks after the data have been shifted one place to the left. Each time a shift takes place, the data in bit 16 are shifted out and lost and a data bit is entered into bit 1. Some shift register operations require words that are greater than 16 bits. In this case, more than one register can be cascaded, as shown in Figure 16-12. Data is shifted into bit 1 of word 1. Instead of the data being lost out of bit 16, it is transferred to bit 1 of the of the second shift register.

Shift registers are used primarily to control a process on a conveyer system. As an object is inspected, it causes data to be entered into the register. The inspection results

Bit numbers
16 15 14 13 12 11 10 9 8 7 6 5 4 3 2 1
(a)
0 0 1 1 1 0 0 1 1 1 0 0 1 1 1 0
Register 105
Initial condition of register
(b)
0 1 1 1 0 0 1 1 1 0 0 1 1 1 0 0
Register 105
Condition register after shift forward (left)

FIGURE 16-11 Forward 16-bit synchronous shift register

Data in
16 15 14 13 12 11 10 9 8 7 6 5 4 3 2 1
Word 1
Word 2
Shifted out
16 15 14 13 12 11 10 9 8 7 6 5 4 3 2 1

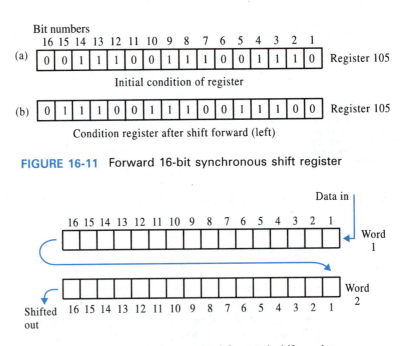

FIGURE 16-12 Cascaded two-word forward shift register

determine which logic state is entered. For example, the presence of an object produces a logic 1 and the absence generates a 0. An example of a practical application for a serial shift register is the filling and crowning of bottles on a conveyer line in a brewery, as shown in Figure 16-13.

The bottles physically touch each other as they travel down the line. They are all stopped simultaneously for two seconds when a bottle is being filled. The conveyer runs only long enough to move the distance of one bottle diameter before it stops to fill the next bottle. Light Sensor 2, placed two bottle positions from the filler, detects if the bottle is full. Each bottle is capped two bottle positions from Sensor 2. Light Sensor 3, placed two bottle positions from the Capper determines if the cap is placed securely on the bottle. A pneumatic ram, two bottle positions from Sensor 3, pushes a bottle off the conveyer if it is rejected because the bottle is not full enough or the cap is not secure.

Figure 16-14 shows a ladder diagram of the logic circuit that is used to inspect and reject the filling and capping processes.

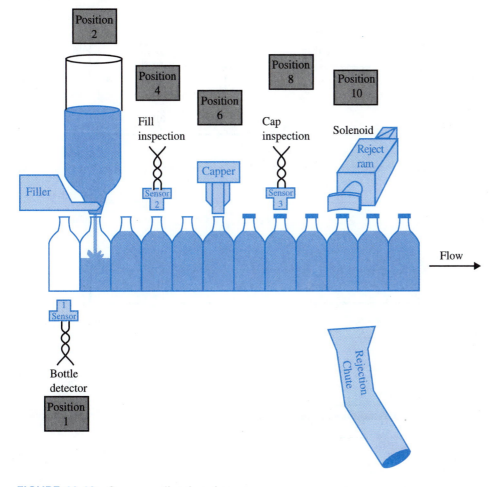

FIGURE 16-13   Conveyer line in a brewery

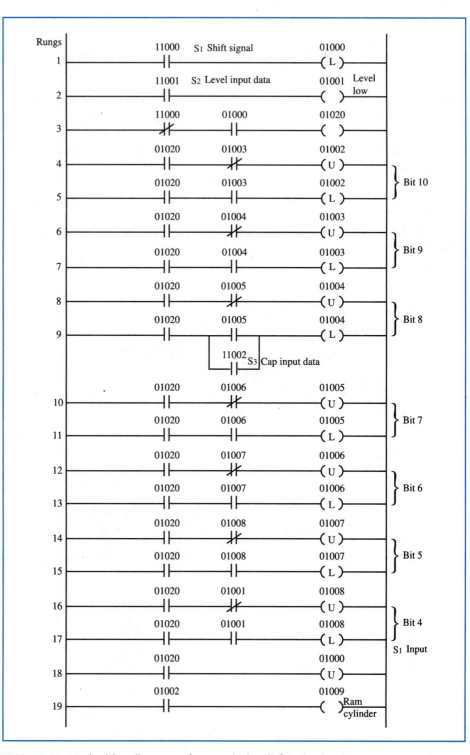

Step 1: Sensor 1 detects when a new bottle passes a reference point on the conveyer line. When the neck of a bottle is sensed, contact 11000 of Rung 1 turns On and latches output 01000. When the bottle is past the sensor, the condition of Rung 3 becomes True and energizes output coil 01020. The energized coil causes input contacts 01020 from Rungs 4 and 18 to turn on during the remainder of the scan cycle. This action is similar to providing a clock pulse to all flip-flops in a synchronous shift register used in digital electronics. The duration of the pulse ends when the Unlatch output is energized in Rung 18. Another pulse will not occur until the next bottle is detected by Sensor 1.

Step 2: The operation of a serial shift register is shown in Figure 16-15(a). Whatever logic level is present at a data input of each flip-flop, it is entered after the clock pulse arrives. Since the outputs of the flip-flop are fed into the data input of the adjacent flip-flops, all data is shifted one position to the left every time a clock pulse arrives.

Each pair of Latch and Unlatch output coils with the same number in Rungs 4 through 17 perform the same function as a flip-flop. When the Latch rung is activated, the flip-flop sets to a 1. When the Unlatch rung is activated, the flip-flop resets to a 0.

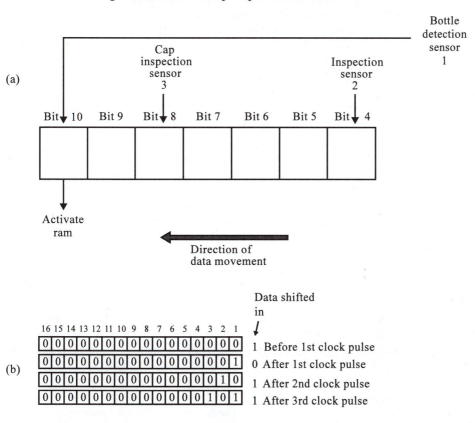

FIGURE 16-15 The operation of a shift register

Step 3: Whenever a bottle that is not full passes Sensor 2, contact 11001 of Rung 2 turns On and energizes output coil 01001. Contact 01001 closes and causes Latch 01008 to activate when the clock pulse reaches Rung 17 during the scan cycle. This condition loads a 1 into Bit 4 of the register shown in Figure 16-15(a). If the bottle is full, a 0 is loaded into Bit 4 when a clock pulse arrives. Any 1 is shifted every time a clock pulse arrives, as shown in Figure 16-15(b).

Step 4: Rung 9 of the ladder is an OR function. Whenever a cap is not properly placed on the bottle, Sensor 3 will set Bit 8 by energizing Latch 01004. If a 1 is already shifted into the bit from Sensor 2, the bit will remain latched.

Step 5: Whenever any 1 state bits are shifted into Bit 10, output 01002 is latched on Rung 5. This latched condition closes contact 01002 of Rung 19 to activate a pulse valve that causes a spring-return ram cylinder suddenly to extend and retract. The extension movement rejects the bottle by pushing it off the conveyer into a chute.

SEQUENCERS

Before PLCs were developed, the control function of an automatic assembly machine was often performed by an electric drum sequencer, shown in Figure 16-16. The drum sequencer is a cylinder with pegs placed in holes at varying horizontal positions around the outside. As a motor turns the drum, switches aligned with the pegs close when they make contact, and switches where no pegs exist remain open.

Pegs are inserted into holes, one row at a time, to program the drum sequencer. Each row provides a step in the sequence. A closed switch represents a 1 and an open switch represents a 0. The operation of the sequencer provides a direct relationship between action and time. Action at the output occurs when the pegs cause the contacts to open or close. The duration of each sequence is controlled by the speed of the motor. Since each step takes the same amount of time, consecutive pegs are installed if the output must be on longer than one step.

Many sequential operations can also be controlled by words in a PLC memory. The binary bits are loaded with 1s and 0s to form the same kind of pattern programmed into the drum by the pegs. The steps in the sequence must be loaded into consecutive memory locations known as *files*. To sequentially control output devices, words from each file are transferred to the output module in consecutive order.

A home clothes washer is a machine that performs an automatically-controlled sequence of operations. Figure 16-17 shows a simplified drawing of the timing mechanism that activates solenoids, relays, pumps, and motors. As the electric motor slowly turns the disc, the fingers make contact with conductor tracks to turn the controlled devices on and off at the proper times and in the proper sequence during the wash cycle.

Many manufacturing processes are sequential. An example is a pick-and-place operation shown in Figure 16-18(a) in which a pneumatic robot takes watch covers from one conveyer belt and inserts them onto the back of watch bodies on another conveyer belt. The six-step operation begins when a suction cup at the end of the arm makes contact with a cover on conveyor belt A. The robot has only two movements: *rotate* and *extension*.

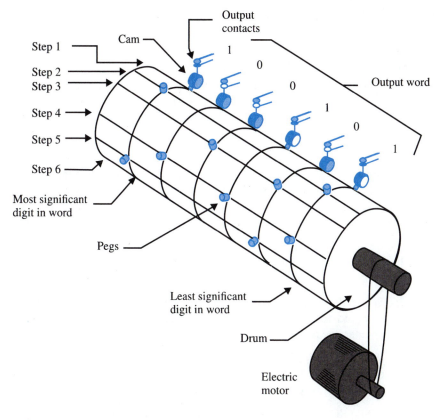

FIGURE 16-16 Sequencer drum

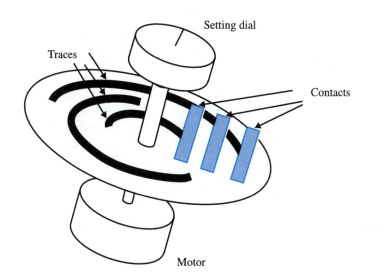

FIGURE 16-17 A timer mechanism that controls the sequence
of steps required to wash a load of laundry

Rotate.   When a 1 is applied to the rotate solenoid, air forces the body to turn counterclockwise 90 degrees. When a 0 is applied, the solenoid deactivates and air pressure stops, allowing a spring to turn the body clockwise 90 degrees.

Extension.   When a 1 is applied to the extension solenoid, air forces a cylinder to extend the arm upward to a height 1mm above the watch covers on conveyer belt A. When a 0 is applied to the solenoid, it deactivates and air pressure stops, allowing a spring to retract the arm to a height equal to the top of the watch bodies on conveyer belt B.

When a 1 is applied to the suction solenoid, a vacuum port is activated, causing suction to occur. The conveyer belts are activated when a 1 is applied to the driver motor. The speed of the motor is set so that the belts travel the correct distance during the sequence steps. Figure 16-18(b) shows the sequencer steps that control the assembly procedure.

Step 1:   The cylinder extends over belt A and the suction at the end of the arm picks up the watch cover.

Step 2:   While the cylinder is extended and the cover is held in place, the solenoid for the robot body is actuated, causing the arm to swing counterclockwise by 90 degrees.

Step 3:   While the cover is held in place and the robot arm is positioned over conveyer belt B, the extension solenoid is deactivated, causing the cylinder to lower until the cover snaps securely onto the watch body.

Step 4:   While the robot body is in position at conveyer belt B, the suction pressure is removed and the extension solenoid is activated, causing the arm to rise.

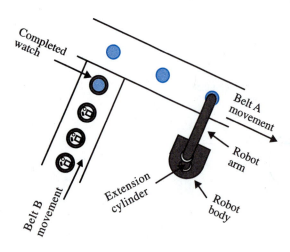

(a)

Step	Suction pick-up	Robot body activated	Extension cylinder	Belt A	Belt B
1	1	0	1	0	0
2	1	1	1	0	0
3	1	1	0	0	0
4	0	1	1	0	0
5	0	0	1	0	0
6	0	0	1	1	1
7	1	0	1	0	0

(b)

FIGURE 16-18 A robotic watch assembler

Step 5: When the cylinder fully extends the arm, the body solenoid is deactivated, causing the arm to rotate clockwise by 90 degrees.

Step 6: While the arm is extended and positioned above conveyer belt A, both conveyer belt motors are activated and run until the end of the sequence step.

Step 7: A cover on conveyer belt A and a body on conveyer belt B are in the required positions to repeat the entire sequence of steps.

16-4 WIRING DISCRETE I/O MODULES

When connecting field devices to a DC I/O module, certain rules must be followed to ensure electrical compatibility. Two common types of field circuits operate in the on/off switching condition:

Contact Circuits.   Relays operate in the on/off condition as their mechanical contacts open or close. Because they can accommodate a large range of voltages, they are popular and useful in control environments with a broad mix of electrical I/O circuit requirements.

Solid State Circuits.   It is common for DC input and output field devices to use solid-state circuitry to operate in either on or off conditions. Specifically, transistors are driven into saturation or cutoff to perform the required switching action.

Before a field device is wired to the PLC system, it must be determined whether the I/O module requires a *sinking* or *sourcing connection.* Sinking and sourcing are terms set to describe the direction *conventional* current flows between the I/O module and the field device when the control signal is at the active state.

The sourcing or sinking specifications are dictated by how the I/O module interprets logic states. Some PLCs recognize a 1 state logic signal as a high-level voltage (near 5 volts, or + 24 volts). Others recognize a 1 state logic signal as a low voltage (near ground). The following circuits show how to connect field devices to the appropriate sinking and sourcing I/O modules.

SINKING INPUT MODULE

Figure 16-19 shows a field device connected to a sinking input module. Turning on the PNP transistor produces a positive DC 1 state voltage at the output lead of the field device. The module sinks current into its input terminal when conventional current flows from the positive potential of the field device's output lead.

SOURCING INPUT MODULE

Figure 16-20 shows a field device connected to a sourcing input module. A low voltage (near zero volts) 1 state is produced at the output lead of the field device when it is activated. The near ground potential of the DC supply is switched to the module when the NPN transistor is turned on. The module sources conventional current from its input terminal to the more negative potential of the field device's output lead.

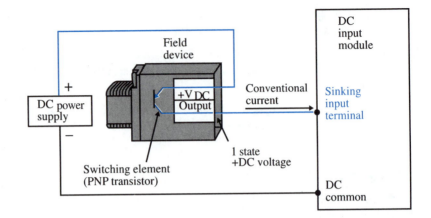

FIGURE 16-19 Sinking input module

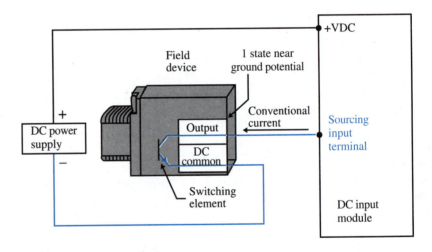

FIGURE 16-20 Sourcing input module

SINKING OUTPUT MODULE

Figure 16-21 shows a field device connected to a sinking output module. The field device must be connected between the positive side of the power supply and the output terminal. When the output module is activated, it produces a low voltage (near zero volts) 1 state. Conventional current sinks into the module's output terminal from the +DC supply through the field device.

SOURCING OUTPUT MODULE

Figure 16-22 shows a field device connected to a sourcing output module. The field device must be connected between the negative side of the power supply and the output terminal. When the output module is activated, it produces a high-level 1 state (near +5v, or +24v) voltage. Conventional current sourced (supplied) from the output module flows through the field device to the negative terminal of the power supply.

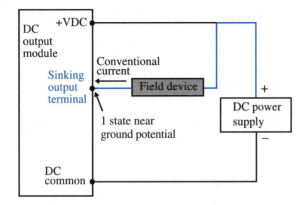

FIGURE 16-21   Sinking output module

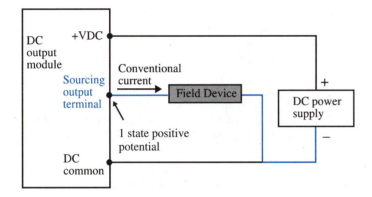

FIGURE 16-22   Sourcing output module

Programmable controllers manufactured in the United States typically recognize a high-level voltage (of +5v or +24v) as a logic 1 state. Therefore, they utilize sinking input modules and sourcing output modules. Most PLCs manufactured in Europe recognize a 1 state as a low-level voltage (near zero volts) signal. Therefore they have sourcing input modules and sinking output modules.

16-5   ANALOG INPUT AND OUTPUT SIGNALS

Programmable controller circuits that are programmed with relay logic instructions operate by using discrete signals. The True or False output signals they produce are based on the ON or OFF conditions applied to the input and the design of the logic circuit.

In many industrial applications, especially process control, the PLC receives analog electrical signals. These signals represent a varying condition anywhere between two limits. An example is a variable voltage that indicates the liquid level of a tank. In contrast, discrete circuits usually indicate the extreme limits, such as the low level or high level of the tank. Field devices that produce analog signals include thermal indicators, pressure transducers, flow rate detectors, and other sensors. When analog output signals are produced by PLCs, they are able to control a variable process. An example is a motor speed application: The motor may be driven at 0 RPM when zero volts are produced, at maximum speed when the largest voltage is generated, or at any speed in between as the PLC analog output voltage varies. In contrast, a discrete output would drive a motor at two extremes, 0 RPM or maximum speed.

ANALOG INPUT MODULE

The analog signal being monitored is fed to an analog input module. Its function is to convert the amplitude of the analog signal into proportional digital data in the form of a numerical value. An analog-to-digital converter similar to the type described in Chapter 2 performs this operation. Each analog module is inserted into a slot of a PLC rack.

The input module must have an electrical range compatible with the sensor or signaling device that feeds it. Standard analog input signals are:

$$4 - 20 \text{ mA}$$
$$1 - 5 \text{VDC}$$
$$0 - 10 \text{VDC}$$
$$-10 - +10 \text{VDC}$$

Most modules have the capability of inputing any one of these analog signals. Some modules have DIP switches for selecting the desired current or voltage range (Figure 16-23(a)). Figure 16-23(b) indicates the required settings of the DIP switches for each selection.

Most modules have the capability of receiving more than one analog signal. Figure 16-24 shows a module with four groups of three input terminals. The plus and minus terminals for the respective channels indicate the polarity of the analog input signal. The SLD terminal is provided for the shield wire of each pair of input cables. All four minus terminals share a common analog ground.

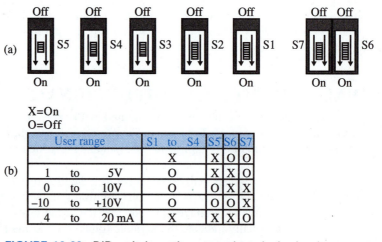

X=On
O=Off

	User range			S1	to	S4	S5	S6	S7
					X		X	O	O
(b)	1	to	5V		O		X	X	O
	0	to	10V		O		O	X	X
	−10	to	+10V		O		O	O	X
	4	to	20 mA		X		X	X	O

FIGURE 16-23 DIP switch settings to select desired voltage range

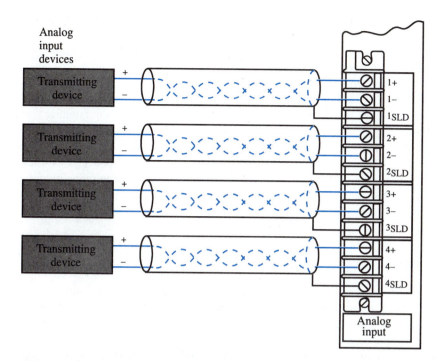

FIGURE 16-24 Field wiring terminations for an analog input module

Each module utilizes a 16-bit register for each of the channels, as shown in Figure 16-25. The analog current or voltage being measured is converted into pure binary or a BCD format before it is loaded into the register.

The most common input signal supplied to the PLC analog module by sensors is 4–20 mA. Suppose that a module is programmed to convert the analog current value into a proportional 16-bit pure binary word. It would be capable of producing 2^{16} (65,536) different numbers to represent the range of 4 to 20 mA. A digital number

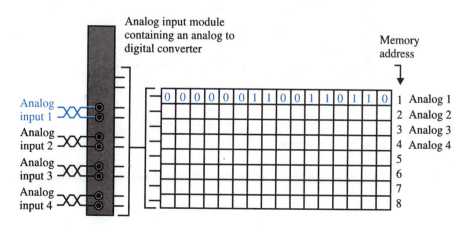

FIGURE 16-25 A binary number that represents an analog input voltage is sorted in a memory register

of 1111 1111 1111 1111 would be loaded into the register by the digital-to-analog converter if 20 mA was fed to the input. The number 9999 would be displayed by a digital readout. If 4 mA are applied to the input, the module converts the analog signal to a number of 0000 0000 0000 0000. The number 0000 would be displayed.

ANALOG OUTPUT MODULE

The digital signal in a register can be converted into an equivalent analog signal by the analog output module. An analog-to-digital (A/D) converter similar to the type described in Chapter 2 is used by the module to transform the digital number to the analog signal. Each analog output module is inserted into a slot of the PLC rack.

The output module must have an electrical range compatible with the output device that it activates. Standard analog output signals are:

$$
\begin{array}{rcl}
4 &-& 20\ \text{mA} \\
1 &-& 5\text{VDC} \\
0 &-& 10\text{VDC} \\
-10 &-& +10\text{VDC}
\end{array}
$$

DIP switches may be used for selecting the desired current or voltage ranges, if the module has the capability of outputing any one of these analog signals. If none of these four signals is compatible with the output device, an optional amplifier may be used to interface the module to the actuator.

The field wiring termination of the output module is shown in Figure 16-26. Each of the four output devices is driven by its own set of connections. The plus and minus terminals for the respective channels indicate the polarity of the analog output signal.

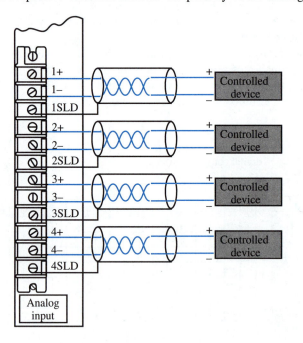

FIGURE 16-26 Field wiring terminations for an analog output module

The four minus terminals share a common analog ground internally. The SLD terminal is used for a shielding connection.

Each module utilizes a 16-bit register for each of the channels, as shown in Figure 16-27. The digital number in the register can be in the pure binary or BCD format. The most common output signal supplied by the PLC analog output module is 4–20 mA. Suppose a pure binary format is used. The module would produce 2^{16}, (65,536) different analog output current levels. A binary value of 1111 1111 1111 1111 would produce 20 mA. A maximum BCD value of 1001 1001 1001 1001 would be connected to 20 mA, and would display 9999 on a numerical readout if groups of four binary bits were used to represent each decimal digit. A binary or BCD value of 0000 0000 0000 0000 would produce 4 mA.

A PLC application using analog modules is shown in Figure 16-28. The modules are used in a closed-loop system that sets and regulates the speed of a DC motor. The operation of the system is as follows:

1. The desired speed of the motor is programmed by the thumbwheel switches. The setpoint signal of 900 is fed into a digital input module that is connected to the PLC memory address 040.

2. The PLC monitors the speed of the motor by receiving a negative feedback signal from the tachometer that is mechanically coupled to the motor. The diagram shows that −1 volt is fed into an analog input module connected to the PLC memory address 041. The module is calibrated to produce a digital output of 100 for every 1 volt that is applied to its input.

3. The multi-bit words in addresses 040 and 041 are manipulated mathematically by the CPU to perform the subtraction function. The result of the CPU calculation is stored in multi-bit word 042.

4. The number 800 produced by the CPU is sent to the analog output module located at address 043. The module is calibrated to produce an analog output of 1 volt for every digital value of 100 applied to its input. An error signal of 8 volts is applied to the amp that drives the DC motor.

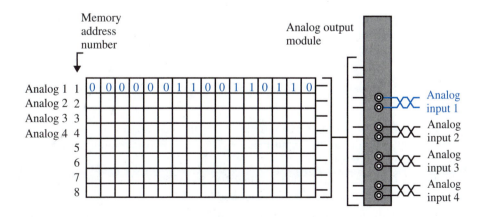

FIGURE 16-27 A binary number that represents an analog output voltage is stored in a memory register

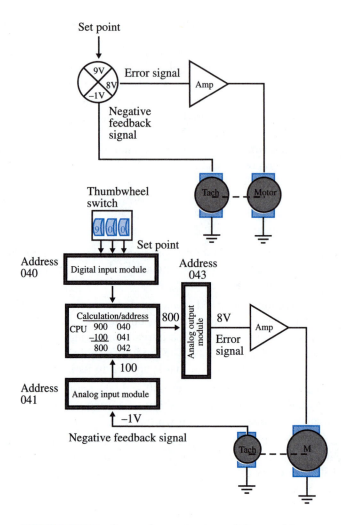

FIGURE 16-28 An analog output module used to control a closed-loop system

The diagram that is programmed into the PLC to control the motor speed is shown in Figure 16-29.

1. The GET instruction 040 fetches 900 from the thumbwheel switch that is coupled to the digital input module. The module feeds digital numbers into memory address 040.

2. The GET instruction 041 fetches 100 from the analog input module that feeds digital numbers into memory address 041.

3. The SUB instruction 042 subtracts the number in address 041 from the number in address 040. The answer displayed beneath the symbol is stored in address 042.

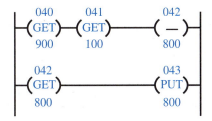

FIGURE 16-29 Ladder instructions programmed into the PLC to control motor speed in a closed-loop system

4. The GET instruction 042 fetches the answer from the SUB instruction in the previous rung.

5. The PUT instruction 043 transfers the digital number from GET instruction 042 to the analog output module connected to address 043.

Suppose that the physical load increases and slows the motor. The tachometer speed reduces to a lower RPM and produces a signal of −0.5 volts. The analog input module converts the voltage to 050. A new calculation is performed by the CPU which subtracts 050 from 900 to produce an answer of 850. The analog output module converts the digital number 850 into 8.5 volts. The increase of 0.5 volts at the amplifier is enough to produce the required torque for the motor so that it returns to its original speed.

16-6   SPECIAL PURPOSE MODULES

Through advancements in integrated circuit technology and software programming, PLCs have evolved into sophisticated equipment capable of controlling most industrial operations. However, some operations cannot be performed by using standard I/O modules. Therefore special I/O modules have been developed to interface the processor unit with field devices in certain applications.

These special I/O interfacers, sometimes called **preprocessing modules**, may incorporate an on-board microprocessor that performs complete processing tasks independent of the CPU or processor scan. By performing control operations by itself, the preprocessing module enables the processor unit to operate more efficiently. Other special purpose modules simply condition input signals, such as low-amplitude voltages or high-frequency data.

Special purpose I/O modules are used by medium- to large-sized PLCs. The following section provides information about some of the common special purpose I/O modules used.

BAR CODE MODULES

Bar code modules are primarily used to gather information about different types of products or parts. Bar codes were originally developed for inventory control by libraries in

England to keep track of books. Soon afterward, they appeared on products in retail stores and supermarkets. Eventually, they were adopted by industry for product identification, production, and inventory control.

Additional bar code applications include:

Item Sorting on high-speed conveyer lines.

Item Tracking to identify the location of a product on an assembly line.

Tool Crib Checking to keep track of the assignment of tools.

Product Verification to determine which products are put into containers.

A bar code is a series of rectangular bars and spaces representing letters, numbers, or symbols. They are usually printed on the outside surface of the object being identified. Figure 16-30(a) shows a block diagram of a bar code reader. To read the bar code, a light source from an optical scanner illuminates the symbol. The series of light bursts reflected back to the scanner indicates the pattern of the symbol. A photodetector converts the light bursts into an equivalent electronic signal, as shown in Figure 16-30(b). After the electronic pulses are amplified and conditioned into a digital signal, this signal is sent to a decoder. The decoder interprets the pulses as letters, numbers, or symbols, and changes them into an ASCII code. This code is processed by a microprocessor after it is received from the scanner.

There are three types of bar code systems: *Unattended, Attended,* and *Scanning.*

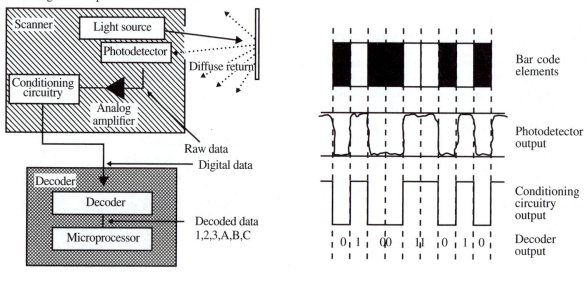

(a)　　　　　　　　　　　　　　　　　　　　　　　(b)

FIGURE 16-30　A bar code system　(Courtesy of Allen-Bradley)

Unattended

Unattended systems are automatic scanners that do not require a human operator. As items move on an assembly line or conveyer, a light beam is reflected off an oscillating mirror. This action causes the beam to sweep back and forth, making the beam appear as a line of light. As the symbol moves into range, the beam sweeps the entire symbol to provide the required information.

Attended

Attended systems require a human operator to trigger the scanner, to move the scanner across the symbol, or to move the symbol into the scanner depth of field. Attended scanners fall into two categories: stationary and portable. A stationary scanner is mounted in place and the operator moves an object marked with the symbol past the beam. A portable scanner can be moved by the operator to cause the beam to pass the label on an object.

Scanning

The scanning system uses a scanner in the form of a wand or gun-type laser designed to be hand-held and pointed at a bar code symbol. A flexible cable connected to the input module provides the electrical interface.

RADIO FREQUENCY MODULE

Some programmable controllers use radio-frequency (RF) technology to perform manufacturing operations. Radio frequencies are used to pass information back and forth between the processor unit and the product. The PLC uses such information as product identification, product tracking, data collection, and product control.

To perform the required communication, small tags are attached to a product. These tags are either *active* or *passive*. An active tag contains a battery which enables it to operate as a transmitter. The signal it sends is received by an RF module connected to the PLC. The active tag can also receive a radio signal from the RF module. The passive tag does not have a battery, and is only capable of receiving radio signals. Each tag contains 100 to 2000 bytes of memory. Information about the product is stored in the tag. For example, a tag might be used on a factory assembly line that makes sofa furniture. Information is read from the tag at each station to ensure that the proper options are installed for a special order. The tag provides instructions on such items as the type of wood, the stain color, the fabric, and the hardware.

VISION SYSTEM MODULES

Some programmable controllers can control manufacturing operations by vision. A vision system typically consists of a camera, lighting, a vision input module, a processor unit, and a video monitor (see Figure 16-31).

Vision systems are used primarily for part presence sensing, measurement, alignment, and other inspection functions. An example of a vision application is the placement of surface-mount integrated circuits onto a printed circuit board by a robotic arm. During machine setup, two steps must be "taught" to the vision system by the programmer. First, a good part is placed in front of the camera. The camera provides a picture of the image, which is then converted into binary data that is stored in memory.

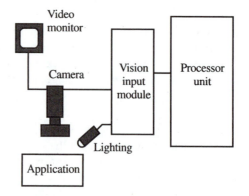

FIGURE 16-31 Block diagram of a typical vision system

Next, the robot arm is physically moved to a precise location where the IC must be placed. This data is also entered into the vision system memory.

During the assembly operation, the vision system controller determines the part orientation by comparing it with data previously taught to the controller. If, for example, the vision controller recognizes that a lead is broken, it will reject the part. A good part is placed on the board precisely at the location that the vision system was taught.

Vision systems can inspect parts at the rate of several hundred a minute very accurately. Because bad parts are rejected before being inserted, not as many completed boards become scrap, resulting in significant cost savings.

PID CONTROL MODULES

Some programmable controllers are capable of performing proportional, integral, and derivative (PID) control. PID systems are used to perform automated closed-loop operations in both motion or process manufacturing operations. The module has an on-board microprocessor. As setpoint and feedback signals are fed into a PID module, the microprocessor performs computer software PID calculations. The appropriate output signal is sent from the PID module to the actuator that controls the desired operation. PID control is explained in Chapter 3.

FUZZY LOGIC MODULES

Some industrial applications require more accurate control than PID closed-loop systems can provide. Fuzzy logic is capable of providing very complex control operations that were previously impossible to perform. Some programmable controllers can perform fuzzy logic control. Setpoint, feedback, and output control signals are processed by an on-board microprocessor in the fuzzy logic module. Fuzzy logic control is explained in Chapter 3.

STEPPER MOTOR MODULE

Stepper motors are used in motion control applications where position locations need to be very precise, such as in the movement of each axis on a robotic arm.

The purpose of a stepper motor output module is to produce a pulse train that is compatible with stepper motor translators. The pulses sent to the translator control the rotation, distance, speed, and direction of the stepper motor. The module receives command data from the CPU that provides the specific values for these three parameters. The module begins to send out signals when it is initialized by a start command. Once the motor is in motion, the module does not receive any further input from the CPU.

The number of pulses dictates the distance the motor will turn. The step rate of the pulses (1 to 60KHz) determines the acceleration, deceleration, and speed of the motor. The direction is controlled by which output line (forward or reverse) transmitted the signals. The operation of stepper motors is described in Chapter 6. Figure 16-32 illustrates a typical interface connection of the stepper motor.

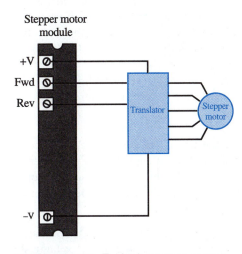

FIGURE 16-32 Typical stepper motor connection diagram

THERMOCOUPLE MODULES

Thermocouples are used as measuring devices in temperature control related applications. Since they produce a very small voltage (approximately 45 mV at maximum temperature), their signals must be conditioned by an input interface module. The thermocouple input module filters, amplifies, and digitizes the signal through an A/D converter so that it can be used by the processor unit. Thermocouple devices are described in Chapter 10.

16-7   TROUBLESHOOTING PROGRAMMABLE CONTROLLERS

When a fault develops in a PLC installation, two methods are used to troubleshoot the system: by trial and error, or by a logical procedure. Trial and error takes too much time because it often involves arbitrarily checking connections and field devices. The best way to proceed is to observe the system in operation and determine the symptoms. Then, using these observations, logically identify the source of trouble.

There are six areas in a PLC system where a fault is likely to occur:

1. I/O field device failure.
2. Incorrect wiring of the I/O field devices.
3. Input Module.
4. CPU or Power Supply.
5. Programming Error.
6. Output Module.

(The most likely cause of a PLC problem is a programming error. These mistakes are usually the result of either loading the wrong address, or an improper programming design scheme that causes the operation to perform incorrectly. This problem is more likely to occur in extremely long programs. To verify the accuracy of a program, the documentation that provides the designed ladder diagram should be compared to the diagram on the display screen.

If the circuit on the monitor is correct, the following procedure provides a logical sequence of steps that will help determine where a fault exists in one of the five remaining problem areas:

Step 1: Observe the LED indicator on the input module when someone activates a field device, such as closing a push button. If the LED does not light, the problem lies in the push button, the power supply, or in incorrect wiring. The fault can be determined by taking voltmeter readings.

Step 2: If a voltage is present at the input terminal when the field device is activated, the corresponding indicator lamp should illuminate and the contact on the display screen should intensify. If the indicator lamp turns on and the display contact does not intensify, the input module probably has failed. It is a good practice to have replacement modules available.

Step 3: If the monitor shows that the input contact symbol becomes brighter, the input field device and input module are functioning properly. By energizing the input, the output symbol should also intensify if the rung is in a True condition. If the symbol does not become brighter, the CPU may be the problem. Some PLCs have a fault register that indicates an internal malfunction. The register holds a binary-based number. Each bit represents a specific type of fault, such as the memory, the timer, the software, or a bus error. Most internal problems require repair work by the manufacturer.

Step 4: If the output on the monitor intensifies and the LED in the output module does not turn on, the problem is probably in the output interface module. To verify the problem, the module should be replaced.

Step 5: If the LED indicator lamp on the output module illuminates and the output field device does not energize, the module is probably functioning properly. To determine if the field device is defective, use a voltmeter to see if a signal is present at the output module terminal. If it is, check the connections to the device to determine whether it is properly wired. If there are no incorrect wiring connections, repair or replace the field device.

The proper operation of output modules and output field devices can also be determined by using a software troubleshooting function that is referred to as *forcing*. This feature enables the technician to energize or de-energize the output and then verify the signal status with a meter. Inputs can also be turned on or off by the forcing function. Input forcing is helpful when many outputs that are controlled by one input need to be checked.

CHAPTER PROBLEMS

1. Write the Boolean expression for the ladder circuit in Figure 16-33.

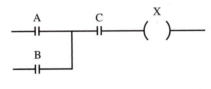

FIGURE 16-33

2. Assuming that a 1 state is applied to the Enable/reset line in Figure 16-34, how long will output line AC ≠ PR be High after a 1 state is applied to the control line.

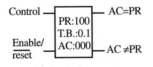

FIGURE 16-34

3. What is the numerical value found at V20 of Figure 16-35 after the condition contact becomes True?

2400 IF (─┤ ├─ 1500) THEN LET V20 = V10 + 2

FIGURE 16-35

4. __ T __ F   If the scan bypasses a rung with a timer instruction due to a jump command, it is necessary to reprogram the timer.

5. __ T __ F   A subroutine is bypassed if the rung condition of a JSR instruction is False.

6. When a forward shift takes place in a shift register, the data moves to the _____ (left, right).

7. __ T __ F   Data can be entered into a shift register only at bit 1.

8. __ T __ F   The method of cascading is limited to only two registers.

9. When several steps are programmed into a sequencer, the data is loaded into consecutive memory locations known as ____.
   a. module groups   c. files
   b. storage blocks   d. all of the above

10. __ T __ F   One advantage of sequencers is that several outputs can be activated simultaneously.

11. Which of the following is the most common method used to extend the time that an output is activated by a sequencer: ____
    a. Connect a gear reducer to the output.
    b. Slow the clock pulse speed of the PLC.
    c. Place a logic state that activates the output in the bit of consecutive memory locations.
    d. Program a timer delay subroutine.

12. A _____ (Low, High)-level signal applied to a sourcing DC input module is recognized as a logic 1 state by the PLC.

13. A PLC that operates by observing high-level voltages as logic 1 states uses a _____ (sourcing, sinking) output module.

14. List four standard signals that are commonly used by analog I/O modules.

15. How many different pure binary number combinations can the 16-bit register in an analog I/O module use to represent analog values?

16. Which of the following function statements describe the reason for using specialty modules? ____
    a. Signals are conditioned to a usable form.
    b. They enable the processor unit to operate more efficiently.
    c. They enable small PLCs to become more powerful.
    d. They can perform control functions independent of the processor unit.

17. Place the appropriate letter next to the following specialty modules to indicate which type of interface function each performs:

    a. Input   ____ Thermocouple
    b. Output   ____ Vision
    c. Input and Output   ____ Bar Code
                                ____ Radio Frequency
                                ____ Stepper Motor
                                ____ PID
                                ____ Fuzzy Logic

SECTION

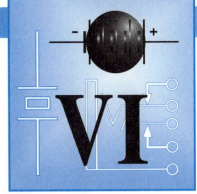

VI

MOTION CONTROL

As industries compete in the world market, they must strive to increase production with lower costs. To be competitive, many manufacturing companies are using motion control automated equipment to make their products. As technology advances, this equipment produces products more quickly and with higher quality. To enable these systems to operate effectively, the technicians who install, repair, and calibrate them must understand their operation.

This section of *Industrial Control Electronics* provides information about motion control equipment. Chapter 17 explains the operation of feedback devices that measure the motion of an output actuator or the load it is moving. Chapter 18 describes the operation of each element in a closed-loop and the parameters controlled in a motion system. Chapter 19 discusses different types of motion control systems and control loops used for various applications.

CHAPTER 17

Motion Control Feedback Devices

CHAPTER OBJECTIVES

At the conclusion of this chapter, you will be able to:

○ Describe the differences among angular velocity, angular displacement, and linear displacement feedback devices.

○ Explain the operation of the following motion control feedback devices:

Tachometers

Potentiometers

Optical Encoders

Resolvers

Inductosyns

LVDTs

Linear Displacement Transducers

○ List the factors that determine the voltage produced by a tachometer.

○ Describe the difference between incremental and absolute optical encoders.

○ Convert Gray code to binary and binary to Gray code.

○ Describe the difference between ratiometric tracking and phase digitizing resolvers.

INTRODUCTION

Automation in motion control applications is only possible if the controller section receives information about conditions in the manufacturing process. These conditions include displacement, position, speed, and acceleration. The devices capable of monitoring these conditions are called *transducers*. The transducer performs the measurement of the condition and produces a feedback that provides information on the results.

Some transducers produce an analog output signal. The amplitude of the signal is proportional to the measured quantity. Other transducers produce a digital output consisting of a group of 1s and 0s called *words*. The words are numerical values that represent the measurement.

Motion control detectors are used to measure conditions in a wide variety of applications. For example, they measure the precise depth of a hole bored by a drill bit; they sense the precise position of the work piece in an automatic milling operation; and they measure the acceleration and speed of a robotic arm performing a welding operation.

419

This chapter explains the operating principles of several types of motion feedback transducers. Each device falls under one of the following feedback transducer categories: angular velocity, angular displacement, or linear displacement.

17-1   ANGULAR VELOCITY FEEDBACK DEVICES

TACHOMETERS

A **tachometer** is a device that measures the angular velocity of a rotating shaft. The most common type of tachometer is the DC generator. It has a permanent magnet stator which produces magnetic flux lines. A conventional DC armature using a commutator and brushes is connected to the rotary object being measured. When the object spins, the armature cuts the flux lines and produces a voltage. The equation for generated voltage is:

$$Vg = KB(RPM)$$

where,

Vg = Generated Voltage

K = A constant which represents the physical construction (armature length, armature diameter, and so forth)

B = Strength of the magnetic field

RPM = Angular Velocity measured in revolutions per minute.

Since the physical size of the armature and the field strength are constant, the generated voltage is proportional to the angular velocity of the shaft. If connected to the armature, a voltmeter with an RPM faceplate can be calibrated to indicate the speed of a shaft's rotation. As the tachometer rotates in one direction, the polarity of the voltage is positive. The polarity reverses if the tachometer turns in the opposite direction. The result is that the tachometer output indicates both direction and speed.

Tachometers are often used in high performance servo applications where they provide velocity feedback for speed control purposes.

17-2   ANGULAR DISPLACEMENT FEEDBACK DEVICES

POTENTIOMETERS

Perhaps the simplest device used as a feedback indicator is the **potentiometer**. It is capable of converting mechanical motion into an electrical voltage variation.

The potentiometer consists of a fixed resistive element with an AC or DC voltage connected across its ends. Ideally, the resistance of a feedback resistor is uniform across its entire length. A conductive metal slide called an *arm* moves across the element. A voltage divider output is developed from one end terminal of the element to a terminal connected to the arm.

The arm is mechanically coupled to the object that is being measured. As the object moves, the resistance between the terminals varies. A voltage proportional to the resistance forms, indicating position.

Figure 17-1 shows two types of potentiometers. The type in Figure 17-1(a) has a straight resistive element. It is used to measure linear displacement. Suppose the object being measured has a linear motion range of two inches. If the resistive element is also two inches long and has 10 volts applied across it, then an output of 5 volts will indicate that the object is halfway through its travel. The potentiometer illustrated in Figure 17-1(b) has a circular resistive element. It produces an output voltage that is proportional to the angular displacement of the shaft. If 10 volts is applied across the element terminals, an output of 5 volts will indicate that the object has turned halfway through its rotation. Since the resistive element does not make a complete circle, rotary measurement is limited to about 330 degrees.

The potentiometer has a reputation for lacking accuracy because as the arm moves across the element, both the arm contact and element wear. They also become dirty as a result of the residue buildup from friction and oxidation. The resistance of the potentiometer therefore changes, reducing its accuracy. Potentiometers are making a comeback because development of materials, such as hard conductive plastics that don't wear or oxidize, are improving their performance.

OPTICAL ENCODERS

An **optical encoder** is an electromechanical device that is used to monitor the direction of rotation, position, or velocity of a rotary or linear operating mechanism. This device typically has four major elements: a light source, a light sensor, an optical disc, and signal conditioning circuitry. They are shown in Figure 17-2.

The light source is usually a light-emitting diode (LED) that transmits infrared light. The light sensor is a phototransistor that is more sensitive to infrared energy than to visible light. The optical disc is connected to the shaft being measured so that they rotate together. The disc is made of plastic, Mylar®, glass, or metal. It has opaque and translucent regions. The disc is placed between the light source and sensor as shown in Figure 17-3(a). As the disc is rotated, light passes through the translucent segments and is blocked by the opaque areas. Figure 17-3(b) shows that when light passes

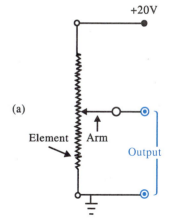

(a)

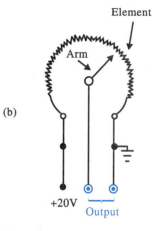

(b)

FIGURE 17-1
Potentiometer symbols

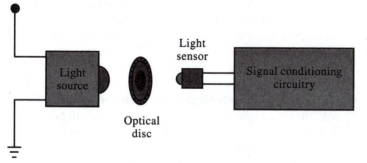

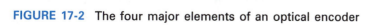

FIGURE 17-2 The four major elements of an optical encoder

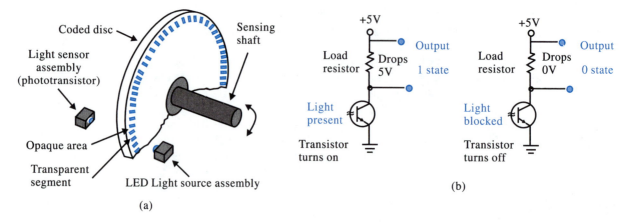

FIGURE 17-3 Operation of an optical encoder

through these segments, it strikes the base of the transistor which conducts. The voltage at the detector output goes High. When light is blocked, the transistor is off and a Low is formed at the detector output. The conditioning circuitry uses Schmitt triggers to convert pulses from the phototransistor into square waves.

Optical encoders are classified as either *incremental* or *absolute*.

Incremental Encoders

Figure 17-4 shows an incremental encoder disc in its simplest form. It has one track of imprinted marks precisely positioned around the outside of the disc. The quantity of marks is equivalent to the number of outputs for every 360 degree revolution of the encoder. As the disc rotates, light passes through the translucent slots, generating a series of electrical pulses. The exact number of pulses produced is determined by the amount of travel.

Single track encoders are typically used as tachometers to measure the rpm of a rotating device. The velocity information is determined in one of two ways; by measuring the time interval between pulses or by counting the number of pulses within a time period. The disadvantage of using one track is that the direction cannot be determined.

FIGURE 17-4 Single track incremental disc

Figure 17-5 shows a partial diagram of a two-track encoder. The two tracks (labeled A and B) provide the pulse train that indicates displacement. The number of pulses is recorded by a digital counter circuit. A separate light source and detector are used for each. The tracks are in quadrature, meaning that they are 90 degrees out of phase with each other. A continuous square wave is produced by each track as the encoder rotates. Track A lags Track B by 90 degrees when the shaft rotates in a CW direction. Track A leads Track B by 90 degrees when the shaft rotates in the CCW direction. Therefore, the direction is indicated by the relationship of which track leads or lags.

Figure 17-6(a) shows a circuit that counts the number of degrees the incremental encoder rotates. It also indicates which direction the encoder is turning. When the encoder rotates clockwise, the count increments. When the encoder turns counterclockwise, the count decrements. The circuit consists of the encoder that produces the A and B outputs that are phase shifted by 90 degrees from each other. The A output of the

CW = Track B goes high 90° before A
CCW = Track A goes high 90° before B

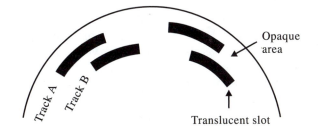

FIGURE 17-5 A two-track incremental encoder

encoder is connected to the input of the one-shot. Each time the A signal goes low-to-high, the one-shot is activated and produces a temporary high-level pulse.

The waveforms produced by each component in the circuit as the encoder turns are shown in Figure 17-6(b). Each time the output from track A goes high, a pulse is applied to the NAND gates by the one-shot. The B output of the encoder is connected directly to the CD (count-down) NAND and the inverter. The output of the inverter is connected to the CU (count-up) NAND gate. Each NAND gate requires that both inputs be high to produce an active output. The direction of rotation determines which of the two NAND gates is active. Before operating the encoder, it is necessary to turn it to the 0-degree indexing position and reset the counter to 0000. The waveforms show that when the encoder rotates clockwise, the CU NAND gate is active. It supplies pulses to counter input CU, causing it to increment. When the encoder rotates counterclockwise, the CD NAND gate feeds pulses to counter input CD, causing it to decrement.

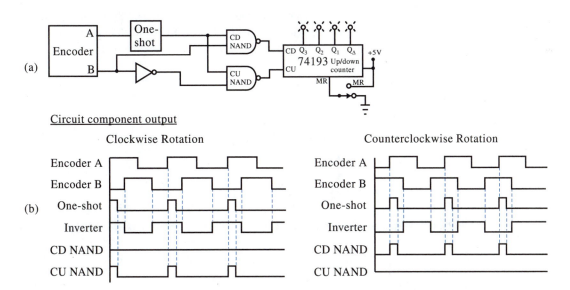

FIGURE 17-6 Incremental encoder conditioning circuitry

To indicate the amount of distance traveled from a reference point, a third track is added to the incremental encoder. Track 3 contains only one translucent segment. The signal produced is called the index pulse, or zero reference pulse. This pulse is generated once per revolution in the CW or CCW direction, as shown in Figure 17-7. It is used to establish a reference or home position. When the index pulse is detected, the counter is reset to zero. As the encoder rotates clockwise the counter will increment. When the encoder rotates counterclockwise, the counter will decrement. The net sum of the increments and decrements is indicated by the number in the counter. The count provides the user with the amount of travel from the reference point at any point in time. For example, suppose Track A of the encoder produces 1000 pulses per revolution. If the counter reads 100, the encoder has traveled 36 degrees; 150 would indicate 54 degrees, and so forth. By feeding the index pulse into a separate counter, the number of revolutions made by the encoder shaft can also be recorded.

The number of degrees per count is called **resolution**. The larger the number of translucent segments on an encoder wheel, the fewer degrees per count, thus, the better the resolution. The number of pulses per revolution is limited by the physical line spacing of the segments and by the quality of light transmission. Higher resolutions are available through various multiplication techniques.

One method uses Track A and Track B and counts in three modes called X1, X2, and X4. The X represents multiplication. Figure 17-8 summarizes the three modes of operation. In the X1 mode, the number of pulses generated is equal to the segments of Track A on the encoder wheel. An encoder with 500 segments will generate 500 counts per revolution. Only the off-to-on transitions in Track A are counted. In the X2 mode, both the off-to-on and the on-to-off transitions in Track A are recognized. Therefore, an encoder with 500 segments will generate 1000 counts per revolution. In the X4 mode, the off-to-on and on-to-off transitions of both tracks are used by the external counter. An encoder with 500 segments on each track will generate 2000 counts per revolution.

Incremental encoders are used for high speed applications. The disadvantage of the incremental encoder is that the contents will be lost if power to the counter is interrupted. Once power is restored, the reference point must be reestablished.

Quadrature encoder output

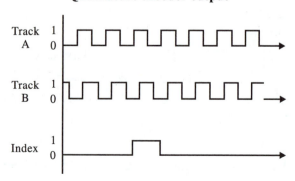

FIGURE 17-7 Quadrature encoder output

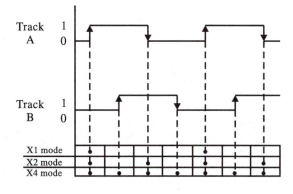

FIGURE 17-8 Quadrature modes

Absolute Encoders

Figure 17-9 shows an absolute encoder disc. It has four concentric tracks that vary in size. They are smaller at the outside edge and become larger toward the center. The tracks on the disc form sections with unique code patterns to represent particular positions. This code is derived from light sensors located at each track. A parallel output from these detectors produces highs and lows to form a 4-bit pure binary *whole word* that indicates position. By using four digits, the binary count of 0000 through 1111 is possible. If each number represents a sector of equal size, the encoder disc is divided into sixteen 22.5 degree sections.

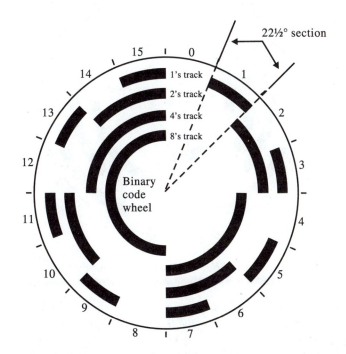

FIGURE 17-9 Four-bit absolute binary encoding disc

Using a pure binary encoder sometimes presents a problem. If all the bits do not change at exactly the same instant as the encoder crosses over from one sector to the next, a false number may temporarily be detected. For example, when going from the seventh section to the eighth, all four bits on the encoder must change at exactly the same time and as the count increments from 0111_2 to 1000_2. However, if the most significant bit changes slightly earlier than the other three bits, the encoder output would temporarily read 1111_2. This represents a 180-degree error of the encoder position. Since it is extremely difficult to construct a disc with enough precision to prevent such slight differences in the bit-switching times, another numbering system, called a **Gray code**, is frequently used.

The Gray code is a system that also uses 1s and 0s. Table 17-1 shows a comparison between the standard binary system and the Gray code. The table shows that when counting is done in the standard binary system, it is not uncommon to have more than one bit change at a time. The Gray code, however, is designed so that only one bit changes at a time. This characteristic makes the Gray code system an ideal choice for encoder discs. Even with a disc that lacks precision, if the bit changes too soon or too late, the temporary error is insignificant. The 4-bit Gray code encoder disc is divided into 16 sectors, as shown in Figure 17-10. To operate as a 4-bit optical encoding device, four separate light sources with four separate phototransistors are used for each of the four tracks.

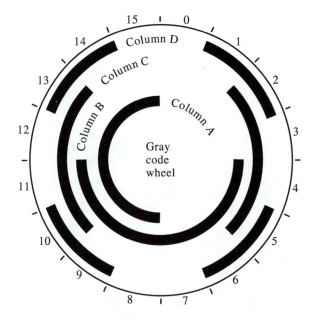

FIGURE 17-10 Four-bit Gray code encoding disc

Gray-Code-to-Binary-Code Conversion   One method used to convert a Gray code number to its equivalent binary value is shown in Figure 17-11. This table is used to convert 4-bit Gray code numbers to 4-bit standard binary numbers. Similar tables can be used to convert numbers that contain greater than or fewer than 4 bits.

TABLE 17-1   GRAY CODE

Decimal	Gray Columns ABCD	Binary $2^3 2^2 2^1 2^0$
0	0000	0000
1	0001	0001
2	0011	0010
3	0010	0011
4	0110	0100
5	0111	0101
6	0101	0110
7	0100	0111
8	1100	1000
9	1101	1001
10	1111	1010
11	1110	1011
12	1010	1100
13	1011	1101
14	1001	1110
15	1000	1111

Binary number system	2^3	2^2	2^1	2^0
Gray number system	Column A	Column B	Column C	Column D
Gray section	1	1	1	0
Binary section	1	0	1	1

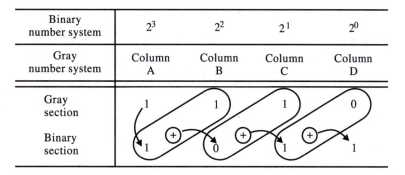

FIGURE 17-11   Gray-code-to-binary-code conversion table

The conversion from Gray code to binary code can be accomplished by following these five steps:

1. Enter the Gray code number to be converted into the Gray section of the table. The number 1110 is entered. From Table 17-1, which compares the standard binary system and the Gray code, note that the most significant bit (MSB) is the same for both.

2. Because the MSB of the Gray code number is 1, place another 1 in column A of the binary section of the table.

3. Exclusive-OR the bit in column A of the binary section with the bit in column B of the Gray section. Place the result in column B of the binary section of the table.

4. Exclusive-OR the bit in column B of the binary section with the bit in column C of the Gray section. Place the result in column C of the binary section of the table.

5. Exclusive-OR the bit in column C of the binary section with the bit in column D of the Gray section. Place the result in column D of the binary section of the table.

The equivalent standard binary number 1011_2 is generated from the Gray code number of 1110.

EXAMPLE

Convert the Gray code value 1001 to an equivalent binary number.

SOLUTION

Gray value   1   0   0   1
⊕ ⊕ ⊕
Binary value   1   1   1   0

It is possible to perform the same conversion process by using logic circuitry. Figure 17-12 shows a combination circuit that consists of three exclusive-OR gates that convert 4-bit Gray code numbers to equivalent 4-bit standard binary numbers. As the Gray code number 1110 is placed into the register on the left, the figure illustrates how the 1s and 0s are manipulated by each gate to generate binary number 1011_2 at the output.

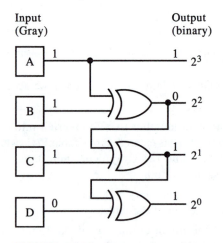

FIGURE 17-12 Gray-code-to-binary-code conversion circuit

Binary-Code-to-Gray-Code Conversion   Some applications require that binary numbers be converted to equivalent Gray code values. Figure 17-13 shows how this process is performed.

Binary number system	2^3	2^2	2^1	2^0
Gray number system	Column A	Column B	Column C	Column D
Binary section	1 (+)	0 (+)	1 (+)	1
Gray section	1	1	1	0

FIGURE 17-13   Binary-code-to-Gray-code conversion table

Follow these steps to convert binary code to Gray code:

1. Enter binary 1011_2 number into the binary section of the table.
2. Since the MSB of the standard binary number is 1, the same value is placed in column A of the Gray section.
3. Exclusive-OR the binary bits of columns A and B, and place the result in column B of the Gray code section.
4. Exclusive-OR the binary bits of columns B and C, and place the result in column C of the Gray code section.
5. Exclusive-OR the binary bits of columns C and D, and place the result in column D of the Gray code section.

The Gray code number 1110 is generated from the standard binary value of 1011_2.

EXAMPLE

Convert the binary number 1010 to its Gray code equivalent.

SOLUTION

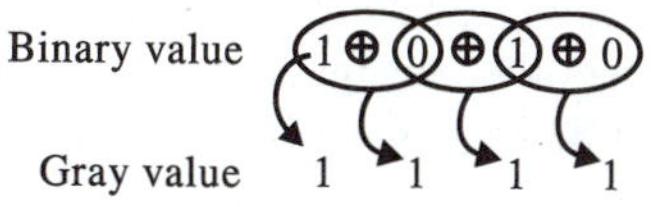

Binary value   1 ⊕ 0 ⊕ 1 ⊕ 0

Gray value   1   1   1   1

The binary-to-Gray conversion function can be performed by a combination circuit consisting of three exclusive-OR gates, as shown in Figure 17-14. The placement of 1s and 0s at the inputs and outputs of the gates shows how the binary number 1011_2 is converted to its equivalent Gray code number (1110) by the circuit.

Absolute encoders are used in applications where devices being measured are inactive for long periods of time or equipment power is turned off. When the

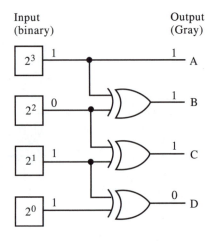

FIGURE 17-14 Binary-code-to-Gray-code conversion circuit

devices are reactivated or power is restored, positional information of the encoder does not have to be obtained from a storage counter. Instead, the position can be read from the disc. Absolute encoders are also used for devices that are relatively slow. Cranes, telescopes, and floodgates are a few examples. The common types of binary codes used by absolute encoders are natural binary, binary coded decimal (BCD), and Gray code.

A typical incremental or absolute encoder disc has six to twenty tracks. The disadvantages of optical encoders are that they can break when subjected to shock or high temperatures and their light sources are not always reliable.

RESOLVERS

The resolver is a rotary transformer capable of making precise position measurements and, in some applications, velocity measurements. The resolver consists of a rotor winding that is rigidly attached to the rotating shaft being measured and to stator windings that are stationary. Depending on the manufacturer, the rotor or the stator coils are used as the primary. The primary winding is typically driven by a reference voltage at a frequency ranging from 400Hz to several KHz. As the shaft rotates, the relative position between the rotor and stator windings changes, which causes the amplitude or the phase shift of the induced signal at the secondary to vary. The amplitude of the voltage, or the phase shift, corresponds to the shaft position. The physical configuration of a resolver is shown in Figure 17-15.

The most popular methods of determining rotational measurements by a resolver are *ratiometric tracking* and *phase digitizing*.

Ratiometric Tracking

The ratiometric tracking resolver is shown in Figure 17-16(a). It consists of one rotor winding and two stator windings wired 90 degrees apart. The rotor winding connected to the shaft is electrically connected to an AC supply through slip rings. One stator coil is labeled *sine,* and the other coil is labeled *cosine.*

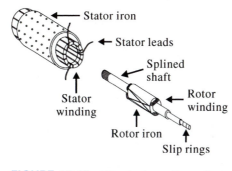

FIGURE 17-15 Physical configuration of a resolver

The waveforms in Figure 17-16(b) depict the operation of the resolver as it rotates from 0 to 270 degrees. When the rotor coil is in the 0-degree position, it is parallel to the cosine stator and perpendicular to the sine stator coil. The result is that maximum voltage is induced into the cosine coil and zero volts into the sine coil. As the rotor moves from the 0-degree position toward the 90-degree position, it induces progressively less AC into the cosine winding and progressively more into the sine winding. At 90 degrees there is no voltage induced into the perpendicular cosine winding, and the AC voltage at the sine coil is maximum. As the rotor turns from 90 degrees to 180 degrees, the sine voltage reduces to zero volts. The cosine voltage increases until it is at the maximum level, but it is now 180 degrees out-of-phase with the rotor signal. Therefore, at any 360-degree position, the angle of the shaft is determined by the unique waveforms produced at the sine and cosine coils.

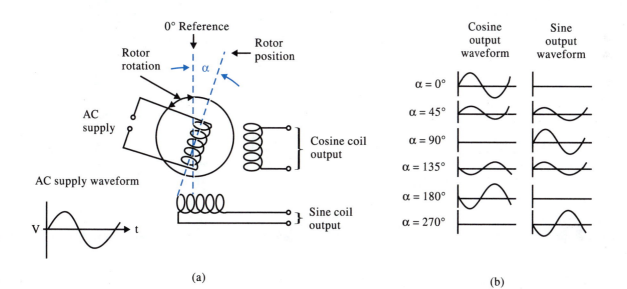

FIGURE 17-16 A ratiometric tracking resolver

By applying the sine and cosine output-stator-winding signals to an analog-to-digital converter, referred to as a ratiometric tracking resolver-to-digital converter (R/D), absolute position can be determined. The R/D calculates the ratio of the sine and cosine voltages by the following formula:

$$\frac{V_{s_1}}{V_{s_2}} = \frac{VR\ Sin\ \theta}{VR\ Cos\ \theta} = Tan$$

The division function finds the resultant tangent value that is used to determine the shaft angle. A 10-bit R/D will convert one resolver revolution into 1024 digital parts.

Phase Digitizing Method

The phase digitizing resolver operates on the principle that a phase shift between a reference signal and the voltage induced into the rotor is a direct measure of the shaft position. A simplified diagram of this resolver is shown in Figure 17-17(a). Two stator windings are mechanically fixed at a 90-degree angle to one another. The rotor winding is free to rotate, and the shaft around which it is wound is connected to the physical object being measured. Both stator windings operate as the primary of a transformer. A 2kHz 10V P-P excitation signal is applied to the stator winding on the left. This signal is referred to as the RPO, which means *reference phase output* signal. Another excitation signal, which lags the RPO by 90 degrees, is applied to the winding

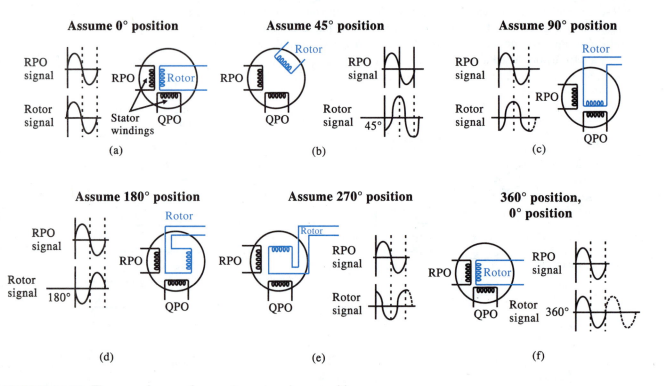

FIGURE 17-17 The waveforms of a resolver at various positions

on the bottom. This signal is referred to as the QPO, or *quadrature phase output,* which means 90 degrees out-of-phase. The rotor winding operates as a transformer secondary. It moves 360 degrees in either direction as the shaft rotates. The voltage induced into it by both stator windings produces a signal that represents position.

Figure 17-17(a) shows the resolver at 0 position. The voltage induced by the RPO winding into the rotor coil is maximum because the coils are parallel to each other. No voltage is induced by the QPO winding into the rotor winding because they are perpendicular to each other. Therefore the rotor output signal will be in-phase with the reference (RPO) signal.

Figure 17-17(b) shows the resolver at the 45-degree position. The rotor coil is at a 45-degree angle to both stator windings. An equal amount of voltage is induced by the RPO and QPO windings into the rotor coil. The rotor output shifts to the right and lags the reference (RPO) signal by 45 degrees.

Figure 17-17(c) shows the resolver at the 90-degree position. The voltage induced by the QPO winding into the rotor coil is maximum because they are in parallel. No voltage is induced into the rotor by the RPO winding because they are perpendicular to each other. Therefore the rotor output signal will be in-phase with QPO or lag the reference (RPO) by 90 degrees.

Figure 17-17(d) shows the resolver at the 180-degree position. The rotor is in parallel with RPO, but it has been rotated 180 degrees. Therefore a phase shift of 180 degrees is developed. Figure 17-17(e) shows that at 270 degrees the rotor coil is in parallel with QPO, but it is 180 degrees out-of-phase, resulting in a 270-degree phase shift with the reference RPO. At 360 degrees or 0 position, RPO and the rotor will be in-phase again, as shown in Figure 17-17(f).

Figure 17-18 shows the resolver connected to two operation modules. The first, the digitizing reference module, generates and synchronizes the excitation signals necessary for phase digitizing. An 8MHz clock is the heart of the system. A frequency divider reduces this high-speed oscillator signal into a lower frequency of 2KHz. This signal becomes the reference phase output, which is connected to the RPO winding. It is also fed into a phase-shift circuit, which produces a signal that lags RPO by 90 degrees. This becomes the 2KHz signal that is applied to the QPO winding.

To determine the phase shift between the reference (RPO) signal and the rotor output signal, a second device, called the position detection module, is used. It consists of a high-speed counter, a zero crossover circuit that monitors the RPO signal, and another zero crossover circuit that monitors the rotor output signal. When RPO crosses the zero reference line in the positive direction, it is detected by the zero crossover circuit which starts the counter. The clock pulses fed to the counter are supplied from the digitizing reference module at a frequency of 8MHz. The counter increments until the rotor output crosses the zero volt reference line in the positive direction. When it does, it is detected by the second zero crossover circuit which stops the counter. The number will be trapped, which represents position information called feedback units. Feedback units are simply measurements of how far the shaft has turned.

The counter is capable of incrementing from 0 to 3999, or 4000 counts. If the rotor is at the 0 position, there will be no phase difference between the reference signal and resolver output. Thus, the number in the counter will be 0. If the rotor turns somewhere between 359 and 360 degrees, the phase shift will be nearly one period of a cycle. 8MHz is 4000 times greater than 2KHz. Therefore nearly 4000 counts, called

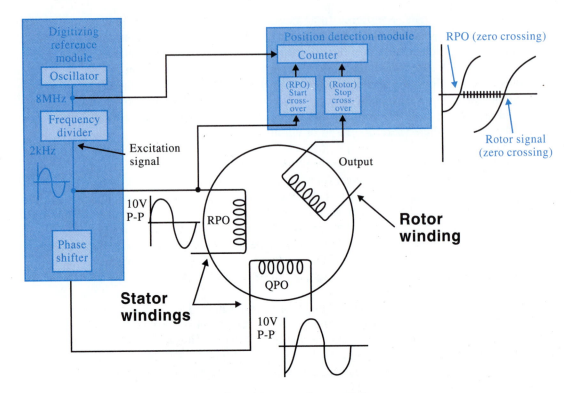

FIGURE 17-18 A resolver connected to the operation modules

feedback units, will increment the counter during one cycle. Thus, the number 3999 represents a shaft revolution of just under 360 degrees.

Figure 17-19 shows that since 4000 feedback units equal one 360-degree revolution, 1000 represents a 0- to 90-degree clockwise rotation; 2000 indicates a 180-degree position; and a 3000 count is a 270-degree rotation. The counter is continuously updated to indicate any change in position. As the resolver is turned CCW, the count becomes smaller.

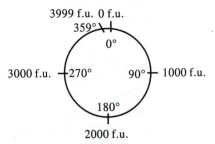

FIGURE 17-19 Resolver units

Figure 17-20 shows how a resolver monitors linear position. A ball screw mechanism is physically coupled to the resolver shaft. The ball, which rides along a groove in the screw, is located inside the machine block. As the shaft rotates, the ball forces the block to move horizontally. The direction that the screw turns determines which way the machine block moves. Every turn of the screw moves the block a horizontal distance of .25 inches. Since 4000 feedback units are generated by every 360-degree turn, the measuring resolution is .0000625 inches per 1 feedback unit.

The advantage of resolvers is that they can operate under environmental conditions that are too hostile for encoders. Resolvers function reliably when exposed to continuous mechanical shock, vibration, extreme temperature and humidity changes, and around oil mists, coolants, and solvents.

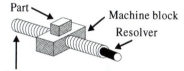

Assume:
Ball screw is geared to resolver 1:1
1 screw rev. moves axis .25 inch
1 screw rev. = 1 resolver rev.
.25 inch = 4000 f.u.
Therefore .0000625 inch = 1 f.u.

FIGURE 17-20 Resolver gearing example

17-3 LINEAR DISPLACEMENT FEEDBACK DEVICES

INDUCTOSYN

Although encoders and resolvers are ideal for measuring rotary position, they can make linear position measurements by employing leadscrews (ball screws). However, ball screw mechanisms lack accuracy due to backlash and are expensive. A device that overcomes these drawbacks by making linear position measurements directly is called an **inductosyn**.

The inductosyn is shown in Figure 17-21. It consists of a ten-inch continuous rectangular pattern called a *scale*. The scale is a conductive printed circuit track covered by an insulating layer. The function of the track can be considered the same as the windings of a coil. The scale remains fixed to a solid axis, such as a machine-tool bed.

Another part of the inductosyn is the *slider*. It has two separate but identical printed circuit tracks bonded to the surface of the plexiglass board that move parallel to the scale. These two tracks have a rectangular pattern identical to the pattern on the scale. The slider tracks are insulated from each other. The patterns are also offset one-quarter of a cycle, or 90 degrees, from each other. The slider is about four inches long and is separated from the scale by a small air gap. The slider moves along the scale with the

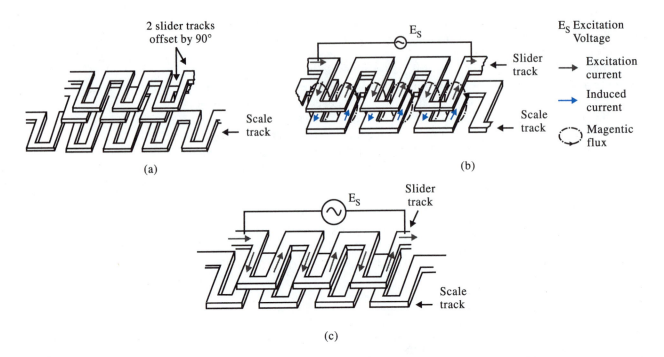

FIGURE 17-21   Linear inductosyn

device that is being measured, such as a machine-tool carrier. Figures 17-21(b) and (c) illustrate the magnetic coupling between one of the slider tracks and the scale track.

An excitation current is supplied to the slider track. When a slider track is located exactly adjacent to the scale track, the inductive coupling is maximum. The result is that the AC current of the slider induces a maximum AC signal into the scale track (Figure 17-21(b)). Figure 17-21(c) shows that when the slider and scale patterns are opposite, no signal is induced into the scale track.

The operating principle of the inductosyn is similar to that of the resolver. The digitizing reference module and position detection module used for the resolver are also used by the inductosyn. The 2KHz reference signal is applied to one slider track, labeled RPO. A 90-degree phase-shifted signal is applied to the other slider track, labeled QPO. The magnetic fields formed around each track cut into the scale track. Like the rotor of a resolver, an AC voltage is induced into the scale track. The relative position of the slider tracks to the scale track determines the phase difference between the RPO signal and the scale output. When the RPO signal crosses the zero reference point in the positive direction, the counter begins to count. The counter is stopped when the scale output crosses the zero reference point in the positive direction.

Suppose the length of each inductosyn section (the same pattern as on square wave) is 2mm. If the slider moves 1mm from the zero degree reference position, the counter will increment to 2000.

LINEAR VARIABLE DIFFERENTIAL TRANSFORMER (LVDT)

A **linear variable differential transformer (LVDT)** is an electromechanical transducer. It produces an AC voltage proportional to the linear physical displacement of an

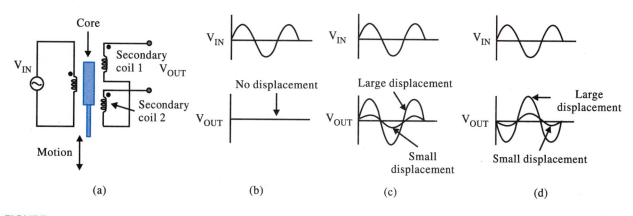

FIGURE 17-22 LVDT

object. In its simplest form, an LVDT consists of one primary and two secondary coils wound around a hollow tube. A ferrous metal cylinder called a *core* is placed inside the tube. It provides a path for the magnetic flux linking the coils. Figure 17-22(a) shows the basic winding configuration of the device.

When the primary coil is energized with alternating current, voltages are induced in the two secondary coils. The secondary windings are connected series opposing. The transducer output is the vector difference between the two voltages induced in the secondaries.

When the magnetic core is perfectly centered, the coupling of the primary coil's magnetic field will be the same to both secondary windings. Both secondary winding voltages will be equal but opposite. Therefore, the voltages cancel, resulting in a net output of zero volts. This is shown in Figure 17-22(b). If the core moves upward, the magnetic coupling will increase at secondary 1 and decrease at secondary 2. Therefore, the voltage at winding 1 becomes larger, and the voltage at winding 2 becomes smaller. The result is that an output voltage develops that is in-phase with the input voltage. The further the core moves the greater V_{OUT} becomes, as shown by the waveforms in Figure 17-22(c).

If the core moves downward below the center position, the voltage will increase at secondary 2 and decrease at secondary 1. The result is that an output voltage forms that is 180 degrees out-of-phase with the input voltage, as shown in Figure 17-22(d). Therefore, the size of the output RMS voltage is proportional to the amount of the core displacement from center, and the phase indicates the direction of displacement.

Figure 17-23 uses equivalent circuits to show why the differential output voltage is equal, in-phase, or out-of-phase with the input voltage. Batteries are used to represent the instantaneous peak voltages of each coil.

By connecting an AC voltmeter across the output terminals, an RMS voltage proportional to the displacement will be displayed. However, the voltmeter can indicate only voltage levels, not direction.

Figure 17-24(a) shows two filtered half-wave rectifiers connected to the secondary coils. They convert the AC output signal of the LVDT to a variable DC voltage, providing an analog representation of the core position. When the voltage is positive, it indicates the positive displacement of the core. A zero-volt output is produced when

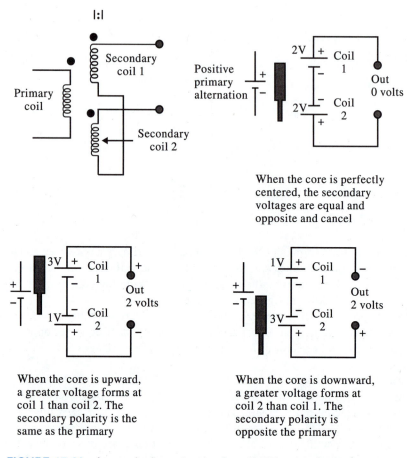

FIGURE 17-23 An equivalent circuit of an LVDT using batteries

the core is centered. A negative voltage indicates the negative displacement of the core. The graph in Figure 17-24(b) shows how the DC output voltages form at three different core positions—core at A, core at center, and core at B.

The frequency of the excitation voltage ranges from 60Hz to 20,000Hz at less than 10 volts. The unfiltered output voltages can range from 0.5 volts AC to 10 volts AC. Most LVDTs have a displacement range of about one inch from center in each direction. If a displacement greater than one inch is required, a ratioing mechanical gear apparatus can be used.

LINEAR DISPLACEMENT TRANSDUCER

A device that applies magnetorestrictive technology for measuring linear position is the **linear displacement transducer**. Figure 17-25 shows a drawing of this device. It has a housing that contains its electronics. A tube, called the *waveguide*, is spring mounted at the end connected to the housing. The other end is hard mounted. Also inside the housing are two strain tapes made of a special nickel alloy. They are welded

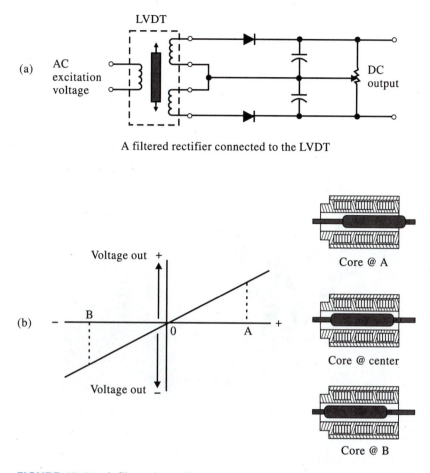

(a)

A filtered rectifier connected to the LVDT

(b)

FIGURE 17-24 A filtered rectifier connected to the LVDT

to the waveguide. A coil wrapped around each tape is placed within the field of two stationary bias magnets. A ring-shaped permanent magnet passes along the outside of the waveguide, generating a magnetic field. The ring (magnet) is attached to the object being measured. A conducting wire inside the waveguide runs the entire length of the transducer tube. The measurements are taken along the tube from an indexing point near the housing to the location of the ring.

To measure linear position, a current pulse is launched from the housing along the wire at the speed of light. An electromagnetic field expands around the entire length of the waveguide. The momentary interaction with the ring's magnetic field produces a helical field that causes a twisting motion, or torsional strain pulse, in the waveguide. This twist action is known as the Wiedemann effect. The strain pulse propagates along the waveguide as an ultrasonic wave at the speed of sound, or over 9000 feet per second. When the pulse arrives at the housing, it is detected by the two tapes. The pulse puts a strain on the tapes which causes their reluctance to momentarily change. The strength of the flux lines that pass from the bias magnets through the tapes fluctuates. This change in intensity causes the flux lines to cut across the pick-up coils. A

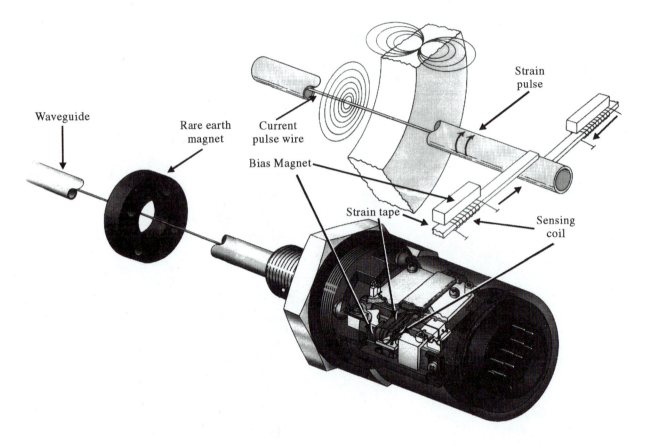

FIGURE 17-25 Linear displacement transducer (Courtesy of Temposonics™)

momentary electrical pulse is induced into the sensing coils and sent to the detector. This is known as the Villari effect.

The output signal from the coils is processed into either an analog voltage of 0 to 10 volts, or a 16-bit binary or BCD digital information.

The position information is determined by measuring the lapsed time between the launching of the current pulse and the arrival of the strain pulse sensed by the coils. Linear position can be measured from one inch up to 25 feet.

CHAPTER PROBLEMS

1. Which of the following parameters are measured by a tachometer? _____
   a. Position   c. Direction
   b. RPM speed   d. Distance

2. __ T __ F   A tachometer operates the same way as a DC generator.

3. __ T __ F   A feedback potentiometer is a linear feedback device.

4. __ T __ F   A potentiometer is considered a precision measuring device.

5. Which of the following parameters are measured by the optical encoder? ____
   a. Direction   c. Velocity
   b. Position   d. All of the above

6. What is the function of the Schmitt trigger in an optical encoder?

7. List two ways an incremental optical encoder can measure velocity.

8. __ T __ F   A one-track incremental encoder is capable of measuring direction.

9. The number of degrees per count in an incremental optical encoder is called

   _____.

10. __ T __ F   The disadvantage of an incremental encoder is that if power to the counter is removed, the count is lost.

11. __ T __ F   An advantage of an absolute optical encoder is that if power is lost and then restored, it is not necessary to reestablish a reference point.

12. The advantage of the _____ code over pure binary numbers is that only one bit changes at a time when its numbers increment or decrement.

13. Convert Gray code number 1000 to an equivalent binary number.

14. A 5-bit Gray code wheel can be divided into _____ sectors of equal size that

    are each _____ degrees.

15. Convert the binary number 0101_2 to an equivalent Gray code number.

16. Logic circuits that convert binary numbers to Gray code, or vice versa,

    use _____ gates.

17. Which of the following parameters are measured by a resolver? ____
    a. Position   c. Velocity
    b. Direction   d. All of the above

18. When a rotor operates as a transformer primary, it will induce _____

    (minimum, maximum) voltage into a parallel stator winding, and _____

    (minimum, maximum) voltage into a perpendicular stator.

19. The best type of position device to use in conditions where there is extreme heat, humidity, and dirt is the ____.
    a. potentiometer   b. encoder   c. resolver

20. __ T __ F   The operation of an inductosyn is similar to that of a resolver.

21. A resolver is typically used to measure ____ movement, and an inductosyn measures ____ movement.
    a. rotary   b. linear

22. __ T __ F   The LVDT uses transformer action to measure position.

23. The Wiedemann effect is a ____ action caused by the interaction between two magnetic fields.
    a. pushing   b. pulling   c. twisting

CHAPTER 18

Elements of Motion Control

CHAPTER OBJECTIVES

At the conclusion of this chapter, you will be able to:

○ Describe the four motion control parameters.

○ Identify the elements of a closed-loop motion control system.

○ List and describe the operation of devices and equipment used to perform the function of each motion control element.

○ Define the following terms associated with motion control:

Following Error	Damping
Indexing	Backlash
Tracking	Bandwidth
Home Position	Holding Torque
End Point	Loop Gain
Traverse Rate	

○ Identify six different mechanical systems that transmit force from a rotary actuator to a load, and describe their operation.

○ List and explain different types of position movements used in motion control applications.

INTRODUCTION

Motion control is the process whereby a system converts input commands into controlled mechanical movements. The movements can be linear, rotary, or a combination of the two. The commands can be provided by mechanical cams or gears, a simple potentiometer, or a complex computer.

The purpose of this chapter is to describe the mechanical movements, the power transmission, and the parameters controlled in a motion system. It also identifies each element of a closed-loop motion control system and explains its function. Many types of equipment and devices covered earlier in the book are used by the elements. Brief explanations of their operation and how they are used by each element are included.

18-1   MOTION CONTROL PARAMETERS

The purpose of a motion control system is to control any one, or any combination, of the following parameters:

Position
Velocity
Acceleration/Deceleration
Torque

POSITION CONTROL

This parameter controls the motion displacement of an object to a specific location. This control includes command controlled actuation and the monitoring of motions.

VELOCITY CONTROL

Speed, or velocity, is the quantity of motion that takes place with respect to time. The particular manufacturing application determines the type of speed that is required to perform the job. One example of a high-speed application is an automated insertion machine that makes traverse movements to place electronic parts onto a printed circuit board. In some applications, it is important to have good speed regulation. Speed regulation is the ability to maintain a constant speed under varying load conditions. A machine tool spindle that requires constant speed under varying load conditions due to changes in the cutting action is an example of a speed-regulated application.

ACCELERATION/DECELERATION CONTROL

Acceleration and deceleration are changes in speed with respect to time. Their rates are affected by inertia, friction, and gravity.

Inertia.   Inertia is a measure of an object's change in velocity. It is also a function of an object's mass. The larger the inertial load, the longer it takes to accelerate and decelerate.

Friction.   Friction is the resistance to motion caused by surfaces rubbing together. Mechanical parts of a machine and the loads that they drive exhibit some frictional forces. Friction is caused by factors such as bearing drag, sliding friction, and system wear.

Gravity.   The gravitational force of an object affects its acceleration and deceleration rates. In horizontal movement applications, gravitational force has no effect. It only applies to upward and downward vertical movements.

TORQUE CONTROL

Force is a quantity that causes motion. Force can exist whether or not it is moving an object. When measuring motion produced by a linear actuator, force as a unit of measurement is referred to as inch-ounces or foot-pounds.

Torque is a twisting type of force that causes an object to rotate. It is used as a unit of measurement for rotary actuation, such as with electric or hydraulic motors. The English unit of measurement is inch-ounces or foot-pounds. In design applications, torque calculations must be made to determine if the size of a motor actuator provides enough acceleration torque, constant velocity torque, or deceleration torque to drive the load. Torque is also the force that can prevent an object from moving by holding it in a fixed position.

18-2   MOTION CONTROL ELEMENTS

Figure 18-1 shows the block diagram of a motion control system. Each block represents an element that performs a specific function. There are many different types of devices available to perform the functions of each element. Most of these devices have been described in previous chapters. This chapter explains the operation of a motion control system, its elements, and the role the devices play in the overall scheme.

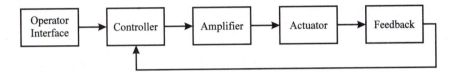

FIGURE 18-1 Elements of a motion control system

OPERATOR INTERFACE
The operator interface is used to communicate with the other elements of a motion control system. Interfaces such as keyboard terminals, hand-held keypads, thumbwheels, and switch panels are used for data entry input. Display terminals such as CRT screens, LCD displays, and indicator lamps provide a way to visually read the programs, messages, or other information useful for the programmer.

CONTROLLER
The controller is the "brain" of the system. It consists of memory where data instructions from the user are stored and microprocessor-based circuitry that processes information. When activated, the controller translates the instructions in the program memory into a series of analog or digital signals that are fed to the amplifier. In a closed-loop system, it processes the input and feedback signals and produces an output, either electronically or through mathematical computations, by utilizing a software program that causes the actuator to produce the desired motions.

AMPLIFIER
The amplifier receives signals from the controller and converts them into power sufficient for the actuator to drive the load.

ACTUATOR
The actuator receives power from the amplifier. It provides the "muscle" required to drive the load so that the desired physical motion is achieved.

FEEDBACK
The feedback element closes the loop in a control system. It provides information about the condition of the mechanical load under control. The information is in the form of an analog signal or digital data.

18-3   TERMINOLOGY

Before describing each block of the motion control system in more detail, an understanding of the following terms is necessary.

Following error is the positional error during motion resulting from the load movement lagging behind the desired movement specified by the command signal.

Tracking is the movement of the load by the actuator as it attempts to follow a changing command signal in a motion control position system.

Home position is a reference position from which movements are measured in a motion control system.

End point is the desired location to which a load is moved by a motion control position system.

Traverse rate is a fast rate of speed at which the load is moved from one position to another in a point-to-point position servo system.

Damping is the prevention of overshoot of the load past the end point in a motion control system. Damping is achieved by reducing the overall gain of the system to slow the movement. The greater the damping, the slower the movement.

Backlash is the movement within the space between mating parts or gears. Normally, gears need some space to accommodate lubrication. However, it is desirable to have minimal backlash, otherwise positioning errors may occur.

Bandwidth is the measure of how quickly the controlled quantity tracks and responds to the command signal. The greater the bandwidth, the slower the system responds.

Holding torque is the peak force that can be applied to the shaft of an energized motor at standstill without causing it to move. Also known as static torque, it is the result of the interaction between magnetic fields of the stator and rotor.

Loop gain is a measurement value based on the ratio of output speed to the following error and is expressed in inches per minute per mil of error. A mil equals 0.001 inches. The loop gain is improved by increasing the gain of the proportional loop of the system.

18-4   OPERATOR INTERFACE BLOCK

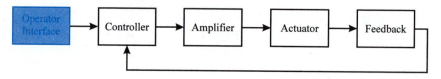

DIAGRAM A

The operator interface block, highlighted in Diagram A, is the equipment used by the operator to communicate with the motion control system. An input device provides a means for the user to enter information. Data can be entered from a keyboard or a thumbwheel switch or downloaded from a host computer. The information entered forms the control parameters for the command signals that cause the system to make the desired motions. The terminal also includes a monitor screen unit. It enables the user to examine and program various types of information. The main parameter menu in Figure 18-2(a) provides an easy way to program the parameter values. The parameters include:

○ The direction, distance, and any other information that specifies the end point position.

○ The velocity data that instructs the system how fast to get to the desired location.

○ The gain setting for the derivative function that controls the acceleration and deceleration rates when starting, stopping, or deviating from a desired speed.

○ The gain setting for the integral function that controls the tolerance for positioning and speed.

The status display in Figure 18-2(b) enables the operator to examine various drive parameters at any time while the system is operating. The screen is constantly updated.

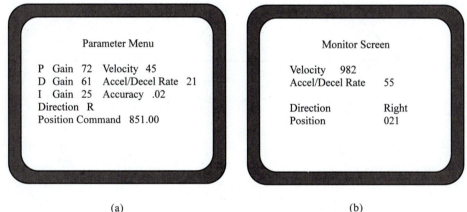

       (a)   (b)

FIGURE 18-2 Operator interface screen

18-5 CONTROLLER BLOCK

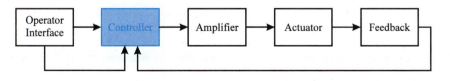

DIAGRAM B

The controller, highlighted in Diagram B, is the brain of the system. Its primary function is to produce electrical signals, either as digital pulses or as analog voltages, that cause the actuator to perform its desired operation. The desired motion profile the controller produces is the result of collecting information like command signals, feedback signals, parameter adjustments (such as gain settings), and other data. This information is processed by the controller before it sends the proper output voltages to the amplifier. The processing operation can be performed by analog electronic circuitry or by digital equipment.

A typical analog motion controller requires the adjustment tuning of many potentiometers by trained technicians during installation. Because of component aging, these circuits must be retuned periodically. A digital motion controller requires the tuning by inputing calibration data from a keyboard or by downloading from a computer. The computer software processes command signals, feedback signals, and parameter settings to produce the desired output. A typical motion profile generated by the output is shown in Figure 18-3. The velocity profile in Figure 18-3(a) resembles a trapezoid. It is characterized by constant acceleration, constant velocity, and constant deceleration.

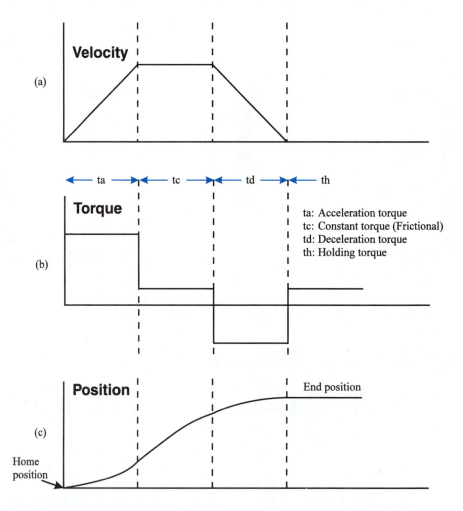

FIGURE 18-3 Typical motion profile

The torque profile in Figure 18-3(b) shows that a positive force causes accelerations, a small force causes constant speed, and a negative force causes deceleration. A small torque exists to hold the load in a fixed position. Figure 18-3(c) shows the resultant positioning of the object being driven. The parameters established at set-up and the size of the load determine the profile.

The profile parameters are normally programmed during set-up. However, they can be established during normal operation or changed under emergency conditions. The profile can be maintained by derivative and integral control or adjusted by fuzzy logic control as load conditions change.

There are various types of motion controllers available. One type is a dedicated controller that is very specialized and designed to perform a specific task. These controllers perform operations such as computerized numerical control, welding, and laser cutting. Another type of controller system, which provides signals to cause positioning movements, is called an **indexer**. These controlled movements are referred to as **indexing**. One specialized controller is a unit that is in chip form or constructed on a single printed circuit board. These controllers are often placed inside the machine enclosure of the equipment they operate. General purpose controllers also exist. These higher level units are very versatile. They are capable of coordinating several control operations simultaneously, performing high-speed mathematical calculations, and communicating with other controllers. The operations they are capable of performing include X-Y positioning, palletizing, and computer integrated manufacturing (CIM). These units are also able to manage faults in the system by shutting down power or activating an alarm if a problem develops. Programmable controllers and computers are considered general purpose controllers.

18-6   AMPLIFIER BLOCK

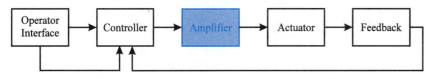

DIAGRAM C

The amplifier, highlighted in Diagram C, converts the low-level output signal from the controller to a level of electrical power sufficient to drive the actuator. In most motion applications, the actuator is a motor. The selection of an amplifier is based on how closely the characteristics of its outputs match the voltage and current requirements of the actuator.

The most common types of amplifiers include:

DC Servo Amplifier.   A DC servo amplifier is a linear amplifier that is capable of producing a bidirectional DC voltage for powering a brush-type DC motor. When it outputs a positive voltage, the motor rotates in one direction. A negative voltage causes the motor to rotate in the opposite direction. DC amplifiers may consist of a simple circuit with a power transistor or power operational amplifier, or a DC drive that consists of complex circuitry.

Brushless DC Amplifier. A brushless DC amplifier consists of electromagnetic stator coils that are arranged in a cylindrical shape. Its rotor is made of a permanent magnet that is free to turn inside the stator. By switching the DC power applied to the coils on and off sequentially, a rotating magnetic field is created. The rotor turns because of its magnetic interaction with the rotating stator field.

AC Servo Amplifier. The AC servo amplifier is used to supply power to a single-phase squirrel-cage type induction motor called an AC servo motor. It provides AC power to two sets of stator coils, the main winding and the auxiliary winding. The power from an AC line is fed directly to the main winding. Power to the auxiliary winding is 90 degrees out of phase from the main power and is produced by an amplifier circuit. Its amplitude is determined by the error signal that results from the difference between a command signal and the feedback signal. By varying the amplitude to the auxiliary winding, the speed and position of the rotor can be controlled.

AC Inverters. AC inverters, also known as AC drives, are used to drive single-phase and three-phase induction motors. They control the speed of the motor by changing the frequency and amplitude of the simulated AC waveforms they supply. The two most popular AC drives are the PWM and vector drives. Both drives utilize a pulse width modulation amplifier that produces a simulated AC wave to drive AC induction motors. Microprocessor-based circuits produce small signals that switch the amplifier's output power on and off to produce squarewave signals. As the duty cycle of the resulting square waves is changed by the switching signals, the output frequency and voltages can be varied to turn the motor at the desired speed and torque and to the desired position.

Stepper Motor Amplifiers. Stepper motors are position motors that are energized by electrical pulses. As the pulses are applied to a series of coils that are positioned in a cylindrical pattern, a permanent magnet rotor is forced to turn. Each incoming pulse causes the rotor to turn a specified angular distance. The rate at which the pulses are applied controls the rotor velocity. The pulses are commonly supplied by programmable controller modules or other peripheral devices that interface digital equipment with the motor. The pulses originate from a microprocessor-based device. Since the pulses are low voltage, they are used to switch amplifier transistors on and off. The output of each transistor is connected to a stator coil in the stepper motor. Stepper motor amplifiers are also referred to as indexers.

18-7   ACTUATOR BLOCK

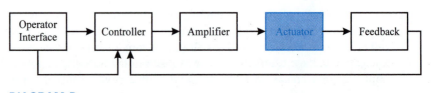

DIAGRAM D

The actuator, highlighted in Diagram D, provides the actual physical linear or rotary motion. Motion control actuators are usually electric powered motors, hydraulic motors, or hydraulic cylinders. The motor's rotary force is transmitted by the actuator transmission system.

ELECTRIC MOTORS

There are many types of electric motors used to drive the load in motion control applications. The most common types are brush-type DC motors, brushless-type DC motors, AC servo motors, induction motors, stepper motors, and hydraulic actuators.

Brush-type DC Motors

The most widely used actuator in servo applications is the DC motor that uses brushes. Among its advantages are proven performance, availability at a variety of specification ranges, and favorable cost. Its primary disadvantage is its higher maintenance requirement due to brush wear.

Brushless-type DC Motors

The rotor of the brush-type DC motor consists of the commutator, an iron core, and wire coils coupled to the same shaft. Because these elements are made of metal and are therefore heavy, there is a high rotor inertia. High starting and stopping inertia is an undesirable characteristic in positioning operations. The brushless DC motor reduces rotor inertia by replacing the mechanical commutation with an electronic commutator and by using permanent magnets instead of heavy electromagnet windings. The stators, which are made of polyphase windings, create a rotating field as they are energized by semiconductor switching circuits. The interaction of the stator field and the permanent magnets causes the rotor to turn. The brushless DC motor is used in applications that require low inertia, high torque, and a wide variable speed range desirable in positioning operations.

AC Servo Motors

The operation of an AC servo motor is similar to a split-phase induction motor. Two phases displaced by 90 degrees applied to two sets of stator coils cause a resultant rotating field. One phase is supplied to the main winding from the AC power source. The second phase is supplied to the auxiliary windings by a servo drive amplifier. The movement of the flux lines causes the rotor to turn. The strength of the magnetic field at the auxiliary winding can be altered by the servo amplifier. As it changes, the speed of the motor can be varied. When the field weakens, the motor runs at a slower speed. If the field strength is reduced to zero, the motor stops rotating. The AC servo motor has linear torque-speed characteristics, which are desirable in some applications for positioning and velocity.

Induction Motors

Until the vector drive was developed, induction motors were not used in servo systems because inductance lag in the rotor circuit caused slow response characteristics. The vector drive controls the speed and position of the AC motor by varying the synchronous frequency and voltage to the stator. For the rotor to turn, current must be induced into its coils from the rotating stator field. The magnetic induction takes place when the rotor coils turn at a rotation speed slower than the synchronous speed. At low synchronous frequencies, the induction action does not work effectively. Therefore, the induction motor is usually limited to high velocities and high-speed positioning applications.

Stepper Motors

A number of motion applications use open-loop systems to perform their operation. One of the most reliable open-loop actuators is the stepper motor. It is an extremely accurate device for positioning and does not require a feedback device to operate.

Power ratings of stepper motors are typically restricted to one horsepower and below, and they turn at low speeds. If the torque requirements of the load exceed the capability of the stepper motor, a position error is introduced which cannot be corrected. Also, its speed performance limits its usefulness in velocity control applications. The clear advantage of an open-loop system is the low cost of the equipment.

For critical applications, the accuracy of the step position can be verified by adding a feedback device to form a closed loop. Stepper motors that use feedback devices, such as encoders, only perform end point verification. They do not verify the position on the fly. For example, suppose the command signal indicates an end point value of 200 and the encoder indicates a position count of 198 when the motor stops. An error signal causes the stepper to move the extra two steps.

Hydraulic Actuators

Hydraulic actuators are playing an increasingly important part in high performance machine tools, steel mill machinery, metal fabrication machinery, and other types of motion control applications. Hydraulic actuation may be linear or rotary depending on the load requirements.

Linear actuation is performed by a cylinder with a piston and a rod. The heart of a hydraulic actuator system is the electrohydraulic servo valve. Figure 18-4 shows the cross-sectional view of a servo valve. The bottom portion of the armature is attached to a push rod that moves a metal spool. When the valve is de-energized, the armature is vertical and the spool is centered. In this position, the spool blocks the flow to the cylinder. When the servo valve is energized, a direct current electrical signal is applied across points A and B of the coil wrapped around the armature. Because the armature is made of iron, it becomes an electromagnet. If point A is more positive than point B, the top of the armature becomes a north pole and the bottom becomes a south pole. The poles of the armature interact with the poles of two permanent magnets mounted inside the body of the servo valve. Because like poles attract and unlike poles repel, the armature moves a few degrees in the clockwise direction and pushes the spool to the left. Fluid from a pump flows into P_4 (inlet port 4) and through P_2 to the cap end of the piston. As fluid is forced out of the cylinder from P_1 through P_3 and to the tank, the cylinder rod extends. If the opposite voltage polarity is applied to points A and B, the change of current direction reverses the poles of the electromagnet. This causes the armature to rotate a few degrees in the counterclockwise direction, thus causing the spool to move to the right. With the spool in this position, fluid flows into P_4, out of P_1, and produces a force against the rod end of the piston head, pushing it to

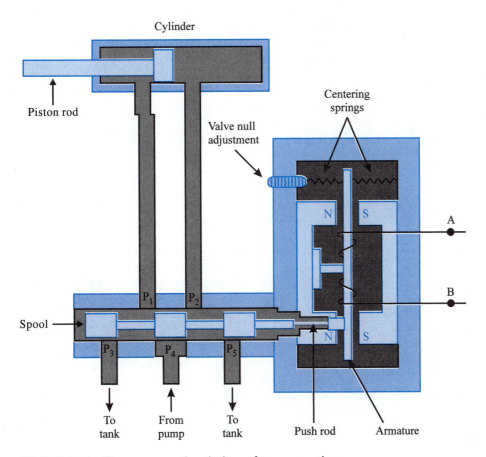

Cylinder

Piston rod

Valve null
adjustment

Centering
springs

P_1 P_2

Spool

P_3 P_4 P_5

N S

A

B

N S

To
tank

From
pump

To
tank

Push rod

Armature

FIGURE 18-4 The cross-sectional view of a servo valve

the right. The fluid on the cap end is forced out of the cylinder from P_2 through P_5 and
to the tank. This causes the rod to retract. By varying the amount of current to the coil,
the servo valve is capable of controlling infinitely variable positioning in both directions and controlling the flow rate to regulate the extension or retraction speed. By
replacing the cylinder with a hydraulic motor, the servo valve is capable of controlling
the speed and direction of rotary actuation.

Figure 18-5 shows the block diagram of a basic electrohydraulic closed-loop servo
system. As the command signal is compared to the feedback signal, an error signal is
generated. The error signal is amplified and becomes a proportional electrical current
that operates the electrohydraulic servo valve. The output of the servo valve is a flow
of hydraulic fluid which is proportional to the applied electrical current. A change in
polarity will cause the direction of fluid flow to reverse. This flow drives the hydraulic
actuator, which in turn drives the load. A transducer measures the load output and develops a proportional electrical signal, which is fed back and compared to the input
signal. If the system is a linear position servo, an LVDT is used as the feedback
device. The core of the LVDT is attached to the piston rod. If the system is a rotary
velocity servo, a tachometer is used. The tachometer is attached to the shaft of the

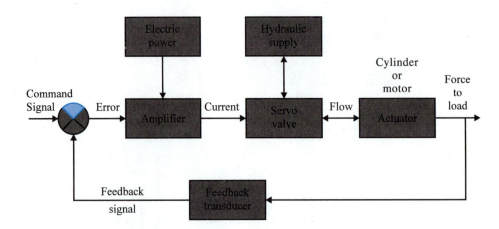

FIGURE 18-5 Block diagram of basic electrohydraulic closed-loop servo system

rotary actuator, such as a hydraulic motor. The error signal will cause corrective action to be taken until the output meets the requirements of the command signal.

There are three basic types of electrohydraulic servo systems: position control, velocity control, and force control. An example of a position control application is a numerically-controlled machine tool that uses a piston to position the cutting head. Here, the cutting head will move to a position where the feedback signal equals the command signal established by a numerical value from a computer. An example of velocity control is where the feedrate of a machine tool application, such as grinding, boring, or drilling, is performed. Hydraulic cylinders are capable of force amplification. An example of force control is the maintenance of a constant tension in a rolling or winding operation.

ACTUATOR TRANSMISSION SYSTEMS

There are several types of mechanical systems used for transmitting the rotary force of the motor to cause the load to move. The following six systems—direct drive, gearbox, leadscrew, rack-and-pinion, worm-and-wheel, and belt drive—are used in the vast majority of applications.

Direct Drive

The simplest method of transmitting motion from the rotary actuator to the load is by using the direct drive. The configuration shown in Figure 18-6 shows how the motor shaft connected to a load pulley transmits the force directly.

Gear Box

The gear box consists of two or more gears. Basically, a gear is a wheel with teeth cut into its circumference. One gear, called the *driver gear,* is directly connected to the shaft of the motor. Another gear, called the *follower gear,* is connected directly to the shaft that drives the load. The gears mesh together because their teeth are of the same size, as shown in Figure 18-7. When the motor turns under controlled conditions, force is transmitted from its shaft, through the gears, and to the follower shaft which causes the load to move at the desired speed and direction.

Motor

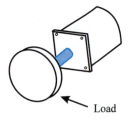

Load

FIGURE 18-6 Direct drive transmission system

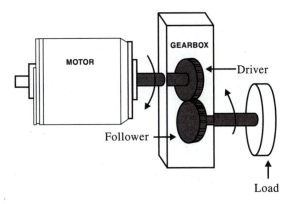

FIGURE 18-7   Gear box transmission system

Leadscrew

The leadscrew is widely used for converting rotary motion into linear motion. The mechanism is shown in Figure 18-8. The screw element is directly coupled to a motor shaft. A block element, which is often referred to as a table, has internal threads that mesh with the screw threads. As the motor runs, the screw turns and causes the table to move laterally. The screw is normally mounted at each end for support, and a guide shaft prevents the table from turning. Each degree the screw turns causes the table to move a precise distance. To eliminate backlash and reduce friction, ball bearings can be placed between the threads of the block and leadscrew. This system is referred to as a *ball screw.*

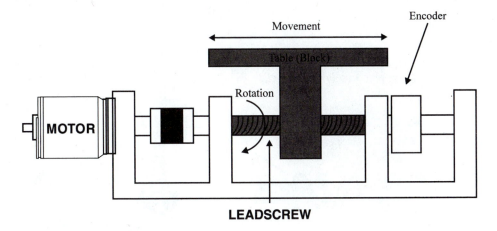

FIGURE 18-8   Leadscrew transmission system

Rack-and-Pinion

Another transmission system that converts rotary motion to linear motion is the rack-and-pinion shown in Figure 18-9. A *rack* is a straight bar with involuted gear teeth cut into one surface. A pinion gear is attached to the shaft of a motor. As the motor

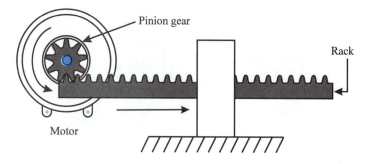

FIGURE 18-9 Rack-and-pinion transmission system

runs, it turns the pinion and forces the rack to slide. The direction of motor rotation determines the direction of linear movement by the rack. The RPM speed of the motor determines how fast the rack moves.

Worm-and-Wheel

The worm-and-wheel assembly mechanism is shown in Figure 18-10. It consists of a worm-pinion cylinder with a helical thread cut on its outer surface. The worm is meshed with a specially designed gear wheel. The wheel is manufactured with curved tooth tips so that it will fit well into the worm. Worm-and-wheel assemblies are able to transmit relatively large loads because the faces of the teeth on each gear are in full contact. They are used in applications where precision movements of heavy loads are required. This mechanism is also self-braking and can be used to hold the load in place. They are not used in high-speed operations because of the high friction that is developed at the contacting surfaces of the teeth.

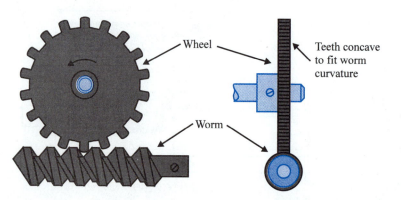

FIGURE 18-10 Worm-and-wheel transmission system

Belt Drive

Belts and pulleys are used in a wide variety of mechanical drive systems. A belt drive system is shown in Figure 18-11. The motor shaft is coupled to a drive pulley. Another pulley, called the follower, is connected to an idler shaft for support. The rotary torque

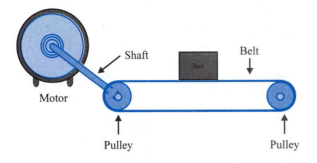

FIGURE 18-11 Belt drive transmission system

from one pulley is transferred to the other pulley by a belt made of durable fabric. A common application of a belt drive is a conveyer system.

POSITIONING MOVEMENTS

Actuators are used to move loads to specified locations. The movement of an object from one position to another is accomplished in one of three ways: single axis, multiple axis, and contouring. The three axes are identified as X, Y, and Z. The X and Y axes are at right angles to each other on the same plane. The Z axis is perpendicular to the plane of the XY axis.

Single Axis Positioning

The simplest and perhaps the most common method of moving a load is by using single-axis motion control. The single-axis system causes objects to make linear movements in both directions. An example is moving a spindle as it rotates into or out of a stock during boring or drilling operations, as shown in Figure 18-12.

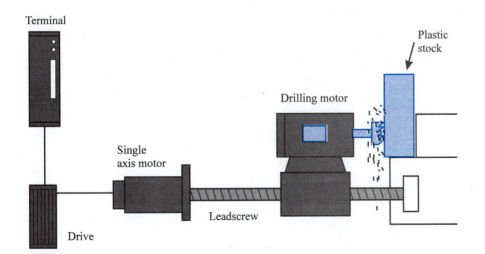

FIGURE 18-12 Single axis positioning

Multiple Axis Positioning

If the single axis system is expanded to more than one axis, more complex operations can be accomplished. An example of multiple axis positioning is the drilling operation shown in Figure 18-13, where holes are bored into a metal disc at precise locations. The X and Y axis movements control horizontal positioning, and the Z axis causes the drill to move vertically into the disc.

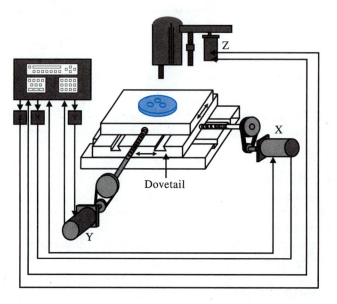

FIGURE 18-13 Multiple axis positioning

Contouring

Multiple axis applications that require continuous path control instead of end positioning use a positioning system known as contouring. To perform contouring, a complex control process is necessary. An example of contouring is the cutting of fabric in a circular arc shown in Figure 18-14. This action requires mathematical computations

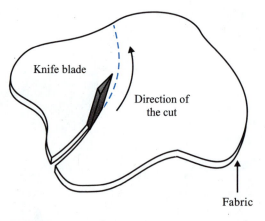

FIGURE 18-14 Contouring movement application

at intermediate positioning points to ensure that one axis corresponds to the position of the other axis.

18-8   FEEDBACK TRANSDUCER BLOCK

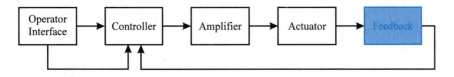

DIAGRAM E

The feedback transducer, highlighted in Diagram E, is used to indicate position, direction, speed, reference locations, and position limits. The selection of which type of device to use is often determined by variables such as environmental considerations, accuracy, and cost. Motion transducers are usually categorized by the type of measurements they make: presence or position.

PRESENCE INDICATORS

The purpose of a presence indicator is to provide a signal that tells the controller when an object being monitored is at a particular location. For example, a home position that is used for a reference point in position measurements must be detected by a sensor. Transducers are also used for limit detection to indicate the end of travel or to prevent an object from moving past a specified location. For safety applications, equipment is disabled to prevent injury when an object is in the wrong location. Presence indicators include limit switches, reed switches, proximity detectors, and opto-electronic sensors.

POSITION TRANSDUCERS

There are two types of transducers used to measure position: rotary and linear.

A popular type of rotary transducer is the optical encoder. By using a light emitter, light is passed through transparent slots in a disc to a light detector. The disc is mounted to a shaft. As the shaft rotates, a squarewave pulse is produced by the detector every time a light shines through a passing slot. Optical encoders are either absolute or incremental. Each type produces a different squarewave signal because the patterns on their discs are not the same. The signals indicate position, direction, and velocity.

The other widely used rotary transducer is the resolver. The resolver is an absolute position sensing device that uses phase- and amplitude-modulated sinewaves to indicate position. Resolver-to-digital (R/D) converters are required to convert the analog signals they produce into position and velocity information.

Linear measurements can also be made by rotary detectors. For example, Figure 18-8 shows a machine table that travels horizontally. As the leadscrew turns, the table travels in one direction. If the rotation is reversed, the table travels in the opposite direction. The encoder rotates when the gear that drives the shaft turns. The measurement system is calibrated to detect the linear distance for every degree of rotary movement.

There are several types of detectors that are designed specifically to make linear positioning measurements. These devices include inductosyns, LVDTs, and linear displacement transducers. An inductosyn has a fixed primary coil energized by an AC signal. The secondary is attached to the object being measured. As the object moves, a resolver type output is produced and is translated into linear information by an R/D converter. An LVDT has fixed primary and secondary coils. The iron core of the transformer is attached to the object being measured. As the object moves, the amount of voltage induced into the secondary from the primary changes. The movement signal is calibrated to indicate position based on the varying secondary output. A linear displacement transducer (making linear measurements) determines the travel time of a strain pulse between the object being measured and an element in the sensor head. Displacement can be determined because it is proportional to time.

CHAPTER PROBLEMS

1. Which of the following parameters primarily maintain speed regulation? ____
   a. Position Control
   b. Velocity Control
   c. Acceleration/Deceleration Control
   d. Torque Control

2. List three factors that impede acceleration or deceleration rates.

3. A twisting force is referred to as _____.

4. List the five major elements of a closed-loop motion control system.

5. Match the following terms with the definitions on the right.

   a. Holding Torque   ____ The prevention of overshooting.
   b. Home Position   ____ A measure of how quickly the actuator
   c. Bandwidth   responds to the command signal.
   d. End Point   ____ The desired location to which a load is
   e. Damping   moved.
   f. Traverse Rate   ____ A reference point from which a load is
                               measured.
                          ____ The force that causes the load to
                               maintain a standstill position.
                          ____ The speed at which a load is moved
                               from one location to another.

6. List three devices that are used to enter data into an operator interface device.

7. Indexers perform the operation of which element in a motion control closed-loop system? ____
   a. Controller   c. Feedback
   b. Amplifier   d. Actuator

8. AC drives perform the operation of which element in a motion control closed-loop system? ____
   a. Controller   c. Feedback
   b. Amplifier   d. Actuator

9. An inductosyn measures _____ (linear, rotary) motion.

10. Stepper motors are primarily ____-loop devices.
    a. open   b. closed

11. A hydraulic actuator is a ____ device.
    a. linear   b. rotary   c. linear or rotary

12. A servo valve controls ____ action.
    a. linear   b. rotary   c. linear or rotary

13. A leadscrew mechanism converts ____ motion into ____ motion.
    a. linear   b. rotary

14. Which of the actuator transmission systems are capable of multiplying rotary speed from the input shaft to the output mechanism? ____
    a. Direct Drive   c. Leadscrew   e. Belt Drive
    b. Gear Box   d. Rack-and-Pinion   f. Belt Actuator

15. In an XYZ system, if two axis movements control horizontal positioning, which axis controls the vertical position? _____

CHAPTER 19

Fundamentals of Servomechanisms

CHAPTER OBJECTIVES

At the conclusion of this chapter, you will be able to:

○ Explain the operation of a closed-loop velocity servo and describe its speed regulating properties.

○ Summarize the characteristics of bang-bang and proportional position servo controls and explain their operation.

○ List and describe the static and dynamic characteristics of a servomechanism.

○ Summarize the characteristics of a point-to-point servo and a contouring position servo.

○ List and explain the function of the multi-loop feedback amplifiers used in position servo systems.

○ Describe the characteristics of PID modes and feedforward.

○ Explain the procedure for tuning a servomechanism.

○ Describe the operation of a master-slave servo system.

INTRODUCTION

Numerous types of motion control systems help manufacture products. Each is uniquely designed to perform the required operation. When developing a system, the designer's objective is to build a machine with the particular characteristics needed for the application.

In this chapter, motion control characteristics are identified. Different types of circuitry and control modes that are capable of providing each characteristic are described. This information is useful when developing velocity and position servomechanisms.

19-1 CLOSED-LOOP VELOCITY SERVO

The control of velocity is very critical in most motion control applications. The purpose of velocity control is to establish the desired rotary or linear speed of a moving object and to provide speed regulation under varying load conditions.

A velocity control system is shown in Figure 19-1. Its angular velocity is proportional to the applied reference input voltage on the left of the diagram. Suppose that a 5V command signal is applied to the input. The system responds by supplying armature current that causes the DC motor to rotate. If the command voltage is increased, the motor runs faster. As the motor runs, a tachometer mechanically coupled to its shaft also turns.

The DC tachometer generates a DC voltage that is proportional to its shaft velocity. For this tachometer, the output is 1 volt per 1250 RPM. This voltage changes polarity when the direction of shaft rotation is reversed. The tachometer provides the feedback signal of the closed-loop system and applies it to the summing junction of the comparator. The 2-volt feedback signal applied to the inverting input of the comparator is subtracted from the 5-volt command signal applied to the non-inverting input of the comparator. The result is that a 3-volt error signal is produced at the output of the comparator. After it is amplified, the error signal is used to drive the motor.

The primary function of the tachometer is to provide speed regulation (which is referred to as *stability*) for the system. Suppose that the motor in Figure 1 turns a drive pulley on a conveyer. The motor RPM is 2500. If a disturbance is introduced, such as a heavy box being placed on the belt, the current supplied to the armature is not sufficient to keep the motor turning at the same speed. As the motor slows down to 1250 RPM, the tachometer voltage decreases to 1 volt, as shown in Figure 19-2. The comparator responds by increasing the error voltage to 4 volts. With an increased error signal a larger current flows through the armature, which produces the higher torque required to meet the increased demand of the load.

Now suppose the disturbance is removed. The elevated current through the armature causes the motor to speed up to 3750 RPM. As the tachometer turns faster, it generates a feedback signal of 3 volts, as shown in Figure 19-3. The comparator responds

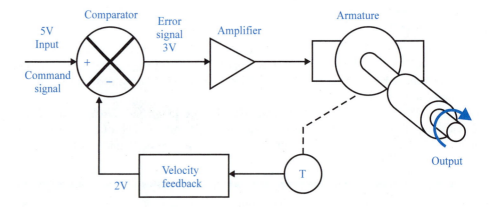

FIGURE 19-1 A closed-loop velocity system operating under normal load conditions

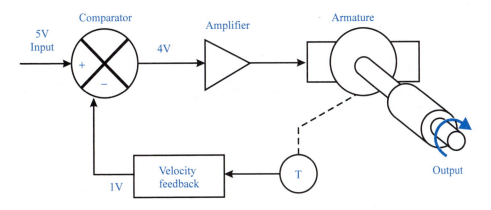

FIGURE 19-2 The velocity system's reaction to an excessive load condition

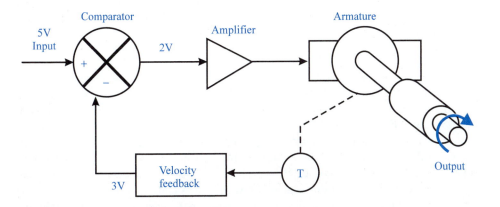

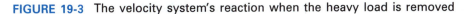

FIGURE 19-3 The velocity system's reaction when the heavy load is removed

by dropping the error signal to 2 volts. The reduced current supplied to the armature causes the motor to slow down. When the system eventually stabilizes, the error signal returns to 3 volts and the motor speed goes back to 2500 RPM.

19-2 BANG-BANG POSITION SERVO

A bang-bang position servo is an inexpensive system that is rather easy to implement. An example of a bang-bang servo is the system in Figure 19-4(a), which uses a cart to transfer grain from one location to another.

The operation begins by moving the cart into the load position so that it can be filled with the incoming material. Once a strain gauge under the track senses that the cart is full, the loading process stops. The cart then immediately moves down the rail as quickly as possible after power is applied to its motor. As the cart hits the "dump" limit switch, power is removed from the motor and it coasts to a stop over a dump chute. A bottom door on the cart opens and dispenses the load as outgoing material.

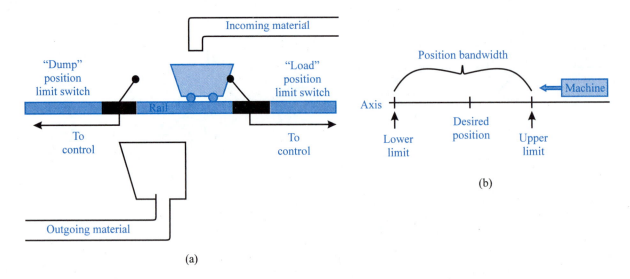

FIGURE 19-4 Bang-bang servo control

Once the strain gauge on the track senses the cart has emptied, the door closes, power is restored to its motor, and the cart moves toward the load position as rapidly as possible. As the cart hits the "load" limit switch, power is removed from the motor and it coasts to a position for reloading.

Figure 19-4(b) shows a graphical representation of the operation when the cart moves to the dump position. The "dump" limit switch represents the upper limit. When it is hit by the cart moving to the left, motor power shuts down. The cart will coast into the desired position somewhere within the bandwidth. If the inertia of the cart is too large due to an excessive load or speed, it will coast past the lower limit and out of range. To correct this problem, it is necessary to lighten the load, reduce the gain of the motor amplifier, or widen the area in which the cart stops. The most convenient option is to lower the amplifier gain. The term "bang-bang" is derived from the manner in which the load moves (at full speed) until it reaches one limit of travel or the other.

The bang-bang servo is used in high-productivity applications that do not require close tolerances. For example, it can be used in a concrete rebar operation where the stock is fed into place as quickly as possible before being cut to size. If the tolerances are within an inch or two, the accuracy of the bang-bang system is adequate.

FIGURE 19-5 A position bandwidth graph of a proportional position servo

19-3 PROPORTIONAL POSITION SERVOMECHANISMS

With the proportional closed-loop system, the bandwidth is reduced to zero. The graph in Figure 19-5 shows that only the desired position exists. There are many applications where proportional position control is used, such as an automated insertion machine that places parts onto a printed circuit board, a robotic arm that makes a weld on an automobile frame, or a numerical control machine tool that cuts patterns into a metal stock.

ANALOG POSITION SERVO

An analog position servo is shown in Figure 19-6. The actuator is a permanent-magnet DC motor. Through a series of pulleys, belts, and gears, the motor drives a rack horizontally in both directions. The rotation of the mechanism moves the wiper on a potentiometer through a 300-degree arc. The potentiometer is the feedback transducer of the system. The voltage on the wiper is an indication of the position. The feedback signal is fed back to a summing amplifier where it cancels the setpoint signal that indicates the desired command position. The summing amplifier operates as the controller and produces the error signal of the system.

When the voltage from the position feedback potentiometer equals the setpoint voltage, there is no error signal. The motor does not move because there is no voltage applied to the power amplifier. To move the rack mechanism to the left, the setpoint voltage is increased. Because the setpoint amplitude becomes greater than the negative feedback voltage, a negative error signal is produced by the summing amplifier. The result is that a positive voltage is produced by the power amplifier. This potential causes the motor to rotate clockwise and drive the pinion gear clockwise, which moves the rack to the left. As the pinion moves clockwise, so does the potentiometer. As the feedback voltage becomes more negative, the error signal becomes smaller and the motor slows down. Eventually, the negative feedback voltage from the potentiometer cancels the positive setpoint voltage and the error signal drops to zero. The rack stops moving.

By reducing the setpoint voltage, the same sequence of events occurs. However, because the polarities and mechanical movements are opposite, the rack moves to the right.

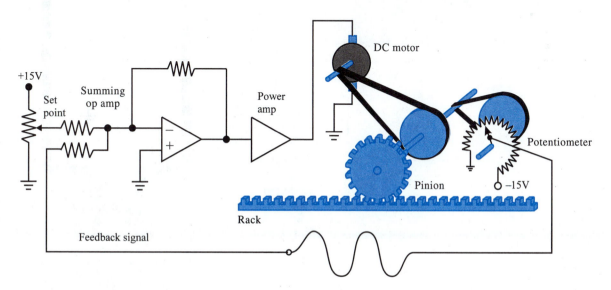

FIGURE 19-6 An analog position servo

ELECTROHYDRAULIC LINEAR POSITION SERVO

The electrohydraulic linear position servo, shown in Figure 19-7, accepts a digital command signal that represents a numerical value. After receiving the command input, the circuit causes a hydraulic piston to move to a linear position which corresponds

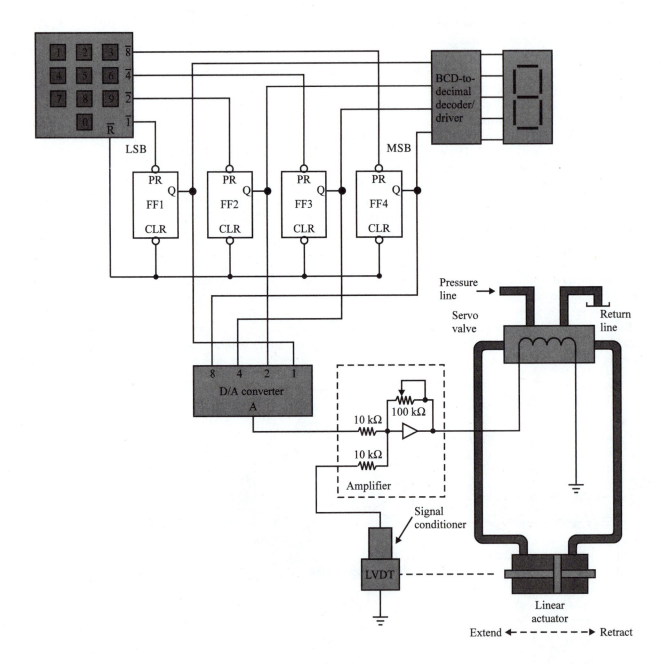

FIGURE 19-7 Electrohydraulic linear position servo

to the number entered on the keypad. The number of different positions is limited to a range from zero to nine inches. Each count causes the piston to move one inch.

Each decimal number entered into the keypad is also loaded into a 4-bit register, consisting of four flip-flops, as a binary-coded-decimal value. The BCD number is then changed to a proportional analog voltage by the D/A converter. The analog voltage, which is either 0 or a positive voltage, is applied to one input of a summing op amp that operates as a comparator. Each time the BCD value increments, the D/A output increases by one volt.

The piston (linear actuator) is mechanically coupled to an LVDT. The LVDT produces an analog voltage, which is either 0 or a negative voltage. This voltage is applied to a second input of the summing op amp comparator. When the piston is retracted at the home position, the LVDT produces a 0-volt output. Each inch it extends from the home position increases the LVDT output by one negative volt.

Operation of the Electrohydraulic Linear Position Servo

Refer to Figure 19-7. Assume that the piston is at the home position and the register contents is 0. Because the output of the D/A converter is 0 volts and the LVDT output is 0 volts, the servo-valve spool is in the center position.

Suppose that the machine operator wants the piston to extend five inches. A description of how the circuit performs this function follows.

1. The operator presses key 5.

2. The push-button closure activates a one-shot multivibrator inside the keypad circuitry that generates a momentary 10-ms 0-state pulse at the $\overline{R}$ output. This negative pulse is sent to the clear inputs of flip-flops FF1–FF4, which causes them to reset simultaneously.

3. The completion of the one-shot pulse during the closure of the number 5 key causes outputs $\overline{1}$ and $\overline{4}$ to go low and outputs $\overline{2}$ and $\overline{8}$ to remain high. Therefore, the parallel-loaded storage register is preset to a BCD 5.

4. The LED readout displays a 5 as the Q outputs of each flip-flop are applied to the BCD-to-decimal decoder driver.

5. The count in the register causes the D/A converter to produce a +5 volt output. This voltage is applied to the top input of the summing op amp, and zero volts from the LVDT is applied to the bottom input lead. The result is the op amp produces a −5 volts and moves the servo-valve spool to the left, which causes the piston to extend. When the piston has moved five inches, the −5 volts from the LVDT cancels the +5 volts of the D/A converter, causing the spool to move to the center position.

6. The operator presses key 2.

7. After the one-shot pulse clears the count, a BCD 2 is loaded into the register. This value causes +2 volts to be applied to the top summing op amp input.

8. A +3 volts are produced by the summing op amp because the +2 volts cancel out some of the -5 volts from the LVDT. The result is that the solenoid action in the servo valve moves the spool to the right and causes the fluid to retract the piston.

9. When the piston is at the 2-inch position, the +2 D/A voltage and the −2 LVDT voltage cancel. As the servo spool centers, the fluid is blocked and the piston rod stops moving.

19-4 CHARACTERISTICS OF A SERVOMECHANISM

The designer attempts to insure that a motion control servomechanism possesses the characteristics necessary to function effectively. These characteristics fall under two general categories: static and dynamic.

The **static** characteristics relate to the steady-state behavior of the servo system. This condition exists when the load is stationary, such as when it is in the home position or has reached the end point. The **dynamic** characteristics relate to the behavior of the system when a dramatic change in the input signal is applied. This condition exists when the load is moving.

The static characteristics of a typical servomechanism are accuracy, resolution, repeatability, and stiffness. They usually relate to a position-type servomechanism.

Accuracy is the degree to which an output will attempt to match the input command.

Resolution is the smallest discrete output change of position that the servo can make.

Repeatability is the range in which the output position of the servo- system will come to rest whenever a given input command signal is repeated.

Stiffness is the reluctance of an actuator to deviate from the desired position specified by the command signal. A system exhibits good stiffness characteristics if it responds quickly by a command signal change or springs back to its stationary position when a load disturbance is introduced.

The dynamic characteristics of a typical servosystem are stability and transient response.

Stability is the maximum amount of allowable overshoot past the desired position. Instability occurs when the load continuously oscillates instead of standing still.

Transient response is the response time of the actuator to sudden changes of speed or position command signals. This value is defined as the response time required to go from 10 percent to 90 percent of the final value. The term *transient* is also known as *bandwidth*.

19-5 DESIGNING A POSITION SERVO

Position control applications typically fall into two basic categories: *point-to-point* and *contouring*.

POINT-TO-POINT

The command signal for many point-to-point control applications is a step voltage applied to the input of the system. A step voltage is characterized by its sudden discrete change in amplitude from one level to another.

Dynamic characteristics such as stability and transient response are of major importance in the design of a servomechanism. To be useful, the system must be stable. Therefore, the amplifier gain must be limited to a value which will give the required degree of stability. To determine stability, a transient response test is performed by using the circuit in Figure 19-8. To perform the test, the transient switch is opened while a new command signal is selected by adjusting the control pot 1 to a new level. The test is performed by observing the motion of the output when the switch is closed. At the moment the step input voltage occurs, the error signal at the feedback potentiometer becomes maximum. The stability criteria is defined as the maximum allowable overshoot. The speed of the response is defined as the time required to move from 10 percent to 90 percent of the final value. Most point-to-point applications require that the traverse movement is swift without causing excessive overshoot.

By connecting a chart recorder across the motor terminals, the transient response test can be displayed graphically. Figure 19-9 shows three examples of the tests. In Figure 19-9(a), the ideal condition is attained by providing the shortest possible rise time with no overshoot. The system is *critically damped* when this condition is achieved. In Figure 19-9(b), an *underdamped* condition occurs as overshoot develops when the rise time is shortened. This is an unstable or oscillatory condition. In Figure 19-9(c), an *overdamped* case occurs where the rise time is lengthened and no overshoot develops. Overdamping is undesirable because the response time is too long.

To achieve a critically damped condition, the amplifier gain must be precisely set to produce the required torque to move the load. If the gain is too high, an underdamped

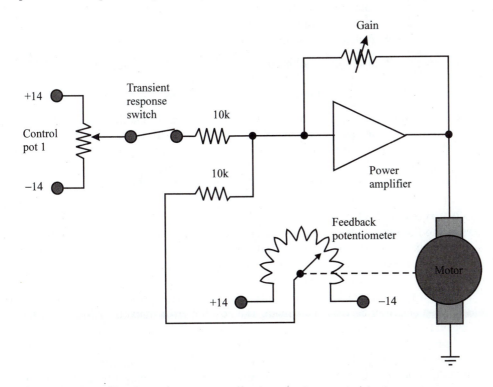

FIGURE 19-8   Circuit to demonstrate the transient response test

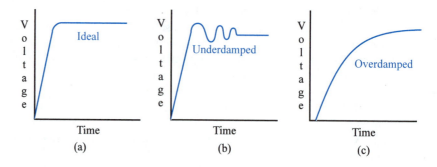

FIGURE 19-9 Examples of transient response tests

condition exists. When it is too low, overdamping occurs. If inertia, friction, or gravitational factors on the load are changed in a critically damped system, amplifier gain adjustments alone might not be sufficient to overcome these conditions. For example, suppose the weight of the load is increased. The increased weight creates a higher static inertia. Consequently, the system will take longer to reach the desired velocity when an input change is introduced because the acceleration is decreased, as verified by the following formula:

$$a = \frac{T}{J^s}$$

where, a = rotational acceleration
              T = Applied Torque
              J^s = system rotation inertia

The system will also require a longer time to travel a given distance:

$$Displacement = Velocity \times Time$$

To achieve a transient response time similar to that of the original load, the gain is increased to overcome the static inertia. However, when the end point position is reached, the dynamic inertia will cause the load to move past the desired position. To overcome various load conditions, system modifications that provide necessary characteristics are made, such as the tachometer feedback described below.

Tachometer Velocity

One method of improving the transient response time for various load conditions is to add a velocity summing amplifier with tachometer feedback. This velocity loop is placed inside the position loop as shown in Figure 19-10. By using two amplifiers, the overall gain of the system is greater than with one amplifier circuit.

At the instant the transient switch is closed, the position error signal produced by the position amplifier becomes maximum. The position error signal is multiplied by the gain of the velocity amp. Since the motor has not started to move, the tachometer output is zero. Because the tachometer signal does not subtract from the position error, the velocity error signal becomes maximum. The large current that results creates a torque that quickly accelerates the motor, providing a fast response time. As soon as the motor speeds up, the tachometer sends a feedback signal to the velocity amplifier.

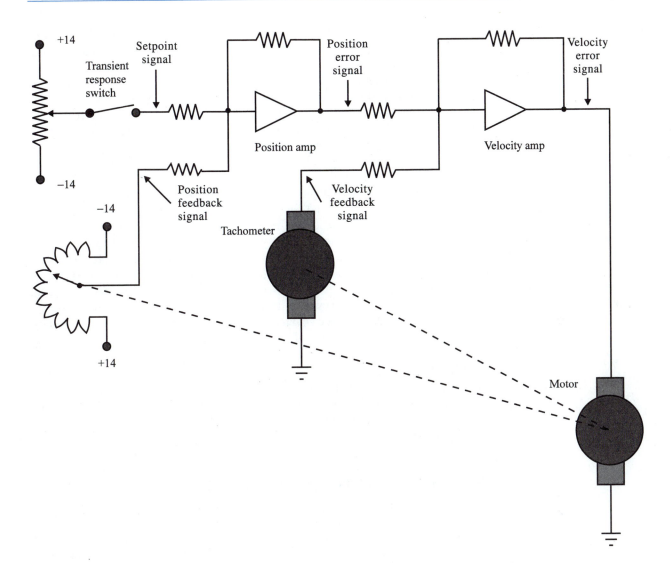

FIGURE 19-10 Tachometer velocity feedback

The combined effect of the position error becoming small and the velocity feedback signal subtracting from it causes the velocity error signal to reduce dramatically. The result is a deceleration braking action called *damping,* which prevents the load from overshooting.

Tachometer feedback provides several additional benefits to a position servo. For example, in a position error system without the tachometer, as the load approaches the desired position, the position feedback voltage changes until it almost equals the command signal voltage but at the opposite polarity. The resulting position error signal is reduced to a level too small to overcome the friction of the load. The result is that the motor slows down until it stops just short of its desired position. Therefore the error signal does not reach zero, leading to a steady-state error. When the tachometer loop is added, the error signal from the position amplifier is multiplied by the amplifier gain of the velocity amplifier to overcome the load friction more quickly. The motor movement rotation continues until the end position is reached. A simple test can be made to demonstrate the impact of a velocity loop. If the shaft of a motor is forced out of the command position without a velocity loop, there is very little resistance. If the velocity loop is added, the shaft of the motor will turn much harder. With increased overall amplification, the system attempts to correct any deviation from the desired position. When the shaft is released, it springs back to the command position. Therefore, the higher system gain provides good stiffness. It also improves accuracy and repeatability characteristics.

In some point-to-point applications, the one-step voltage command signal is not used. Instead, the command signal consists of a series of incremental voltage steps, as shown in Figure 19-11. Its shape resembles a trapezoid profile and is referred to as a *position profile*. The incline slope portion of the profile indicates acceleration. The acceleration stops when the desired speed is reached. The speed is decelerated as the desired position is approached. The acceleration and deceleration rates are typically the same. The voltage reduces to zero when the end point has been reached. As the command steps change, the actuator tracks behind. The advantage of the trapezoid profile signal over a step voltage signal is that less stress is put on the actuator because it can accelerate gradually instead of abruptly.

FIGURE 19-11   A multistep trapezoidal command signal

CONTOURING

The command signal for contouring control is a series of smaller steps than the trapezoidal command supplies to point-to-point servos. The contouring application requires that the actual position precisely tracks the command position steps with a minimal error signal throughout the movement.

19-6   DIGITAL CONTROLLER

Digital controllers that use complex computer programs are required to provide multistep command signals. These computer systems can provide the required control characteristics for various load applications.

A common structure of a high performance digital motion controller is illustrated in Figure 19-12. This cascaded controller consists of:

1. An innermost current loop that provides torque to the motor.
2. A velocity loop around the current loop to control speed.
3. A position loop around the velocity loop that monitors the location of the load.

The speed with which the control functions are performed matches the structure of the circuit. For example, the fastest response time is performed by the current loop function, then the velocity loop event, and then the position loop. The current loop has the highest bandwidth, followed by the velocity loop; the position loop has the lowest bandwidth. Therefore, when tuning the system, it is necessary to start with the innermost loop and work outward.

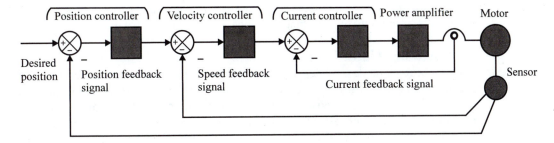

FIGURE 19-12 Digital multi-loop control for high-performance motion control

CURRENT LOOP

The torque power to the armature of a DC motor is supplied by a power amplifier. The signal that regulates the power is provided by the current controller. In many applications, the motor speed must be held constant under varying load conditions. When an operation requires an extremely fast response to a load change, a current inner-loop is placed in the system to keep the speed stable. It consists of an extra summing junction, an amplifier, and a sensor that produces a feedback signal by detecting the motor's armature current. This current inner-loop circuitry is usually contained within a DC drive unit when a DC motor is the actuator, and in an AC drive when an AC motor is used. For example, if a DC motor is loaded down, the CEMF reduces and armature current increases. The current sensor may be a fractional-ohm resistor placed in series with the armature. As the current changes, a proportional voltage change occurs, which is used to produce the current feedback signal. Another current sensor may be a Hall effect device placed next to the armature conductor. A magnetic field proportional to the armature current is present around the conductor. As the field changes, the sensor produces a proportional feedback voltage.

The current feedback signal is compared to the velocity error signal at the summing junction of the current comparator. The output of the current junction, called the current error, is applied to the power amplifier that drives the motor.

The advantage of the current loop is its quick response time. The response time of the two outerloops is a little longer because they are related to the load. Since the response time of the current loop is so quick, it can replace tachometers, which are used for fast reduction applications.

VELOCITY LOOP

For position applications that require good stability characteristics, tachometers usually provide adequate speed feedback. To create a stable position servo loop without using an analog tachometer, software programming can digitally simulate a tachometer's damping effect. The velocity feedback signal is provided by a feedback device such as an incremental encoder. The speed is determined by calculating the rate of change of the position. The actual velocity is subtracted from the position error signal at a comparator to generate the velocity error. The velocity error signal is applied to the current amplifier as a current command input. The comparing takes place at the junction of a summing op amp that functions as a proportional amplifier.

Proportional Gain

The proportional mode provides the primary control function in the velocity loop. The proportional gain setting determines the amount of torque produced in response to the position error signal. A position controller with only proportional gain is very common in contouring applications. A high gain setting provides high accuracy and stiffness. In point-to-point applications, a high gain setting is usually desirable because it provides fast transient responses, high accuracy, good repeatability, and stiffness. The drawback of high gain is that it causes an unstable condition by overshooting and oscillations.

Integral Gain

Another type of velocity control loop commonly used is the proportional-plus-integral (PI) regulator, as shown in Figure 19-13.

If an integral mode control network is added to the system, the output of its amplifier accumulates as long as there is a steady-state error signal applied to its input. Eventually, its output amplitude becomes great enough for the motor to produce the torque required to move the load the remaining distance, eliminating the steady-state error.

When the gain setting of the integral parameters is adjusted properly, the overall positioning accuracy is improved. It also provides stiffness to load torque disturbances. If the gain setting is too high, a low frequency oscillation may occur around the command position. Also, the integral function results in a longer settling time due to a phase shift that it introduces into the system. To reduce the phase shifting in point-to-point applications where the movements are very rapid, the integral gain is adjusted to a low level. In contouring applications where the movements are slower, the phase shifting is not a factor. Therefore the gain can be adjusted to a significant value to provide high stiffness during the travel.

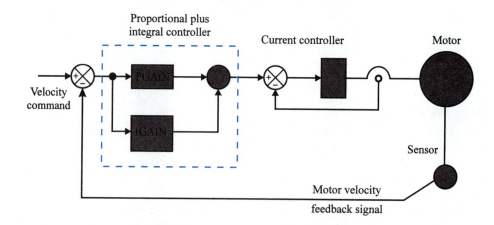

FIGURE 19-13 A velocity control proportional-plus-integral closed-loop regulator

POSITION LOOP

In contouring applications, it is common to use only the proportional mode in the position loop. As the command signal increments through a series of step changes, it is constantly compared to the position feedback signal by the summing junction of the proportional op amp. The resulting output that the op amp produces is the position error which becomes the command signal for the velocity loop.

When a position loop is used for point-to-point applications, it is not designed to follow paths. Instead, it must provide fast transient response time, minimal overshoot, and a good velocity profile. In general, a point-to-point regulator is more complicated than the simple proportional gain-only configuration used for contouring movements.

The multi-loop point-to-point position loop structure shown in Figure 19-14 uses proportional-integral-derivative (PID) mode and feedforward. Each mode and feedforward has unique operational characteristics. Together, they provide maximum performance.

PID is a relatively sophisticated compensation technique to improve the performance of a servo. A servo with a good internal velocity loop will provide many of the benefits of PID. If no velocity (tachometer) loop exists, PID should be considered. It is always best to keep designs simple and introduce complexity only when necessary.

A brief explanation of the purpose for each control mode in a PID with feedforward position loop follows.

Proportional Mode

The proportional mode is the primary control function. It is the simplest of the four functions. The proportional amplifier generates a velocity command proportional to position error. For example, high velocity results when the following error (the amount that the load lags behind the command signal) is large. By increasing the proportional gain, the system can allow the following error to be reduced while still achieving a

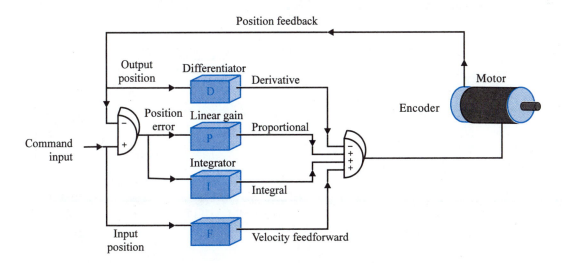

FIGURE 19-14 A PID and feedforward point-to-point position loop

fast transient response. However, the higher gain may cause the actual position to overshoot the commanded position.

Feedforward

When the command signal is introduced in a position servo, an error signal develops. The error signal results in a proportional velocity command that causes the load to move to the end point position. In some applications, the system operation is more responsive if the following error is reduced to near zero during the entire time the load is being moved. By adding a feedforward mode into the system, this requirement is achieved.

Velocity feedforward is produced by altering the command signal with the addition of a boost signal. The larger command signal causes the motor to speed up and therefore minimize the following error. Figure 19-15(a) illustrates how a feedforward command signal is developed.

$$
\begin{aligned}
C &= \text{Command Signal} \\
FF &= \text{Feedforward Signal} \\
F &= \text{Feedback Signal} \\
E &= \text{Error Signal}
\end{aligned}
$$

To keep the error signal at zero, C must equal F, and FF must equal E. Because the command velocity and the gain are known, the error can be computed and added to the command to generate a modified command signal, which will cause C and F to be the same. The diagram shows an ideal system where the feedforward provides the exact velocity command without the need for any position error. In practice, a conservative approach is taken to setting a feedforward value so that it is not quite high enough to completely eliminate an error signal. (If the feedforward gain setting is too high, position overshooting will result.) Therefore, a minimal following error exists. To provide stability, a proper gain setting of the proportional control loop is required.

An example of the effect a feedforward velocity command has on a position following error when making a trapezoidal velocity profile move is shown in Figure 19-15(b).

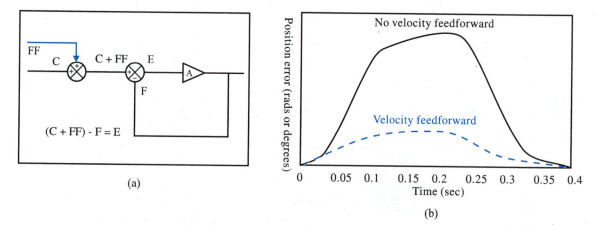

(a)

(b)

FIGURE 19-15   Feedforward control

Derivative Mode

The derivative mode performs an operation similar to that of the velocity feedback tachometer. Recall that as a one-step input command signal is applied, the position error is maximum. Since the position error signal is maximum and the tachometer is not moving, the multiplication factor of the position and velocity amplifier is at the highest level. The result is that the acceleration rate is very high, which improves the transient response time of the system. Once the motor is turning at full speed, the tachometer feedback cancels the velocity command signal developed from position error. This action dampens the overall gain of the system and causes the load to decelerate. As the load approaches the end point, the error reduces and the motor slows down, stopping when it reaches the command position. The disadvantage of tachometer velocity feedback is that the deceleration action begins early during the travel time of the movement.

The derivative amplifier refines the system by replacing the tachometer. The input of the derivative amplifier is the actual position feedback signal. A system that utilizes the derivative mode requires a trapezoidal command signal. As the command signal is ramped upward, the error signal rapidly increases. During the ramping, the output of the derivative amplifier is maximum and it adds to the proportional output to accelerate the motor. When the command signal is constant and at the top of the trapezoid, the derivative is zero and only the proportional output is applied to the motor. When the command signal is ramped downward, the derivative amplifier produces a maximum output but at the opposite polarity. As the derivative voltage cancels the proportional voltage, the motor decelerates quickly. The overall effect of the derivative action is that it quickly boosts the motor speed at the beginning of travel and provides a braking action at the end of travel.

The derivative mode is used for damping in two different situations. The first occurs when the velocity loop is replaced by a current servo. The second situation occurs when it is necessary to reduce overshooting because the proportional gain must be set high.

Integral Mode

The purpose of the integral mode is to provide a velocity command signal that reduces static position error to zero and to provide stiffness for the system. Since integral control is provided by the velocity loop, it is normally not used in the position loop.

19-7   TUNING A SERVOMECHANISM

When developing a servomechanism, it is necessary to tune the system so that it will function effectively. The tuning involves making gain adjustments of various control mode parameters. By making proper gain settings, the system matches the unique conditions of the load and overcomes load disturbances and friction. The choice of which control modes to use in a particular case is made based upon the application's static and dynamic characteristic requirements.

In practice, system tuning is rarely determined by calculations. Instead, a trial-and-error technique is used with the motor connected to the actual or simulated load. The practical approach is to make gain adjustments, excite the system, and then view the response on an oscilloscope or chart recorder. Each parameter is tuned by starting with a low value and increasing the gain until the desired response is achieved. The gain may be adjusted by a potentiometer or by selecting a parameter value on a set-up menu of a computer software program. Achieving accuracy for desired movements often becomes an exercise in matching several gains.

19-8   MASTER-SLAVE SERVO SYSTEM

Master-slave is a form of coordinated motion. In some applications, a number of axes need to be synchronized to enable the entire system to operate properly. The most precise way of achieving this type of motion control is to monitor the primary axis, or master axis, and to slave all the other axes' motions to it. Whenever the operator changes speed or position of the master axis, each slave axis will maintain its relationship to the master by making the necessary adjustments. Historically, gears and line shafts were mechanically coupled to start, move, and stop in precise synchronism to perform the required movements. These control functions are now coordinated digitally by the controller element of a closed-loop system.

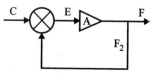

FIGURE 19-16 A basic closed-loop motion control

The concept of master-slave operation is based on a closed-loop motion control system shown in Figure 19-16. The loop's function is to make the output (feedback) follow the command. When the feedback F coincides with the command C, the error E is zero. Figure 19-17 shows a closed-loop slave network that receives its command signal (C_2) from the output (F_1) of a master closed-loop network. Assume that A_2 is a typical drive and motor controller. The error (E_2) that develops causes the motor to adjust its speed proportional to the error amplitude. In this system, output (F_2) lags behind the command signal (C_2) by the amount of the following error (E_2) that develops. The greater the following error, the higher speed the motor will run. A master-slave system does not operate in this fashion. Instead, F_2 must precisely follow the master F_1.

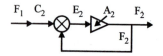

FIGURE 19-17 Slave fed from master feedback

Figure 19-18 shows how the closed-loop system in Figure 19-17 can be modified to operate as a master-slave network. A software module (S) is inserted between F_1 and C_2.

A_2 = A typical drive and motor combination
F_1 = Output of a master network fed into the slave network
C_2 = The modified command signal produced by the software module
E_2 = Error signal
F_2 = Output of the slave network

By supplying the module with the known gain of A_2, its software calculations can predict the E_2 value at any velocity value supplied by F_1. Data entered for computation into the software are the gain of A_2 and the output signal F_1 from the master.

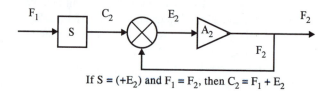

If S = ($+E_2$) and $F_1 = F_2$, then $C_2 = F_1 + E_2$

FIGURE 19-18 Master-slave network

The program predicts the error signal that would develop based on these two input values. By adding the predicted error signal E_2 to F_1, C_2 is boosted so that F_2 coincides with F_1.

A rewinder machine that does not use master-slave control is shown in Figure 19-19. The purpose of the machine is to transfer paper from a large supply roll to make smaller rolls. The supply roll is identified as the feed reel and the smaller roll is labeled the take-up reel. The motor that drives the feed reel runs at a constant RPM and the take-up motor runs at a variable speed. A tension roller rides on the paper and turns as the paper passes under it. As the tension on the paper changes, the roller responds by moving upward if it tightens, or downward if the tension slackens. Mechanically coupled to the roller is the wiper of a potentiometer. Together, the tension roller and the potentiometer form a tension transducer that produces a DC feedback voltage proportional to the tension on the paper.

Suppose that at the beginning of a winding operation, a full supply of paper is mounted on the shaft of the feed reel, and a cardboard tube is placed on the take-up reel. As the rewinding process starts, the wiper arm is positioned at the bottom of potentiometer body (P_1). Since the wiper arm does not feed any cancelling voltage into the summing op amp, the take-up motor will run at a high speed. As paper builds up on the take-up reel, the tension will begin to increase. The roller begins to move upward, causing the potentiometer's wiper to slide in the same direction. As the wiper

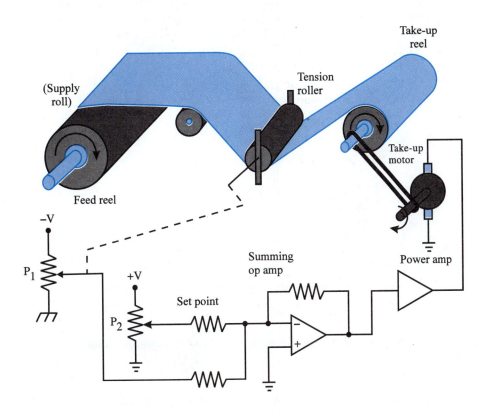

FIGURE 19-19 A rewinder paper machine

moves, the negative voltage it supplies to the summing op amp becomes greater. By canceling some of the setpoint voltage, it causes the take-up motor's speed to reduce and slow down the rate at which the tension changes. The reason the motor reduces speed is to insure that the take-up reel winds the same amount of paper that is unwound by the feed reel. When the take-up roll is full, it is replaced with an empty tube and the process is repeated.

As the supply roll unwinds, its diameter reduces. The drawback of this action is that the rate at which it feeds the paper slows down because the RPM of the feed reel remains constant. A more desirable condition exists when the feedrate remains constant as the diameter of the supply roll changes. A velocity control system called a master-slave network (also known as a master-follower) is capable of performing this function. In a master-slave network, several motors and controllers are linked together to perform coordinated operations.

In the master-slave rewinding machine in Figure 19-20, one motor makes the required speed adjustments by tracking the speed of the other. By convention, the motor that is tracking is called the slave and the motor being tracked is the master. In the rewinding operation, for example, the operator programs the master controller to run the wind motor at a speed that causes paper to be wound at a constant feed rate. An RPM signal is sent to the master controller by a tachometer. The controller uses the data to make complex mathematical ratio calculations. The result of the calculations determines the speed rate at which the slave motor must turn to unwind the paper. This information is sent to the slave controller, which converts the data to the appropriate electrical signals supplied to the feeder motor. To provide better speed and torque

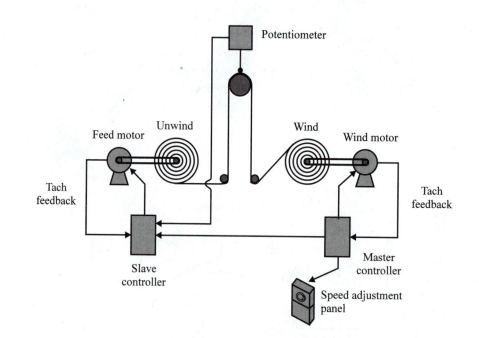

FIGURE 19-20 A master-slave network used to control two motors in a rewinding machine

accuracy, a roller that rides on the paper is connected to a potentiometer. The voltage from the potentiometer is used to produce a feedback signal that indicates paper tension.

CHAPTER PROBLEMS

1. In a closed-loop velocity servo system with a tachometer, if the load slows the motor down, the error signal _____ (increases, decreases).

2. A bang-bang position servo is used in _____ (high speed, precision) applications.

3. A (bang-bang, proportional) _____ position servo has a zero bandwidth.

4. The feedback tachometer in a closed-loop servomechanism performs which of the following functions: ____
   a. Provides speed regulation.   c. Provides stability.
   b. Provides stiffness.   d. All of the above.

5. At the moment a transient command signal is applied to a closed-loop position servo, the error signal is _____ (minimum, maximum).

6. The linear motion control feedback transducer used by the electrohydraulic system in Figure 19-7 is an _____.

7. The _____ (static, dynamic) characteristics of a servomechanism pertain to the condition of the load when it is moving.

8. List the four types of static characteristics.

9. In a transient response test, the system is ____ when it provides the shortest possible rise time without overshoot.
   a. underdamped   c. critically damped
   b. overdamped

10. It is possible to correct an underdamped condition of a position servo by _____ (increasing, decreasing) the system gain.

11. Assuming that the system gain of a position servo remains unchanged, the transient response time _____ (increases, decreases) if the load inertia is decreased.

12. A decelerating braking action called _____ prevents the load in a position servo from overshooting its end point.

13. The following error in a contouring position control application is _____ (minimal, large).

14. A current loop in a position control application has a _____ (low, high) bandwidth.

15. The ____ mode control eliminates steady-state error.
    a. proportional   c. derivative
    b. integral

16. In a slow moving contouring application, the ____ mode provides high stiff-ness during travel.
    a. proportional   c. derivative
    b. integral

17. A ____ provides a better braking action in a point-to-point position system.
    a. tachometer   b. derivative amplifier

18. Feed-forward _____ (increases, decreases) the following error signal in a position servo system.

19. List the two ways in which gain adjustments can be made when tuning a servo system.

20. If a master axis slows down in a master-slave velocity servo, it is likely that the slave axis will ____.
    a. slow down   b. speed up

CHAPTER 20

Functional Industrial Systems

CHAPTER OBJECTIVES

At the conclusion of this chapter, you will be able to:

○ Explain the operation of a digital motion control position system.

○ Describe the operation of a lubricating drilling machine.

○ Explain the operation of a package sorting machine.

○ Explain the operation of an injection molding machine.

○ Describe the operation of an in-line bottle filling machine.

○ Explain the operation of a paint can filling machine.

○ Describe the operation of an X-Y axis medicine dispenser.

○ Explain the operation of a plastic profile extrusion machine.

INTRODUCTION

The materials throughout this book have been organized into topic areas that describe the operation of various types of components, circuits, and equipment used in industry. This chapter illustrates how the devices discussed in the previous chapters can be combined into functional systems. Practical application examples for both motion and process control systems are provided.

20-1 DIGITAL POSITION CONTROL

Figure 20-1 illustrates how the positioning of a linear-motion device can be controlled by a digital command that is entered by a keypad input with numbers 0-9. A bar mechanism positioned by a rack-and-pinion assembly has 10 position regions with mechanical stops at both ends of travel. The bar is shown at position 0 and its location is detected by a Gray code wheel mounted on the same motor shaft as the gear. When a number greater than 0 is entered into the keypad, the gear turns in a clockwise direction, causing the bar to travel to the left. When the bar reaches the region that is the same as the number entered into the keypad, the gear stops and a light turns on. If a second number is entered greater than the first, the gear will again rotate clockwise until the bar reaches the new location. However, if the second number entered is less than the first entry, the gear will turn counterclockwise, pushing the bar to the right until its position equals the number entered. At that moment the gear will stop and the light will go on.

To better explain the operation of positioning, functional descriptions of the various sections are provided.

Input Section (Keypad).   To move the bar mechanism to any one of 10 locations, a command is entered by a keypad that has 10 different keys numbered 0–9. As the desired value is inserted by pressing one of the keys, a 10-ms negative reset pulse ($\overline{R}$) clears the flip-flops. Then the active-low numerical outputs are activated when an equivalent binary number is parallel loaded into a 4-bit storage register.

Decision Section (Magnitude Comparator).   The control portion of the positioner is a digital circuit called a magnitude comparator. As described in Chapter 2, it is made up of two separate 4-bit inputs and three single-lead outputs. The function of the circuit is to compare 4-bit input A to 4-bit input B. If the binary-coded decimal numerical value applied to input A is larger than B, the A > B output line will go high while the other two outputs go low. If the numerical value of input B is larger than A, the A < B line will go high while lows are generated at the other two outputs. If the binary numerical values of inputs A and B are the same, the A = B line will go high while the other outputs go low.

Amplifier Section.   The amplifier section consists of two separate inputs and one output line. One of the inputs is connected to the A > B output of the magnitude comparator. The other input is connected to the A < B output. The output line of the amplifier section provides an amplified potential that is sufficient to drive the DC positioning motor of the circuit.

When the A < B output of the comparator goes high, it is applied to the input of the operational amplifier. Since the op amp is set at a gain of 1, its only function is to invert the +5-volt signal to a −5 volts. As the negative 5 volts are applied to the summing power amplifier input, it is amplified and inverted to a positive potential.

When output A > B of the comparator goes to a +5-volt high, it is applied directly to the power amplifier, which amplifies and inverts the signal to a negative potential.

Actuator Section.   The device that causes the actual physical positioning in the system is the DC motor. The potential used to activate the motor is provided by the summing power amplifier. When the power amplifier output is positive, electron current flows from ground through the motor to the output terminal. When the power section output goes negative, electron current flows from its output through the motor and then to ground. Since current can flow through the motor in two different directions, it can rotate either clockwise or counterclockwise.

Sensing Section.   Attached to the shaft of the motor is a sensing device that detects the positioning of the system. Figure 20-1 shows that the sensor consists of a disc, four lights, and four optical light sensors. Slots on the disc are

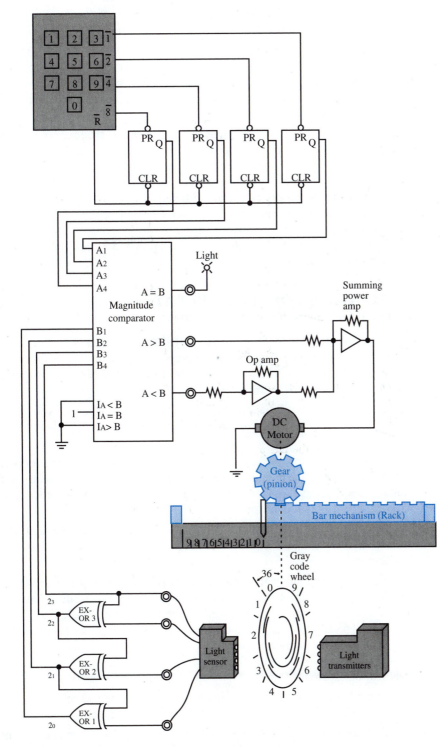

FIGURE 20-1 Digital control of physical position

arranged in a pattern of 1s and 0s in a Gray code format. (A diagram of the disc is shown in Figure 17-10.) It reveals that the Gray code is divided into 10 pie-shaped sectors of equal size, each representing a 36-degree region.

Rack-and-Pinion.   Attached to the end of the motor shaft is a pinion gear that meshes with the teeth on the bar mechanism. When the gear rotates clockwise, the bar moves to the left; when the gear rotates counterclockwise, the bar moves to the right.

Conversion Circuit.   The Gray code numbers generated by the four sensors represent the 10 different physical locations of the digital positioner. However, before the four-digit Gray code numbers are applied to input B of the magnitude comparator, they are converted into their equivalent pure binary values. Conversion is necessary because a binary number is applied to input A of the magnitude comparator and the comparator must compare values from the same number system to operate properly. The conversion process is accomplished by the three exclusive-OR gates (1, 2, and 3) that act as a Gray-code-to-binary encoder.

DIGITAL POSITION CONTROL OPERATION

Refer to Figure 20-1 during the following explanation.

1. Suppose that the bar mechanism is located at position 0. The Gray code wheel mounted on the motor shaft generates a 4-bit output of 0000 that is applied to the encoder, which also generates an output of 0000. Also assume that the number stored in the register is 0000.

2. Since the register output applied to input A of the comparator is the same as the encoder output applied to input B, the A = B output is high and turns on the light.

3. If key 6 is pressed on the keypad, number 0110 is loaded into the register. With the bar mechanism still at location 0, the A input is greater than the B input. Therefore, the A > B output of the magnitude comparator goes high, and the A = B output goes low.

4. The +5-volt signal from the A > B output is applied to the power amplifier, becomes inverted, and is amplified to produce an adequate potential to drive the motor.

5. As electron current flows through the motor from the negative output of the power amplifier to ground, the motor rotates in a clockwise direction. Because the disc and gear are mounted on the same motor shaft, they rotate together. The length of each region at the outer portion of the gear and disc is the same size as the corresponding region on the bar mechanism. Therefore, as the gear and disc turn 36 degrees, the bar mechanism moves the distance of one region.

6. When the bar mechanism reaches region 6, the Gray code wheel generates a 0101, which is converted into a binary 0110 by the encoder.

7. As the register supplies a 0110 to input A of the magnitude comparator, and the encoder output supplies a 0110 to input B, the comparator's A > B output

goes low and stops the motor. Because the gear-and-bar mechanism also stops, the A = B output goes high and turns on the light.

20-2   LUBRICATING DRILLING MACHINE _____

Figure 20-2 shows a drill press with a coolant pump. After the metal stock is secured in a holding fixture, the coolant pump is started by a push button. To ensure that the coolant is flowing where the drill bit makes contact with the metal part, a pressure switch detects the coolant flow.

The drilling process begins by pressing a second push button. The drill bit then turns and a solenoid is energized, which causes a pneumatic cylinder to extend. The extension of the cylinder causes the drill bit to advance toward the stock. As the drill body approaches the stock, it makes contact with a limit switch that activates a five-second timer. During the five seconds, the drill bit bores a hole into the stock. After the timer has timed out, the solenoid de-energizes, causing the spring-return cylinder to retract and remove the drill bit from the stock.

An orange light turns on when the coolant pump is running, a green light is on during the drill-and-feed sequence, and a red light turns on when the equipment is turned off. A stop button turns off the coolant motor, or shuts off the equipment in an emergency situation.

Figure 20-3 shows a hardwired ladder logic diagram that controls the operation of the machine. An explanation of the operational sequence follows:

Step 1:   When the start button is pressed, Rung 1 becomes True, and Control Relay 1 (CR1) energizes. By CR1 energizing, the coolant pump motor

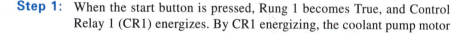

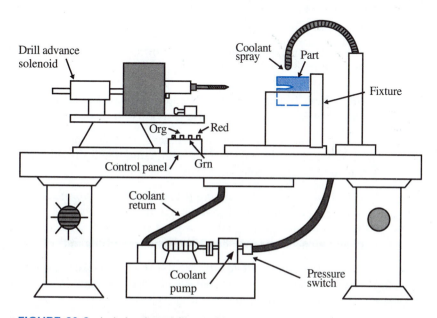

FIGURE 20-2 Lubricating drill machine

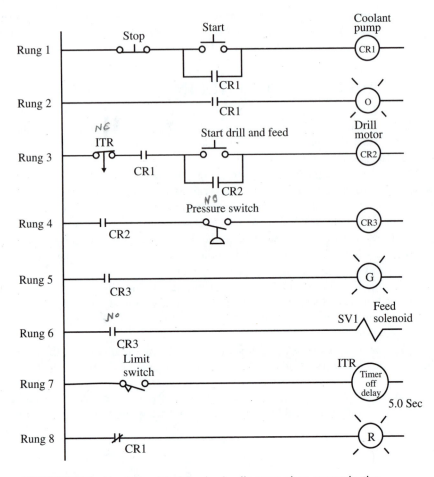

FIGURE 20-3 Hardwired ladder logic diagram that controls the lubricating drill machine

runs, Rung 1 latches, Rung 2 becomes True and turns on the orange light, and Rung 8 becomes False and turns off the red light.

Step 2: If CR1 is energized, Rung 3 becomes True when the drill-and-feed push button is pressed. By CR2 energizing, the drill motor runs and the corresponding contact in Rung 4 closes.

Step 3: If the coolant pump is running properly, oil pressure rises and closes the pressure switch to cause Rung 4 to become True. By CR3 energizing, the green light in Rung 5 turns on and the solenoid in Rung 6 activates a pneumatic cylinder that extends the piston rod.

Step 4: As the drill body extends toward the stock, it makes contact with a limit switch that causes its contacts to close. As Rung 7 becomes True, the timer starts to count.

Step 5: During the five seconds the timer is counting, the drill bit bores a hole into the stock. When the timer times out after five seconds, the N.C. timer contact in Rung 3 becomes False. As the condition of Rung 3 becomes False, the drill motor de-energizes and the spring-return piston rod retracts.

Step 6: As the drill body reverses direction, it breaks contact with the limit switch. Because the limit switch opens and Rung 7 becomes False, the timer resets and the timer contact in Rung 3 becomes True.

Step 7: After the completed stock is replaced with an unfinished stock, the cycle is repeated when the drill-and-feed button is pressed.

20-3 PACKAGE SORTING MACHINE

A company produces bars of soap that are packaged in boxes. Box A, the smallest package, holds 12 bars, box B holds 24 bars, and box C holds 36 bars. Figure 20-4(a) shows a materials handling machine that sorts the boxes from a main conveyer belt according to size. The A boxes are directed to conveyer 1, the B boxes go to conveyer 2, and the C boxes go to conveyer 3.

Figure 20-4(b) shows the three different packages traveling down the main conveyer belt toward the reader. As a package passes through a series of optical sensors called a *light curtain,* its size is detected according to the number of light beams that it blocks. The size A boxes block three beams, size B boxes block six beams, and size C boxes block eight beams. If box A breaks the light beams and is detected by proximity sensor X, solenoid 1 becomes activated. The sensor causes the first arm to swing clockwise by 75 degrees, which deflects the box and steers it toward conveyer 1. When solenoid 1 deactivates after nine seconds, the arm returns to its original position. If six beams are broken, solenoid 2 is activated when box B is detected by proximity sensor Y. The box is steered to conveyer 2 by the arm as it swings out over the main conveyer belt for nine seconds. Box C is deflected down conveyer 3 when eight beams are broken and its presence is detected by sensor Z.

Figure 20-5 shows the ladder diagram used to control the conveyer system that sorts these packages. The start/stop circuit in Rung 1 is used to run or deactivate the motor that drives the four conveyer belts. The row of optical sensors is connected to memory register 030. The BCD number that is present in the register indicates the number of light beams that are broken by the box. The following steps occur if box A interrupts three light beams at the row of optical sensors:

Step 1: When three lights of the optical sensors are broken, a 3 is fetched by the GET instruction in Rung 2. Since the 003 is the same as the 003 in the Equal instruction, output 01001 latches.

Step 2: The latched output 01001 makes Examine-On contact 01001 in Rung 5 True. When the box is detected by sensor X, the Rung is True and latching output 01004 energizes.

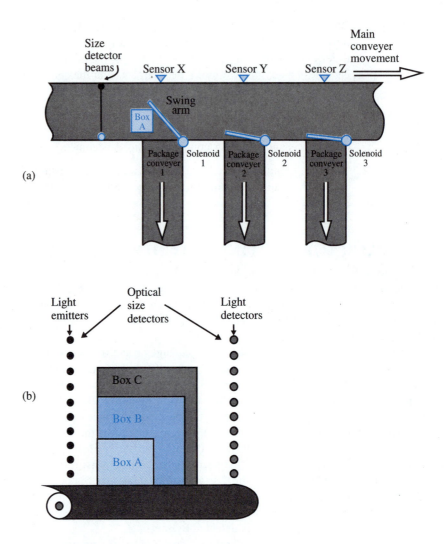

FIGURE 20-4 Package sorting machine

Step 3:   When output 01004 is latched, Examine-On contact 01004 in Rung 8
             becomes True, causing Timer-On output 034 to energize.

Step 4:   When timer 034 begins, Bit 17 of its register sets and turns contact
             03417 of Rung 11 on. The True condition of the rung energizes output
             01007, which activates solenoid 1 to cause arm A to swing open.

Step 5:   When the timer times out after nine seconds, AC = PR and Bit 15 of
             the register for timer 034 sets, causing contact 03415 of Rungs 14 and
             17 to become True. As the condition of both Rungs becomes True, the
             Unlatch outputs 01001 and 01004 energize to create a False condition
             of Rungs 2 and 5.

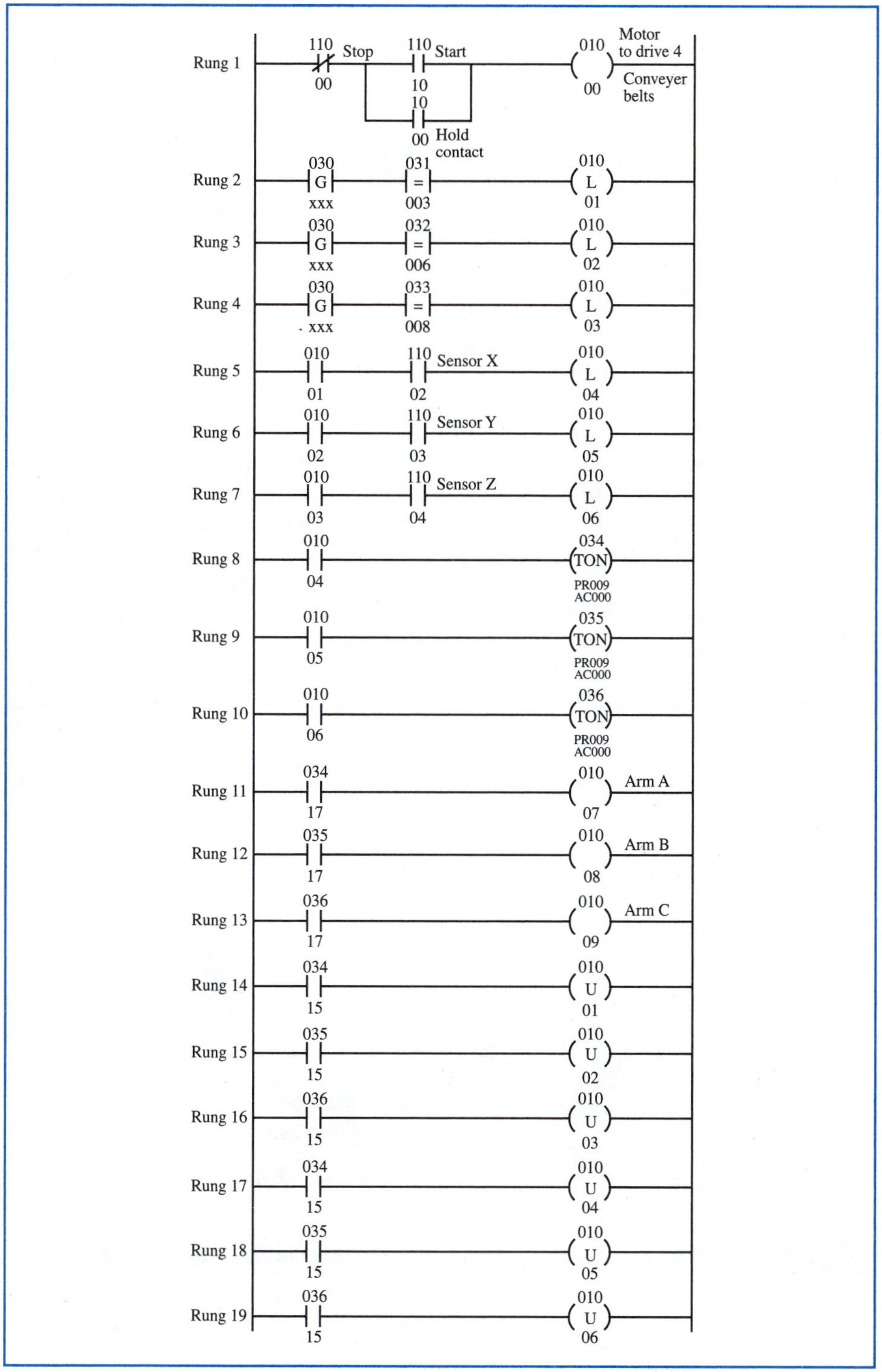

FIGURE 20-5 Ladder logic circuit that controls the package sorting machine

Step 6: When output 01004 unlatches, Rung 8 becomes False and the timer
034 resets. Bit 17 of timer 034 goes to a 0 and deactivates solenoid 1
which causes the arm to swing back to its resting position. Bit 15 then
goes to a 0 to deactivate Unlatch outputs 01001 and 01004.

Steps 1 through 6 are repeated if the detector of box B loads a 006 into the GET instruction of Rung 3. The sequence continues at Rungs 6, 9, 12, 15, and 18. The detection of box C causes Rungs 4, 7, 10, 13, 16, and 19 to be used in the same sequence.

20-4   INJECTION MOLDING MACHINE

An injection molding machine is shown in Figure 20-6. The function of this machine is to form plastic parts, such as toys, computer keys, radio faceplates, or cooking
utensils.

The machine consists of a set of die plates, a cylinder, a storage barrel, and a safety
gate cover. The mechanical movements are activated by a hydraulic motor and various
solenoids. The purpose of the die plate is to create a mold to form the plastic part into
the desired shape. A storage tank with a pre-heater holds plastic pellets. When activated, the pre-heated pellets are gravity fed into the cylinder. The cylinder has two
functions. First, as the piston extends, it compresses the pellets. The compression action creates heat and transforms the pellets into a liquid state. Next, it injects the molten plastic into the mold after the dies are closed. The cylinder piston is extended by a
powerful spring and is retracted by a hydraulic motor that is connected to a rack-and-
pinion mechanism. Before the injection process can be performed, a safety gate which
protects the operator from injury must be in place.

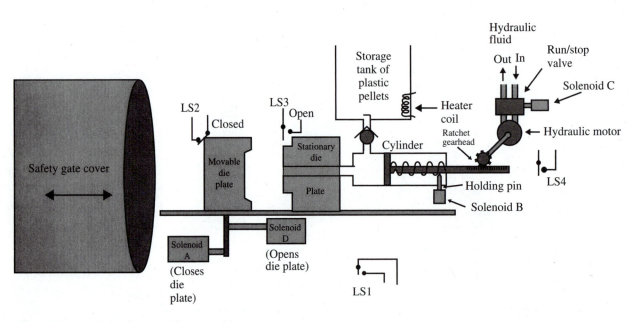

FIGURE 20-6 Injection molding machine

INJECTION MOLDING MACHINE OPERATION

Refer to Figure 20-6. The machine action starts when the operator closes the safety gate. Next, the manual push button is pressed, which energizes a solenoid actuator and closes the moveable die. When fully closed, the plates form the two halves of the mold. Solenoid B is activated and pulls a holding pin free from a notch on the cylinder rod. This action enables the spring to extend the cylinder. Its plunger action injects the liquid plastic into the die. The die remains closed while the plastic cools into a solid state. When the plastic has cured sufficiently, solenoid C is actuated and opens a valve that allows hydraulic fluid to pass through to the motor. The fluid causes the motor to turn a shaft that is connected to the rack-and-pinion mechanism. As the pinion gear turns, it causes the cylinder to retract in preparation for another injection stroke. Finally, the safety gate is retracted and solenoid D opens the movable die to allow the operator to remove the molded part.

This operation requires the use of limit switches, timers, and solenoid action that are controlled by a programmable controller. The PLC program is shown in Figure 20-7. A detailed description of the operation sequence performed by the PLC program follows:

Step 1: With limit switch LS1 closed by shifting the gate and pushing the momentary start button, Rung 1 becomes True.

Step 2: As output 01000 is energized, Rung 2 becomes True and latches. This action activates solenoid A, which causes the die plate to close.

Step 3: When the die reaches the fully closed position, it closes limit switch LS3, which causes Rung 3 to latch. As output 01002 becomes True, it causes Examine-Off contact 01002 to make the condition of Rung 2 False. With Rung 3 True, solenoid B becomes activated and pulls the holding pin out of the notch of the piston rod, allowing the spring to extend the cylinder.

Step 4: As output 01002 becomes energized, Examine-On contact 01002 in Rung 4 becomes True and causes timer 030 to start. While the timer is counting, the molten plastic is allowed to cool and solidify.

Step 5: After 10 seconds, the timer times out, activates Bit 15, and puts Rung 5 into a True condition. With output 01003 energized, solenoid C is activated and starts the hydraulic screw motor, which retracts the piston.

Step 6: When the piston becomes fully retracted, it closes limit switch LS4. With Examine-On contacts 01003 and 11004 True, Rung 6 becomes True. By energizing output 01004, Rung 6 becomes latched, the timer in Rung 4 resets, and Rung 3 becomes False. Solenoid B is de-energized, which causes the holding pin to be reinserted into the notch of the piston rod and Rung 7 to latch.

Step 7: As output 01005 becomes True, solenoid D is activated and the die plate retracts. Contact 01005 in Rung 8 also becomes True.

Step 8: The retraction also causes limit switch LS3 to open, making Examine-On contact 11003 of Rung 6 False, causing Rung 6 to be in a False condition.

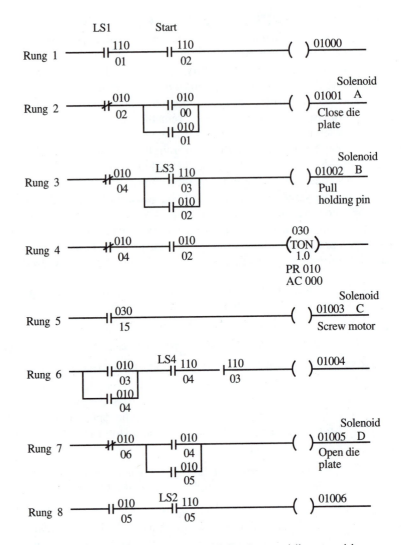

FIGURE 20-7 Ladder diagram of injection molding machine

Step 9: When the die plate is fully retracted, it closes limit switch LS2 and makes Rung 8 True. By energizing output 01006, Rung 7 becomes unlatched.

Step 10: The gate is opened by a machine operator, and the part is removed. The next cycle begins when the gate is closed and the push button is pressed.

20-5   IN-LINE BOTTLE FILLING

Figure 20-8 shows a machine that fills bottles with 16 ounces of medicine. Empty bottles are guided through a chute and into a screw conveyer. A filler carriage with five

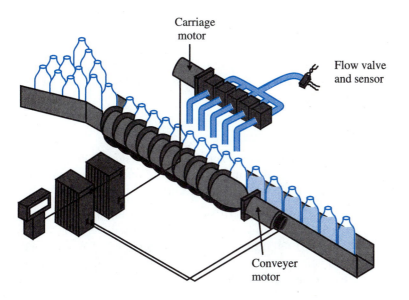

FIGURE 20-8 Pictorial drawing of in-line bottle filler

nozzles is driven by a leadscrew mechanism above the conveyer. When five empty bottles are detected beneath the nozzle heads, the fill operation begins. To speed up the bottle filling operation, the bottles are filled while they are moving. As the conveyer screw is turning, the leadscrew also turns in synchronism so that the carriage moves with the bottles during the fill operation. When each bottle contains 16 ounces, the nozzles are turned off, the leadscrew reverses direction, and the carriage returns at a fast rate of speed to the home position to fill the next batch of bottles.

The two motors that drive the conveyer and carriage are both stepper motors. For each revolution the conveyer screw rotates, the bottles move a horizontal distance equivalent to the width of one bottle. While the nozzles are filling the bottles, the conveyer screw makes five revolutions. The carriage must be synchronized with the conveyer screw so that the nozzles and bottles move together at the same horizontal speed. To achieve this action, the leadscrew motor must turn 14.7 times faster than the conveyer screw motor. Each motor requires 60 steps to make one revolution. The pulses applied to the conveyer stepper motor are supplied at a rate of 60 per second. Therefore, the conveyer makes one revolution per second. Since the stepper motor that drives the leadscrew must turn 14.7 times faster, 882 ($14.7 \times 60 = 882$) pulses must be applied each second during the filling operation.

To supply the stepper motors with the desired pulse rate, thumbwheel switches on the clock pulse generator are set to the proper number. For example, a 060 is preset at the conveyer motor generator so that 60 pulses per second are produced. The number 882 is preset at the carriage leadscrew generator so that 882 pulses per second are produced. The pulses are also supplied to inputs of rungs in the ladder diagram to increment counters. These counters are used to control the number of pulses that are supplied to each stepper motor so that they travel the correct distance.

There is one sensor used by the system. A turbine flowmeter is used to measure how much medicine passes through the flexible tube that feeds the nozzles. Each time the turbine rotates one revolution, a magnetic pulse from a Hall effect sensor is produced. For every ounce that flows through the sensor, the turbine makes one revolution. Therefore, 80 revolutions are produced during the five-second period as 80 ounces (five bottles × 16 ounces) flow through the supply tube. To ensure that the bottles are filled within five seconds, the pump is calibrated to transfer 16 ounces of medicine per second to the tube.

IN-LINE BOTTLE FILLER OPERATION

The programmable controller ladder diagram shown in Figure 20-9 controls the operation of the bottle filler. The following statements describe the sequence of steps that takes place to fill a batch of bottles with medicine.

Step 1: The system is activated by pressing start button 11001 in Rung 1. The push button closure causes the rung to become latched. By energizing output 01000, Rungs 3, 4, 8, and 15 are enabled to drive the motors and pump when other input conditions are met. The rung is unlatched by pressing the stop button.

Step 2: Rung 3 becomes True and causes the conveyer motor to start. Rung 4 becomes True and causes the leadscrew motor to run forward. The True condition of Rungs 3 and 4 also causes Rung 15 to become True and energize the flow valve.

Step 3: As the conveyer motor turns, the stepper pulses are sent to Rung 2 and cause the up-counter to increment. As the leadscrew turns and causes the carriage to move forward, the stepper pulses are supplied to Rung 5 and cause the 031 up-counter to increment. Every time the accumulated value reaches the count of 441, Bit 15 sets. This action increments the counter 032 in Rung 6 and activates the counter reset output in Rung 7. This action recycles the count of CTU 031. The counters in Rungs 5 and 6 are cascaded to obtain a total count of 4410, which is the number of stepper pulses necessary to cause the carriage to travel the required distance.

Step 4: As outputs 01001 and 01002 energize the two motors, they also close the corresponding contacts in Rung 15. This action causes the flow valve to open and start filling the bottles. The Hall effect turbine flow-meter sends pulses to Rung 17 until a count of 80 is reached by CTU 035. As AC = PR, Bit 15 sets, makes contact 03515 in Rung 15 False, and de-energizes the flow valve to stop the medicine.

Step 5: When the stepper pulses have indicated the conveyer and carriage have completed the desired movements, Bit 15 of each counter sets. This condition causes Rungs 3 and 4 to become False and de-energize the motor. It also causes Rung 8 to become True and reverse the direction of the carriage motor toward the home position.

Step 6: As the carriage motor rotates in the reverse direction, stepper pulses increment up-counter 033 in Rung 9. Every time the accumulated

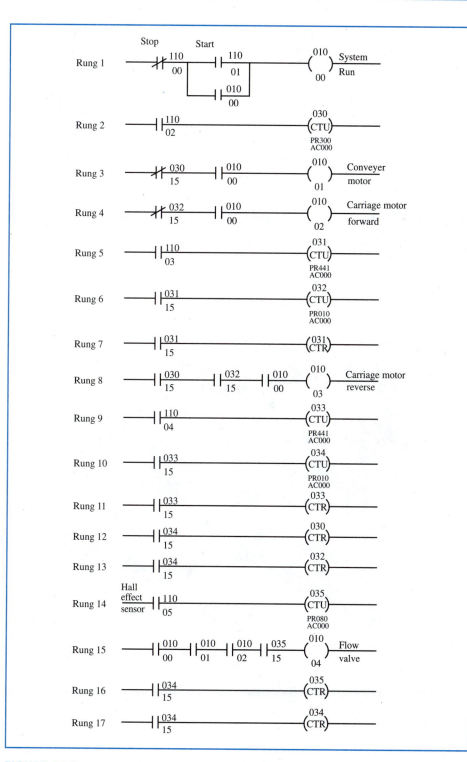

FIGURE 20-9 Ladder diagram of in-line bottle filler

value increases to 441, Bit 15 sets. When Bit 15 goes High, it increments the cascaded up-counter 034 in Rung 10 and activates the CTR output in Rung 11, causing the up-counter 033 to reset back to 000.

Step 7:   When the cascaded up-counters have received 4410 encoder pulses, the carriage is back at the home position. This count is reached when Bit 15 of the 034 CTU output sets. The 1 state of this bit puts Rungs 12, 13, 16, and 17 in the True condition, which causes counters CTU 030, 032, 034, and 035 to reset and begin filling the next batch of bottles.

20-6   BATCH PROCESS PAINT FILLING MACHINE

Figure 20-10 shows a machine that performs a batch process application. The process involves four major steps:

1. Empty cans are put on a rotary table from a conveyer.

2. The cans are filled with paint.

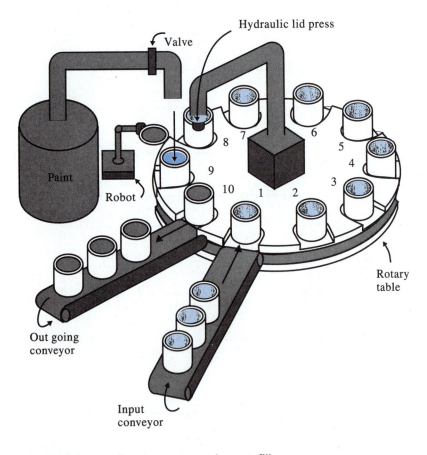

FIGURE 20-10 Batch process paint can filler

3. Lids are placed on the full cans.

4. The filled cans are placed on an outgoing conveyer.

The machine consists of a rotary table with 10 equal sections around its perimeter. The first section (referred to as section 1) accepts empty cans from the input conveyer. The last section (section 10) places sealed cans full of paint onto the outgoing conveyer. A programmable controller is used to perform the controller function of the process. The machine is started by pressing a start button and can be stopped at any time by pressing a stop button.

BATCH PROCESS PAINT FILLING MACHINE OPERATION

When the system is started, the input conveyer inserts an empty can into the slot of section 1 at the table. A limit switch causes the conveyer to turn off when the can is placed in the proper position. The in-position signal opens the valve that causes paint to flow into the can in section 9. While being filled, the can depresses a weight sensitive transducer that produces a voltage proportional to the weight of the can. This voltage is sent to an analog module in the programmable controller. An A/D converter in the module produces a digital signal proportional to the analog signal. When the weight reaches 128 ounces, the valve shuts off and a robotic arm places a lid on the can. When an optical sensor detects reflected light from the lid, a motor turns on to rotate the table. An incremental encoder coupled to the table detects that the table has rotated 36 degrees, the motor turns off, and a signal is sent to a hydraulic press, which will secure the lid onto the can. When a pressure sensor detects that the lid is sealed, it switches the conveyer motors on. As the outgoing conveyor removes the filled can from the table, the input conveyer transfers an empty can into section 1 and the batch process cycle is repeated.

The programmable controller ladder diagram in Figure 20-11 is used to control the operation of the paint filling machine. The following statements describe the sequence of steps that takes place to fill the paint cans:

Step 1:   The sequence begins by pressing the start button 11001 in Rung 1. As output 01000 becomes energized, the rung becomes latched and corresponding contacts in Rungs 2, 4, 7, 9, and 12 close. The system becomes de-energized by pressing the stop button 11000 in Rung 1.

Step 2:   The closure of the start button also causes Rung 2 to latch. By energizing output 01001 in Rung 2, both conveyer belts begin running. When an empty can from the input conveyer enters the slot in position 1, it activates a limit switch that turns Rung 3 True. As the output of Rung 3 energizes, it creates a False condition at Examine-Off contact 01002, Rung 2 unlatches, and the conveyer belts stop. Rung 4 is now in the True condition.

Step 3:   When Rung 4 becomes True, its output causes a valve to open and paint pours into the empty can at position 9. As the paint fills the can, a strain gauge measures the weight and sends an incrementing signal to an analog module. An A/D converter produces a proportional digital signal that is stored in register address 110. A GET and PUT

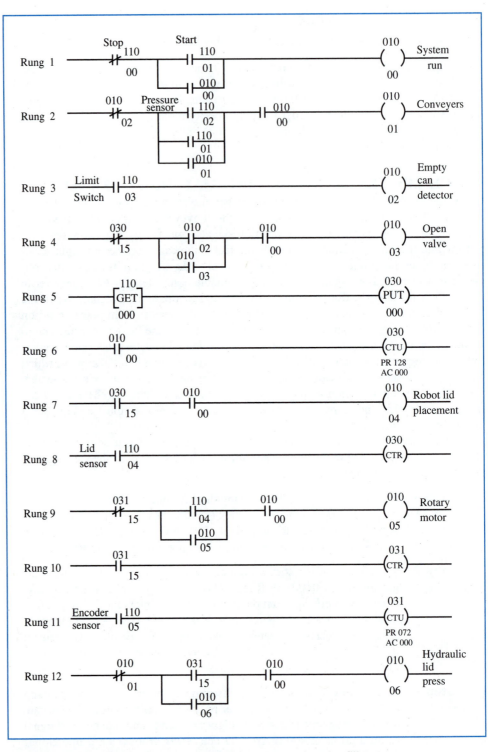

FIGURE 20-11 Ladder diagram of batch process paint can filler

instruction in Rung 5 transfers the signal from the register to the accumulator of up-counter 030 in Rung 6.

Step 4: When the accumulated value equals the preset number of 128, Bit 15 of CTU 030 causes the corresponding Examine-Off contact in Rung 4 to become False and the valve to shut off. It also activates Rung 7, causing a robotic arm to place a lid on the can. When sensor 11004 in Rung 8 detects the placement of the lid, it causes counter 030 to reset and Rung 9 to latch, starting a motor that rotates the table.

Step 5: As the table and incremental encoder rotate, the up-counter 031 in Rung 11 increments. After 72 pulses, the accumulator equals the preset value. Bit 15 of the counter then sets, causing the corresponding Examine-Off contact in Rung 9 to become False. The result is that the Rung unlatches and the rotary motor stops.

Step 6: The setting of 03115 also causes the corresponding contact in Rung 12 to become True. As output 01006 energizes, the rung latches and causes a hydraulic press to seal the lid on the can. The rung also becomes latched.

Step 7: As the hydraulic ram presses down on the lid, the pressure reaches a certain level when the lid is sealed. When this level is detected by a pressure sensor, it puts the Examine-On contact 11002 in Rung 2 in a True condition. The batch process begins the next cycle as both conveyers are activated. The outgoing conveyer transfers a filled can from the slot in section 10, and the input conveyer inserts an empty can into the slot of position 1.

20-7   X-Y AXIS MEDICINE DISPENSER

Figure 20-12 shows a machine that fills a tray of vials with medicine. A gantry table is used to move the X and Y axes, and a dispenser injects the proper amount of liquid into each bottle. The X-Y movement of the gantry is powered by using two permanent magnet DC motors to provide a full range of motion. The position of each axis is monitored by an incremental encoder. The dispenser is located above the gantry in a fixed position slightly higher than the top of each vial. The system is designed to provide high speed, accuracy, and smooth motion that prevents spillage.

X-Y AXIS MEDICINE DISPENSER OPERATION

The process begins with the gantry in the home position, where the dispenser is directly above the farthest bottle shown in the diagram. Next, a valve opens and the medicine is gravity fed into the bottle. By using a flowmeter in the valve body of the dispenser, a volume of 140 grams is measured to ensure that the valve is turned off at the correct time. The next step is to energize the X-axis motor until 25 pulses are detected by the encoder. When the 25 pulses are read, the motor shuts off and the dispenser is positioned above the adjacent bottle to repeat the filling process. After the

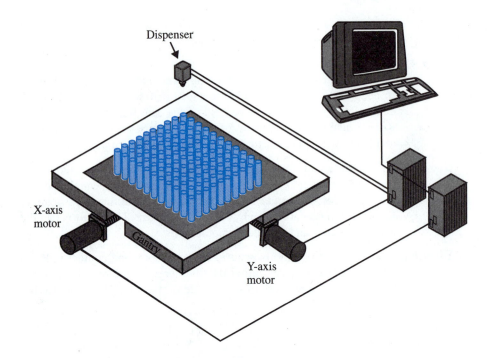

FIGURE 20-12 X-Y axis medicine dispenser

second bottle is filled, the procedure is repeated until all of the bottles in the first row are filled. The X-axis motor then reverses direction until the dispenser is located above the first bottle. Next, the Y-axis motor turns on and moves the table horizontally. When 25 clock pulses are read by the Y-axis encoder, the dispenser is located above the second row of bottles. The dispensing process is repeated until all bottles in the remaining rows are filled. When the closest bottle in the illustration is finished, the gantry is moved to the home position and the system automatically turns off. After the completed tray is replaced by an empty tray of bottles, the start button is pressed and the filling process is repeated.

The programmable controller ladder diagram shown in Figure 20-13 is used to control the operation of the medicine dispenser. The following statements describe the sequence of steps that takes place to fill the vials on the tray.

Step 1:   Figure 20-12 shows the gantry in the home position. The filling operation is started by pressing the start button in Rung 1 of Figure 20-13. By pressing the stop button at any time during the operation, output 01000 is de-energized and the system becomes disabled.

Step 2:   Closing the start button causes the valve to open and dispense medicine into the bottle. A turbine Hall effect flow sensor is used to measure the amount of liquid that is dispensed into the bottle. It generates one pulse for every gram that flows. As these pulses are sent to input 11002 of Rung 4, up-counter 030 increments until it reaches a count of 140.

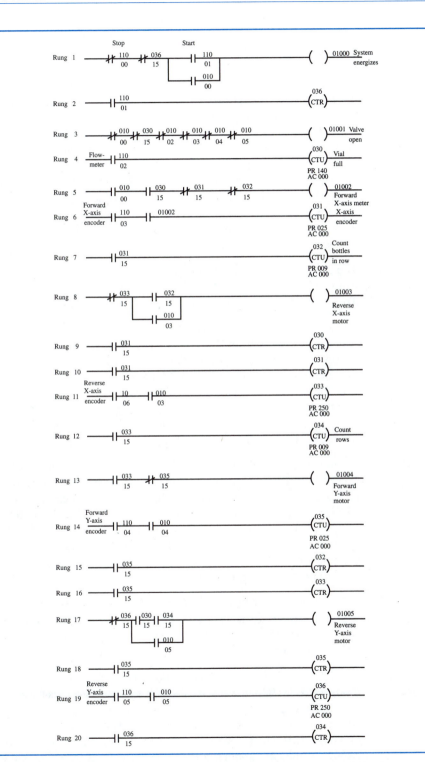

FIGURE 20-13 Ladder logic circuit that controls the medicine dispenser

Step 3: When the accumulated value of CTU 030 reaches 140, its Bit 15 becomes set, causing the corresponding Examine-On contact in Rung 5 to become True. As output 01002 in Rung 5 is energized, it causes the corresponding Examine-Off contact in Rung 3 to become False and the flow valve to shut off. It also causes the forward motor to start and run. Coupled to the motor shaft is an optical encoder that sends pulses to input 11003 of Rung 6. Examine-On contact 01002 in Rung 5 becomes True only when the forward motor is energized. When the accumulated value of CTU 031 reaches a count of 25, Bit 15 sets and causes the corresponding Examine-Off contact in Rung 5 to become False. As a result, the motor turns off. It also causes the Examine-On contact in Rung 7 to become True and increment the accumulated value of CTU 032, indicating the first row is completed. The adjacent bottle is now located under the dispenser.

Step 4: The set condition of Bit 15 in CTU 031 also causes Rungs 9 and 10 to become True. As counters 030 and 031 are reset, Rung 3 becomes True and the valve opens to fill the next bottle.

Step 5: Steps 2 through 4 are repeated until the accumulated value of 9 is reached by CTU 032 in Rung 7, at which time its Bit 15 sets. Bit 15 causes corresponding contact 03215 in Rung 8 to become True. After the tenth bottle is full and CTU 030 reaches a count of 140, Rung 8 is True and the X-axis motor causes the gantry to reverse direction. As output 01003 energizes, the corresponding Examine-On contact in Rung 11 becomes True, which enables the encoder pulse to increment the count of CTU 033.

Step 6: When the count of CTU 033 reaches 250, Bit 15 sets and increments CTU 034 in Rung 12 to indicate that a row of bottles has been completed. It also causes the corresponding Examine-On contact in Rung 13 to become True and energizes the Y-axis motor. Coupled to the motor shaft is an incremental encoder. As output 01004 in Rung 13 energizes, it causes the corresponding Examine-On contact in Rung 14 to become True. As the encoder pulses are fed to input 11004, the counter in Rung 14 increments.

Step 7: When the accumulated value of CTU 035 increments to 025, its Bit 15 sets. The setting of Bit 15 causes the corresponding Examine-On contacts in Rungs 15, 16, and 18 to become True and makes counters 032, 033, and 035 reset.

Step 8: Steps 2 through 7 are repeated until the accumulated value of 009 is reached at CTU 034 in Rung 12. As its Bit 15 sets, it causes corresponding Examine-On contact 03415 in Rung 17 to become True. When the count in CTU 030 of Rung 4 reaches 140, Examine-On contact 03015 in Rung 17 becomes True and output 01005 energizes to reverse the motor. As the corresponding Examine-On contact 01005 in Rung 19 becomes True, it enables the Y-encoder to increment CTU 036.

Step 9: When the accumulated value of CTU 036 reaches the count of 250, its Bit 15 sets. The result is that the corresponding Examine-Off contact in Rung 17 becomes False and causes the motor to stop. Bit 15 also causes Examine-Off contact 03615 in Rung 1 to become False. The system is de-energized until the start button is pressed to fill the next tray. The start button closure also causes the condition of Rung 2 to become True and resets CTU 036.

20-8   PLASTIC PROFILE EXTRUSION MACHINE

Figure 20-14 shows a profile extrusion machine that performs a continuous process control function. This machine operates by feeding plastic pellets into a barrel that contains a fluted auger. As the auger turns, it moves the pellets through heat zones

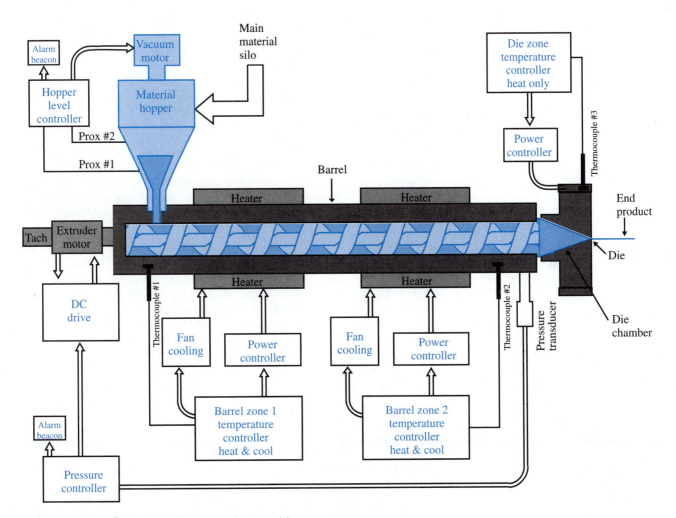

FIGURE 20-14 Plastic profile extrusion machine

until they are transformed into a homogeneous melted core. The melted plastic is forced out the end of the auger into a chamber. Because the chamber shape is tapered, the melted plastic is forced through a die located at the narrow end under high pressure. The shape of the die determines the type of product that is made. For example, the finished product could be flexible vinyl tubing, door weather-stripping material, edging for flower gardens, or a host of other plastic items.

Four functions of the machine are controlled: the hopper level, the barrel temperature, the pressure of the material in the chamber, and the temperature of the chamber.

1. *Hopper Level Control.* The purpose of the hopper level control system is to maintain the proper level of raw plastic in the hopper at all times. Two capacitive proximity sensors are located at different levels within the hopper. Proximity switch 1 (prox #1) monitors the lowest point and proximity switch 2 (prox #2) monitors the highest level. When prox #1 detects no material, the hopper level controller energizes the hopper motor. The motor produces a vacuum in the hopper, which draws the pellets inside through a piping system from the main plastic storage silo (usually located outside the plant). When prox #2 senses material, the hopper level control turns off the hopper vacuum motor. When enough material has been used to lower the material below prox #1, the cycle is repeated. The reason for using two proximity sensors is to introduce hysteresis into the system. Using only one sensor would cause constant cycling of the vacuum system. A warning beacon is used to provide a visual indication to the operator if raw material is not replenished within a predetermined time period.

2. *Barrel Zone Temperature Control.* The barrel of the extruder is basically a length of high-pressure pipe that has been machined to a very close tolerance to house a fluted auger. As the pellets are fed into the rear of the barrel, the auger moves the material through heated zones toward the die chamber. When completely melted, the plastic is forced through the die to produce a continuous length of finished product. Heating zones gradually raise the temperature of the plastic as it passes through the bore. (Raising the temperature abruptly would prevent the plastic from reaching the proper hardness when it solidifies after passing through the die.) In the machine shown in Figure 20-14, there are two zones, each consisting of a temperature sensor, a temperature controller, a power controller, a fan blower, and an electric heater. The temperature controller uses a thermocouple sensor located in the barrel assembly to measure the temperature. The temperature controller then compares the measured temperature with the process setpoint temperature. If the measured temperature is less than the setpoint value, the temperature controller feeds a signal into the power controller to cause the power applied to the heaters to increase. If the measured value is higher than the setpoint value, the output to the power controller is lowered. If the measured value is much higher, the cooling output of the temperature controller is activated to energize a cooling fan.

3. *Pressure Control Loop.* The purpose of the pressure control loop is to provide a constant amount of pressurized material at the head (die outlet) of the extruder. If a constant pressure is maintained at the die outlet, the product

that is extruded will be uniform. A closed-loop pressure controller is used to keep the pressure within the required tolerance range. Typically, the pressures will run between 500 and 3000 PSI, depending on the product size and composition of the raw material. If the pressure sensed by the pressure transducer falls below the operating set point, the pressure controller will automatically increase its output to the DC drive unit. The result is that the DC motor that drives the auger will speed up. As the material is forced into the chamber at a faster rate, the pressure increases. If the pressure rises above the operating setpoint value, the pressure controller's output is lowered. As the auger slows down, the pressure in the chamber decreases.

The components of the pressure control loop follow:

○ *Pressure Transducer.*   This sensing device converts the chamber pressure into a proportional signal and sends it to the pressure controller.

○ *Pressure Controller.*   This equipment compares the set point to the input it receives from the pressure sensor and provides a control signal to the DC drive. It also provides an output to the alarm beacon if the pressure drops below a minimum level.

○ *DC Drive.*   This unit receives a control signal from the pressure controller and provides power to the DC motor that drives the auger.

4. *Die Chamber Temperature Control.*   The die chamber controls the temperature of the plastic in a manner similar to that used by the barrel zone temperature control. It does not, however, use a fan motor to cool the chamber if the temperature rises above a certain level.

HOPPER LEVEL CONTROL CIRCUITRY

The ladder logic diagram of a programmable controller shown in Figure 20-15 is used to control the level of the material in the hopper. The following statements describe the sequence of steps required to refill the hopper each time the material drops to a certain level.

Step 1:   When the material in the hopper is above sensor 1, the proximity switch is in the closed condition and the Examine-Off contact 11000 is False. Therefore, the rung is False.

Step 2:   When the material in the hopper drops below the level of sensor 1, the proximity switch opens and the Examine-Off contact is True, which causes output 01000 to energize. As corresponding Examine-On contact 01000 in Rung 2 becomes True, Rung 2 latches and the vacuum motor begins to run.

Step 3:   When the material in the hopper reaches the level of sensor 2, the proximity switch closes, Examine-On contact 11001 becomes True, and Rung 3 becomes True. As output 01002 energizes, it causes the corresponding Examine-Off contact in Rung 2 to become False. Rung 2 unlatches, and the vacuum motor turns off.

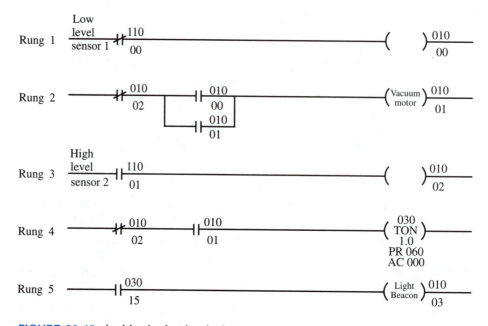

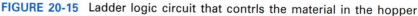

FIGURE 20-15 Ladder logic circuit that contrls the material in the hopper

Step 4: While the motor is running, contact 01001 in Rung 4 is True. Until the material reaches the level of proximity switch 2, Examine-Off contact 01002 in Rung 4 remains True. During this time, the continuity of Rung 4 is True and timer TON 030 increments. If the material in the hopper is replenished within 60 seconds, the high level sensor closes and causes Examine-Off contact 01002 to become False to stop the timer from timing out. If the piping becomes plugged and the material hopper is not replenished within 60 seconds, the timer times out, causing Examine-On contact 03015 in Rung 5 to be True and the warning beacon to energize.

BARREL ZONE TEMPERATURE CONTROL CIRCUITRY

The temperature of each barrel zone is controlled by a digital computer. The digital system shown in Figure 20-16(a) describes one method of controlling the temperature digitally. An opto-triac is used as the load control device that supplies current to the heater coils. By changing the triggering time of the triac, the amount of current supplied to the coils can be varied. The triggering time is controlled with an 8-bit binary code. Recall that an 8-bit binary ranges from 0 to 255. By loading one binary number into a storage register, the triggering pulses can be supplied at 256 different times during an alternation, thus providing full control.

Each time V_A crosses the zero volt X-axis, the zero crossing detector circuit generates a pulse (P_1). For 60-hertz AC power, P_1 is produced every 8.33 milliseconds. As a pulse occurs, it provides a signal to input PL of the storage register, which enables an 8-bit binary number to be parallel loaded from the computer. P_1 also clears the 8-bit counter by applying a temporary pulse to its CLR input line. The counter immediately increments from 0 as 256 clock pulses are supplied every 8.33 milliseconds. When the counter reaches the same value as the storage register, output A = B of the magnitude comparator produces a trigger pulse to the opto-coupler triac. The amount of time that the triac conducts during each alternation is determined by the 8-bit number loaded into the storage register.

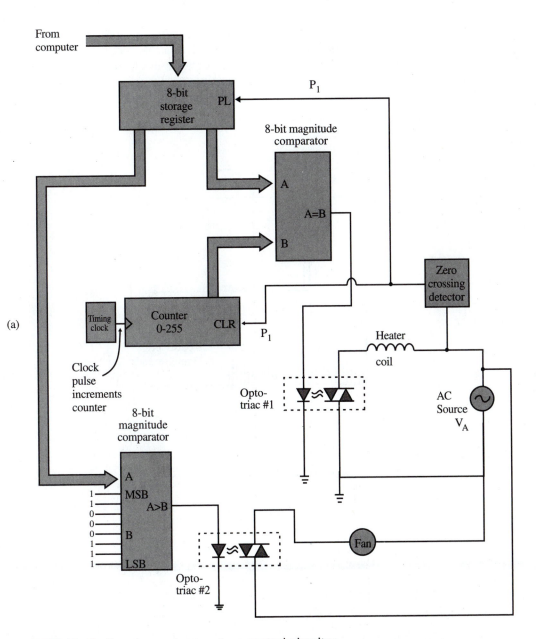

(a)

FIGURE 20-16 Barrel zone temperature control circuitry

Suppose that the storage register is loaded with a binary number equivalent to the decimal value of 1 when P_1 occurs. Simultaneously, the counter is cleared to 0. Only one clock pulse will be required to increment the counter value to 1 and cause the magnitude comparator to generate a trigger pulse. The triac will conduct for essentially the full 8.33 milliseconds and supply maximum current to the heater coils. Suppose that when P_1 occurs, the storage register is loaded with a binary value equivalent to the decimal value of 255 and the counter is cleared. It will require 255 clock pulses before the trigger pulse is generated by the magnitude comparator. This means that the triac will be turned on very late during the alternation, with little or no current being passed to the heater coil. Thus, the coils will generate little heat.

A waveform timing diagram of the circuit is shown in Figure 20-16(b). T_1 is the varying time between the occurrence of P_1 and the triggering pulse. T_2 is the amount of time that the triac conducts and provides load current to the heater coils.

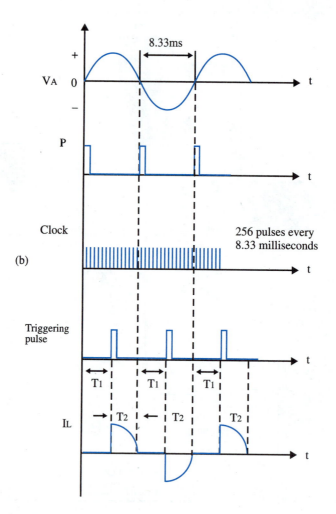

FIGURE 20-16   *(continued)*

The number fed into the storage register is called a control code and is supplied by the computer. Figure 20-17 shows how the control code is determined. A temperature sensor continually outputs an analog voltage representing the temperature. The analog value is transformed into a proportional 8-bit digital number by an A/D converter. This digital value is called the *temperature code* because it represents 256 different temperature values, ranging from 900 to 1156 degrees. When the temperature code is received by the computer, a decoder converts the value into an address memory location. Each location is the address of an 8-bit memory register. There are 256 different memory locations that store the control codes. This block of registers is referred to as a *look-up table*. Each time the crossover circuit in Figure 20-16(a) produces pulse P_1, the memory register that is addressed parallel loads its contents into the storage register. When the counter contents increment to the same number as the storage register, opto-triac #1 is triggered.

Each temperature code has a specific control code associated with it to control the current supplied to the heater coil for that temperature. The computer continuously

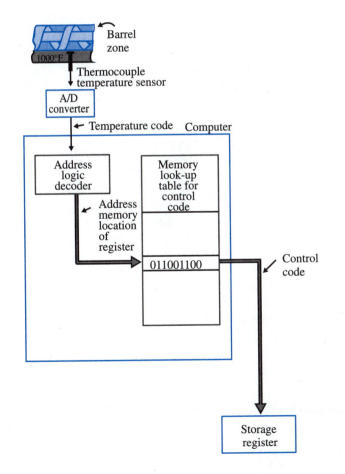

FIGURE 20-17 Motor controlled by computer

samples the zone temperature at a rate of 120 times a second. Suppose the desired temperature in a zone is 1000 degrees F. The binary control code loaded into the storage register is 01100100 (100 in decimal represents one hundred degrees above the minimum range number of 900). If the temperature lowers, a new control code with a lower number is parallel loaded into the storage register. Because the counter reaches the value of the storage register faster, opto-triac #1 is triggered sooner during the alternation and supplies more current to the coil. The larger current will then cause the coil temperature to increase toward the required level. Likewise, if the temperature rises above 1000 degrees, the control code loaded into the storage register becomes larger. Because the counter does not reach the register value until late in the alternation, the triac will supply less current to the coil after it fires. Whenever the temperature rises to 1100 degrees, the value in the storage register will be 11001000 (200 in decimal represents two hundred degrees above the minimum range number of 900) or greater. This situation causes the second magnitude comparator to produce a trigger pulse that fires opto-triac #2. When this occurs, a fan is activated so that the cooling of the coil is accelerated. The same results occur if the temperature rises above 1100 degrees.

PRESSURE CONTROL LOOP CIRCUITRY

The auger that moves the raw material through the barrel is driven by a DC motor. The speed of the motor is determined by the amount of current that flows through its armature. The greater the current, the faster the motor turns. The current is controlled by a trigger circuit similar to the one used to control the current flowing through the coils at the temperature zones. However, instead of using an opto-triac, an opto-SCR is used to supply a varying DC voltage in a DC drive. The trigger signal fed to the SCR is supplied by a computer. The internal structure of this computer is different than the one shown in Figure 20-17 because two feedback signals—pressure and speed—are used to determine when the trigger signal fires.

Figure 20-18 shows the architecture of the motor speed computer and how it processes the control signal. An incremental encoder measures the motor speed by counting the number of pulses it samples within a precise fraction of a second. This number is fed to the computer as an 8-bit digital number. A sensor located at the inlet of the chamber produces an analog voltage that is proportional to the pressure. An A/D converter transforms the voltage into an 8-bit binary value compatible with the computer. The speed and pressure digital codes are fed to a logic decoder that selects an appropriate register address just as it does for decoding the temperature digital code. The following sequence of steps is performed by the processor once every 1/120 of a second:

1. The computer decodes the pressure code and addresses the appropriate control code in the first half of the look-up memory table.

2. The pressure control code is stored in register B and also temporarily stored in register C.

3. The computer decodes the 8-bit sample of the motor speed and addresses the appropriate control code corresponding to the speed. The speed control codes are located in the second half of the look-up table.

4. The speed control code is stored in register A.

5. The computer compares register A with register B. If B (pressure) is greater than A (speed), the register C value is increased by the difference.

   (Refer to Figure 20-16 to follow step 6. The DC drive control circuitry operates in a similar manner.)

6. The computer loads the new code value from register C into the 8-bit storage register, the counter begins to increment, and the triac triggering pulse occurs when the numbers are equal.

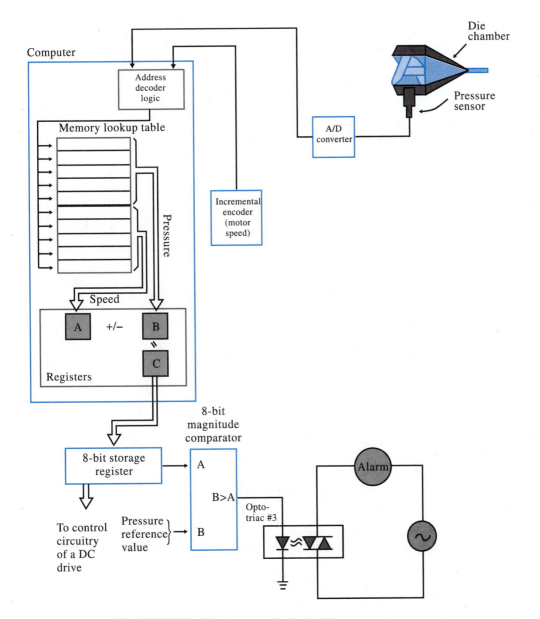

FIGURE 20-18 Pressure control circuitry

If the motor speed is less than desired, register A would be greater than register B, and register C would be decreased by the difference to cause the motor speed to increase. If the pressure drops below a certain level, the control code in register C decreases and causes the B > A output of the magnitude comparator to go High. The result is that opto-triac #3 turns on, and an alarm beacon is activated. When the desired speed is reached, register A equals register B and register C is not changed. The motor remains at the same speed until a new speed is determined because the pressure has changed.

DIE CHAMBER TEMPERATURE CONTROL CIRCUITRY

The die chamber temperature control operates the same way as the barrel zone temperature control except it does not have an over temperature cooling fan.

CHAPTER PROBLEMS

(Refer to Figure 20-1 to answer problems 1 and 2.)

1. What is the purpose of the three exclusive-OR gates?

2. When input A of the magnitude comparator is less than input B, the motor

   turns _____ (clockwise, counterclockwise, nowhere).

(Refer to Figures 20-2 and 20-3 to answer problems 3 and 4.)

3. If the oil pressure drops and opens the pressure switch, the ____.
   a. drill motor will stop turning
   b. solenoid will de-energize and the drill bit will retract

4. When the timer times out, ____.
   a. the drill motor will stop turning
   b. the solenoid will de-energize and the drill bit will retract
   c. both a and b will occur.

(Refer to Figures 20-4 and 20-5 to answer problems 5 and 6.)

5. When six beams are broken, an arm will swing out if sensor ____ detects the presence of a box.
   a. X   b. Y   c. Z

6. How long is each swing arm activated to deflect a box down the desired conveyer belt?

(Refer to Figures 20-6 and 20-7 to answer problems 7 through 9.)

7. The closure of LS4 will affect which solenoid?

8. Which of the following two conditions must be met to activate a cycle on the injection molding machine? ____
   a. LS1 closed   c. LS3 closed
   b. LS2 open   d. Push start button

9. The timer is used to ____.
   a. delay the beginning of the cycle by 10 seconds after the start button is pressed
   b. delay the separating of the dies by 10 seconds to allow the plastic to solidify
   c. delay the beginning of the cycle by 10 seconds after the gate has been closed

(Refer to Figures 20-8 and 20-9 to answer problems 10 and 11.)

10. How many ounces does the flowmeter measure during each cycle before the flow valve closes? ____
    a. 80   b. 16   c. 5

11. How many stepper pulses are applied to the carriage motor to develop movement from the extended position to the home position? _____

(Refer to Figures 20-10 and 20-11 to answer problems 12 and 13.)

12. A ____ is used to determine when a paint can is full.
    a. level detector   c. flowmeter
    b. pressure sensor

13. How many encoder pulses are counted when the rotary table moves from one position to the next? _____

(Refer to Figures 20-12 and 20-13 to answer problems 14 and 15.)

14. How many encoder pulses are counted as the gantry moves from one vial to an adjacent vial? ____
    a. 9   b. 25   c. 140

15. A (An) ____ is used to measure when a vial is full.
    a. level detector   c. pressure sensor
    b. flowmeter   d. optical sensor

(Refer to Figures 20-14 to 20-16 to answer problems 16 to 18.)

16. If the measured pressure in the chamber is too low, the auger motor speed ____.
    a. increases   b. decreases

17. Pellets that enter the hopper from the silo are ____.
    a. gravity fed   c. transfered by an auger
    b. sucked in by a vacuum pressure

18. If the temperature in a heated zone becomes too high, the ____.
    a. system shuts down   c. cooling fan is energized
    b. auger moves faster

ANSWERS TO ODD-NUMBERED PROBLEMS

CHAPTER 1

1. Motion, process or open-loop, closed-loop

3. Negative

5.
Motion Control	*Process Control*
Hall effect speed sensor	Float

7. The prescribed input value that indicates the desired operating point of the manufacturing process.

9. True

11. disturbance

13. (a) Excessive time lag,
(b) Large disturbances

CHAPTER 2

1. increasing

3. $R_F/R_{IN} = 5K/1K = 5$

5. +2.4 volts

7. square wave

9. (a) < (b) = (c) >

11. It determines if one binary number is greater than, less than, or equal to the other binary number.

13. SCR

15. True

17. reverse

19. $2^5 = 32 - 1 = 31$ 10V/31 = .3225V

21. 8

23. high

25. $f = \dfrac{1.44}{(R_A + 2R_B)\,C}$ $\dfrac{1.44}{120k \times 10 \text{ ufd}} = 1.2\text{Hz}$

CHAPTER 3

1. Control section

3. — Process disturbances
— The controller cannot adjust the output to match the process demand.

5. % Differential Gap $= \dfrac{\text{Differential Gap}}{\text{Total Control Range}}$

$= \dfrac{8}{80}$

$= .1 \times 100$

$= 10\%$

7. True

9. increasing

11. (b) integral

13. rate

15. adds to

17. True

19. True

21. If

SECTION II

CHAPTER 4

1. torque, opposite
3. (b) at right angles
5. True
7. increases
9. (a) 864 oz-in
11. (a) armature connections
13. (a) increases armature current
15. (b) a constant speed rating
17. (d) decrease, increase
19. True
21. (c) high starting torque
23. (c) a higher starting torque than a shunt motor
25. (c) more constant speed
27. True

CHAPTER 5

1. (b) Brushes and commutator
3. stator, rotor
5. synchronous speed
7. Percent Slip $= \dfrac{\text{Synchronous Speed} - \text{Rotor Speed}}{\text{Synchronous Speed}}$

$$\times 100 = \frac{200}{3600} \times 100 = 5.56\%$$

9. (b) reverse the main and auxiliary windings with respect to each other.
11. lags
13. two
15. universal
17. — Blown fuse
    — Broken connection
19. increase
21. synchronous
23. overexcited

CHAPTER 6

1. True
3. — Tape transport systems
    — Computer peripheral devices
5. stepper
7. (b) counterclockwise

9. $n = \dfrac{Y \times S}{6}$

$$= \frac{12 \times 360}{6}$$

$$= 720$$

11. (b) the auxiliary winding

SECTION III

CHAPTER 7

1. — High starting torque
    — Higher horsepower
3. (e) All of the above
5. (d) All of the above
7. — They rectify the AC supply, which causes current to flow in one direction through the armature.
    — By turning on at any time during the alternation, they control the amount of current that flows through the armature, thus controlling motor speed.
9. increases, negative, harder, sooner, more
11. (c) maximum
13. Constant Load; (c, e) Hoist, Extruder
    Constant Horsepower Load; (a) Machine Tool Lathe
    Variable Torque Load; (b, d) Pump, Centrifugal Fan

CHAPTER 8

1. False
3. $N = \dfrac{60 \times 60}{4(\text{pole pairs})} = \dfrac{3600}{4} = 900\,\text{RPM}$
5. inverters
7. (a) Rectifies AC into pulsating DC.
    (b) Filters pulsating DC signals into a pure DC voltage.
    (c) Converts a DC voltage into a simulated AC waveform.
9. (a) 1

11. — Phase control rectifier
    — A chopper circuit

13. True

15. (b) PWM and (c) Vector Drive

17. (b) PWM

19. | *Speed* | *Current* |
     |---|---|
     | Resolver | Clamp-on ammeter |
     | Tachometer | Hall effect transducer |
     | Encoder | |

SECTION IV

CHAPTER 9

1. (e) All of the above

3. (b) decreased

5. less

7. (b) decrease

9. (c) stay the same

11. decreases

13. 64.7 − 14.7 = 50psig

15. (b) highest

17. Trim pots to balance the bridge network, if the resistor values change due to component aging.

19. Pump

21. out of

23. Boiler

CHAPTER 10

1. thermal energy

3. Conduction, Convection, Radiation

5. (b) gives off heat

7. $5 \times 10 = 50$

9. Expansion

11. (a) thermoelectric

13. Measures oven and furnace temperatures, molten plastic, and nuclear reactors.

15. linear

17. negative

19. Emitted, reflected, transmitted

21. Several thermocouples connected in series.

23. Ratio pyrometer

CHAPTER 11

1. Cubic feet, gallons, liters

3. $F = \dfrac{WS}{L} = \dfrac{100lb \times 100ft/min}{5\ feet} = 2000lb/min$

5. Temperature, pressure

7. To identify the type of flow currents that are likely to occur so that the proper flowmeter can be selected.

9. (c) Venturi tube

11. is

13. is

15. 20

17. Pressure

CHAPTER 12

1. Feet, meters

3. — To determine if there is enough material to complete a job.
   — Determine inventory.
   — To prevent a container from underfilling.

5. a direct

7. (b) a non-invasive

9. (a) an invasive

11. decreases

13. At the bottom

15. (b) liquids

17. (b) non-invasive

CHAPTER 13

1. (c) target

3. (b) axial

5. (a) loses

7. (a) increases, (b) decreases

9. (d) All of the above

11. (b) infrared

13. True

15. Opposed Sensing Method

17. Convergent Sensing Method

19. Specular Sensing Method

21. 2

23. $R = \dfrac{E}{10 \times (1X)}$　　　$X = \dfrac{(+50)}{200V}$

$= \dfrac{200V}{10 \times 1.25}$　　　$= .25$

$= \dfrac{200V}{10 \times 1.25}$　　　$C = 100 \ mA^2$

$= 16 \ ohms$　　　$C = .001 \ ufd$

SECTION V

CHAPTER 14

1. rails, rungs

3. (c) left, (d) right, (a) top, (b) bottom

5. (c) both latching and interlocking

7. Hand-held programmer
Dedicated terminal
Microcomputer

9. 2

11. False

13. (e) None of the above.

15. (c) Executive
(d) Scratch pad
(b) Application memory
(a) Data table

17. words

19.

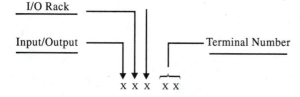

CHAPTER 15

1. program

3. 100 scans

5. (a) True　　　(c) False
(b) False　　　(d) True

7. (a) create a parallel circuit

9. (d) All of the above.

11. True

13. Address　　　**15.** Address
Time Base　　　　Preset Value
Preset Value　　　Accumulated Value
Accumulated Value

17. Data Transfer
Data Compare

19. 0–9

21. (e) All of the above.

CHAPTER 16

1. (A + B)C = X　　　**7.** False

3. 12　　　　　　　**9.** (c) files

5. True

11. (c) Place a logic state that activates the output
in the bit of consecutive memory locations.

13. sourcing

15. 65,536

17. (a) Thermocouple　　　(b) Stepper Motor
(a) Vision　　　　　　(c) PID
(a) Bar Code　　　　　(c) Fuzzy Logic
(c) Radio Frequency

SECTION VI

CHAPTER 17

1. (b) RPM speed and (c) direction

3. True

5. (d) All of the above

7. — By measuring the time interval between
pulses.
— By counting the number of pulses within a
timer period.

9. resolution　　　　**13.** 1111

11. True　　　　　　**15.** 0111

17. (d) All of the above   **21.** (a) rotary, (b) linear

19. (c) resolver   **23.** (c) twisting

CHAPTER 18

1. (b) Velocity Control   **11.** (c) linear or rotary

3. torque   **13.** (b) rotary, (a) linear

5. (e) Damping   **15.** Z axis
   (c) Bandwidth
   (d) End Point
   (b) Home Position
   (a) Holding Torque
   (f) Traverse Rate

7. (a) Controller

9. linear

CHAPTER 19

1. increases   **7.** dynamic

3. proportional   **9.** (c) critically damped

5. maximum

11. increases

13. minimal

15. (b) integral

17. (b) derivative amplifier

19. — Potentiometer
   — Computer set-up menu

CHAPTER 20

1. They convert Gray code to pure binary code.

3. (b) solenoid will de-energize and the drill bit will retract

5. (b) Y

7. Solenoid B

9. (b) delay the separating of the dies by 10 seconds to allow the plastic to solidify

11. 4410

13. 72

15. (b) flowmeter

17. (b) sucked in by a vacuum pressure

GLOSSARY

absolute optical encoder an optical rotary encoder which produces binary data that represents an angular position.

absolute pressure the gas pressure above a perfect vacuum.

accuracy the degree to which an output will attempt to match the input command signal.

AC motor a device that converts adjustable alternating current frequency and voltage to rotating mechanical energy.

actuator an element of a control system which converts electrical, hydraulic, or pneumatic energy into work.

Analog-to-Digital Converter (ADC, A/D) a device which converts an analog voltage applied to its input into a proportional digital output.

armature a laminated iron core with wire wrapped around it in which an emf is induced as it rotates inside the stator field.

armature reaction the distortion of the main field flux lines due to the interaction with the magnetic field around the armature.

astable multivibrator a circuit which generates a continuous square wave output.

auto boost a circuit in the electronic AC motor drive which causes current and torque to increase when the motor is running at a low RPM.

backlash the movement within the space between mating parts or gears.

bandwidth the measure of how quickly the controlled quantity tracks and responds to the command signal.

bang-bang position servo a position servo which moves the load from one end position to another end position very rapidly and without the need for extreme accuracy.

base the region of a transistor which controls the amount of and variation in the current the collector receives from the emitter.

base driver the section of a variable voltage inverter (VVI) which amplifies and isolates control circuit signals.

base speed the rated speed at which the motor operates.

batch process a sequence of timed operations executed on a product being manufactured.

binary code a base-2 number system that contains two characters, 0 and 1, which are grouped together to form words used by computer-based circuitry.

bubbler a tube through which air is forced to determine level by counting the number of bubbles emitted within a specified period of time.

bus line the output terminal of the intermediate section in an AC electronic motor drive.

C

chopper control a method in which the output of the intermediate circuit section in a VVI AC drive is pulsated to vary the average DC voltage.

closed-loop a method of control in which feedback is used by a system to produce a controlled process dictated by a command signal.

collector the region of a transistor which receives current from the emitter and sends (outputs) it as voltage.

color mark sensor a photoelectric sensor which detects the contrast between two colors.

commutation the switching action of the brushes and commutator which causes the armature to rotate.

commutator 1. part of a DC motor that is constructed as a split ring, each segment of which is connected to an end of a corresponding armature coil. 2. a device that employs an opening and closing switch action that occurs when the brushes and commutator segments make and break contact with each other. This process always causes current to flow through the armature in the proper direction.

comparator a device which produces various output signals by comparing the signals applied to its inputs.

compensating windings small windings wired in series with the armature windings which cancel the magnetic field of the armature windings and eliminate armature reaction.

compression 1. the process of storing additional gas into a confined container. 2. reducing the size of the confined container that holds a fixed quantity of gas.

conductance the process by which heat is transferred by a solid.

continuous process one or more operations performed simultaneously as a product is produced during a manufacturing process.

contouring a motion control method which causes the load to make continuous movements along curved or straight lines.

control circuit 1. a circuit in an AC drive that controls when the inverter's switching devices will turn on or off. 2. the section of an electronic AC or DC motor drive which provides a way for the operator to preset operational parameters and control speeds.

controlled variable the actual process that is being controlled by an open- or closed-loop system, such as temperature or pressure.

controller an element which is considered the "brain" of a closed-loop system.

convection the transfer of heat through fluids such as liquids and gases.

convergent sensing method a photoelectric sensor in which the light source and receiver are mounted next to each other at the same angle from the vertical axis. The object to be detected is used to reflect light at one set distance.

converter the section of an electronic AC drive which converts AC line voltage to a pulsating DC voltage.

Counterelectromotive force (CEMF) a voltage generated inside an electric motor which is always of opposite polarity to the applied voltage.

critically damped a system which is tuned so accurately that it causes the controlled variable to reach its end position or desired state very quickly without overshoot.

current limiting a circuit in an electronic DC drive that prevents excessive current from flowing throughout the components when the motor is loaded down.

D

damping the prevention of overshoot of the load past the end point in a motion control system or the desired state in a process control system.

data compare the process of comparing the numerical contents of two PLC registers and making decisions based on their value and the type of instructions used.

data manipulation a category of instructions that enables words to be moved within the memory of a PLC.

data transfer   the process of moving the contents stored in one memory register to another memory location in a PLC.

DC bus line   the output terminal of the intermediate section of a DC electronic motor drive.

deadband   see **differential gap**.

density   1. the weight of a certain volume of liquid. 2. the weight per unit volume of a fluid.

depth-of-field   the distance on either side of a photoelectric sensor's focus point.

derivative   a control scheme whereby the controller produces an output that is proportional to the rate that the error signal changes. This function is also called rate control.

diac   a bidirectional, two-terminal, solid-state device that is used to trigger a triac.

differential gap   the range above and below the set point reached by the controlled variable before the controller element turns an actuator on or off.

differential pressure   the difference in gas pressure between any two points in a system.

differentiator   an operational amplifier circuit that performs the derivative function.

diffuse sensing method   a photoelectric sensor in which the light source and receiver are mounted next to each other. The object to be detected is used to reflect light from the emitter back to the receiver.

Digital-to-Analog Converter (DAC, D/A)   a device that translates digital data into an analog voltage.

displacement   the amount of material replaced by a sensor probe as it measures level.

disturbance   a factor that upsets the manufacturing process, causing a change in the controlled variable.

drive   an electronic device that controls the speed, position, acceleration, deceleration, and torque of electric motors.

drive controller   a device that converts the fixed source voltage and frequency of a power line to an adjustable voltage and frequency.

duty cycle   the ratio of time a squarewave signal is high to the total time period of one cycle.

dynamic   the state in which the controlled variable is moving or changing.

dynamic braking   1. a method in which a resistor is used to absorb kinetic energy to stop a motor quickly. 2. a motor braking action which uses a resistor to absorb kinetic energy from the rotating armature and load.

E

efficiency   the ratio of the power produced by a motor's shaft to the power supplied by the electrical source.

emitter   the region of a transistor which sends (inputs) current toward the collector.

end point   the desired location to which a load is moved by a motion control position system.

error detector or (comparator)   1. the element of a closed-loop control system which produces an error signal by comparing the set point to the feedback signal. 2. a device that produces various output signals by comparing the signals applied to its inputs.

error signal   the difference between the desired response and the actual response.

excess gain   a measure of the amount of light energy that falls on the receiver beyond the minimal amount of light required to operate a photoelectric amplifier.

F

feedback signal   the signal or data fed to the comparator of a closed-loop system from an actuator or processor to indicate the response to the command signal.

feedforward   the process of providing information to the controller element device which indicates that a change is going to occur.

fiber optic   transparent strands of glass or plastic that transfer light for photoelectric sensing.

field of view   the dispersion angle at which a photo-electric sensor can effectively detect light from the emitter.

field poles   the electromagnets in a motor that are stationary.

float   a spherical element that rides on the surface of the material it is measuring to determine its level.

flow   the transfer of material from one location to another.

flow rate   the measurement of flow which is determined by how fast a material is moving past a given point.

fluid   a liquid or gas commonly used in a process control system.

following error   the proportional error during motion resulting from the load movement lagging behind the desired movement specified by the command signal.

full load   the maximum power a motor can provide to drive its rated mechanical load.

Fuzzy Logic   a form of artificial intelligence that enables a computer to simulate human reasoning.

gage pressure   the gas pressure above or below atmospheric pressure.

Gray code   a number system which uses multibits of 0s and 1s, with only one of the bits changing when incrementing or decrementing the count.

H _____

head   a term commonly used to describe the height of a liquid above the measurement point.

High/Low speed adjustment   the adjustment made to prevent a motor from attaining full maximum speed or absolute minimum speed, either of which may cause the motor to short out.

holding torque   the amount of force required to keep the rotor of a motor from falling out of a stationary position.

home position   the reference position from which movements are measured in a motion control system.

hydrostatic   pressure exerted equally in all directions at points within a confined fluid.

hydrostatic pressure   the resultant pressure obtained from multiplying the height times the density of a liquid.

hysteresis   see **process lag time**.

I _____

incremental optical encoder   an optical rotary encoder which has tracks of equally spaced slots that indicate position or speed. Position is determined by counting the number of slots that pass by a photo sensor. Speed is determined by counting the number of slots that pass by a photo sensor within a period of time.

indexing   controlled positioning movements caused by signals sent from a controller.

inductosyn   a linear feedback device which produces a specific code or pulse for each position.

industrial controls   the automated equipment that monitors and controls the operation of a manufacturing process.

inferred measurement   a condition wherin one type of measurement is taken to find the value of another type of measurement.

input/output module   a type of module which interfaces the internal circuitry of a PLC to outside equipment.

instability   an action which occurs in a motion control operation when the device being positioned oscillates because of overshoot.

instrumentation   a term commonly used to describe process control; it refers to the instruments that control and monitor the condition of the process.

integral   a control scheme whereby the controller produces an output that is proportional to the length of time an input signal has been applied. This function is also called reset control.

integrator   an operational amplifier circuit which performs the integral function.

interface   1. the boundary between two media, such as water and air. 2. the connecting together of two different circuits.

intermediate circuit   the section of an electronic AC drive that filters the pulsating DC voltage from the converter and varies the DC voltage supplied to the bus line.

interpoles   small windings wired in series with the armature which counteract armature reaction by producing a local field that restores the main flux lines to the original neutral plane.

inverter   a term commonly used to identify an electronic AC motor drive. It is also the section of the drive that provides high currents to power the motor.

IR compensation   a method in which an electronic DC drive detects armature current to use as a feedback control signal.

J

jump   a change in the normal sequence of program execution in a PLC.

L

level   the height at which a material fills a container.

light sensor   the element of a photoelectric sensor which detects the absence or presence of an object. It is also referred to as a detector or a receiver.

light source   the element of a photoelectric sensor which supplies the light beam to a light sensor. It is also referred to as an emitter or transmitter.

limit switch   a mechanical switch activated by physical contact with a moving object.

linear displacement transducer   a type of linear-motion position sensor which detects the distance from a reference point to a measured object by counting the time interval between a launching pulse and a return pulse.

Linear Variable Differential Transformer (LVDT)   a type of linear-motion position sensor which uses transformer action to produce a signal that is proportional to distance.

load   1. the type of device or equipment to which the sensor output signal is applied. 2. the demand on an actuator by the device to which power is delivered.

loop gain   the ratio of output speed to the following error.

M

main field   the magnetic field that forms between two poles of a magnet which interacts with the magnetic field of the armature.

manufacturing process   the operation performed by an actuator to control a physical variable, such as motion or a process.

mass flow rate   the measurement of flow which is determined by the weight of materials that move during a specific time period.

master-slave   a servo system in which coordinated motion is achieved between two control systems by one (slave) system monitoring the condition of the other (master) system.

measurement device   an element in a closed-loop control system which detects a controlled variable and produces an output signal that represents its status. Other terms used are detector, transducer, and sensor.

methods of detection   the various physical arrangements of a light source and its receiving elements which allow an object to be detected by a sensor.

monostable multivibrator   a circuit which produces a temporary logic level voltage after an activating signal is applied to its input. Also known as a one-shot.

motion control   an industrial control system that controls the physical motion or position of an object.

motor action   the conversion of electrical energy to mechanical energy resulting from the interaction of two or more magnetic fields.

multiple axis positioning   a motion control method that causes a load to make many movements horizontally and vertically.

N

negative temperature coefficient the characteristic of a sensor in which its resistance decreases when the ambient temperature to which it is exposed increases.

neutral plane 1. the axis that is at a right angle to the main field flux lines. 2. the plane that is perpendicular to the flux lines of the motor's main field.

no load a condition during which the motor operates when the physical load is disconnected from the motor shaft.

O

offset the error that remains between the set point and the desired output condition after the controller element has caused a transient response. See **steady-state error**.

On-Off control the most basic type of control system in which the actuator is either fully on or off.

open-loop a method of control where there is no feedback to initiate self-correcting action for the error of the desired operational conditions.

operational amplifier an integrated circuit that performs several types of linear circuit operations.

operator control an electronic device which provides the operator a means of controlling the start, stop, directional, and speed functions of a motor.

opposed sensing method a photoelectric sensor in which the light emitter and detector are positioned opposite each other. The target is detected when it blocks the light beam.

optical encoder a rotary feedback device which produces a specific code or series of pulses for each position.

optoelectronic device an interface device that uses light energy to pass signals from one circuit to another circuit.

overdamped a system that is sluggish and responds to a changed command signal by causing the controlled variable to reach its end position or desired state very slowly and without overshoot.

overload the condition in which an electric motor stalls because it does not have enough power to move a load it is driving.

P

partial-load a condition in which the physical load a motor is driving is reduced from the full-load condition.

PID a controller that uses proportional, integral, and derivative control in one unit.

pipe size the diameter of a pipe that carries fluid.

positive temperature coefficient the characteristic of a sensor in which its resistance increases when the ambient temperature to which it is exposed increases.

potentiometer a variable resistor capable of converting mechanical motion into an electrical voltage variation.

power supply a module in a PLC which provides voltages that are necessary to operate the circuitry throughout the system.

pressure the force exerted over a surface area.

process control an industrial control system that regulates one or more variables during the manufacturing of a product.

process lag time the response lag time of a control system to a set point change or a disturbance.

processor scan cycle a sequence of steps performed by the microprocessor in a PLC during which the input conditions are examined, logic decisions are performed, and appropriate signals are applied to the output.

processor unit the "brain" of the PLC which coordinates and controls the operation of the entire system.

program a list of instructions that guides the PLC through a desired operation.

Programmable Logic Controller (PLC) a solid-state control system which has a user programmable memory for the storage of instructions to implement specific functions such as logic operations, timing, counting, arithmetic, and data manipulation.

programming unit a module in a PLC which provides a way for the user to enter data, edit, and monitor the program stored in the processor unit.

proportional band the range at which a controller produces an output signal proportional to the error signal applied to its input.

proportional control a control scheme whereby the controller produces a signal that is proportional to the error signal.

proximity switch a sensor which detects the absence or presence of an object without making physical contact.

pull-out torque a condition in which the physical load connected to a synchronous motor causes the rotor to fall out of synchronism with the stator field.

Pulse-Width Modulation (PWM) a method in which the output of the inverter section of a PWM or AC Vector drive is pulsated to produce a simulated AC waveform.

R

radiation the transfer of thermal energy through a vacuum.

rate a control scheme whereby the controller produces an output that is proportional to the rate that the error signal changes. This function is also called the derivative control.

rated load torque the torque developed by a motor when driving its rated mechanical load.

regenerative braking a motor braking action which returns power from kinetic energy to the power supply.

regenerative feedback a method which regulates speed or stops a motor quickly by generating a voltage that opposes the supply voltage.

relay ladder logic a programming language used by programmable controllers.

repeatability the range in which the output position of the servosystem will come to rest whenever a given input command signal is repeated.

reset a control scheme whereby the controller produces an output that is proportional to the length of time an input signal has been applied. This function is also known as integral control.

Resistance Temperature Detector (RTD) a type of temperature sensor which exhibits a change in resistance when subjected to a change in temperature.

resolution 1. the number of degrees per count produced by an optical encoder. 2. the number of equal divisions into which a digital-to-analog divides the reference voltage.

resolver a transducer which converts rotary or linear position into an electrical signal by the interaction of electromagnetic fields between movable and stationary coils.

retroreflective sensing mode a photoelectric sensor in which the light source and receiver are mounted next to each other. The light source shines into a reflector that returns the light beam to the receiver. The target is detected when it blocks the light beam.

Reynolds number (R number) a numerical scheme that assigns values to express fluidity of a moving fluid. It represents the ratio of the liquid's inertial force to its drag (viscous) forces.

rodgauge a dipstick inserted into the material being measured to determine level.

rotor the armature assembly of the motor which physically rotates to drive the shaft.

run-away a condition in which the physical load driven by a DC motor is disconnected, causing the motor to accelerate until it breaks apart.

S

Schmitt trigger a device which produces sharply defined square waves from distorted signals or sinewaves.

sensitivity a measure of the amount of change in light intensity that is required by a photoelectric sensor to cause a switching action at its output.

sensor circuitry the circuitry of a photoelectric sensor.

sensor response time   the maximum amount of time that elapses from an input detection signal until the output switches.

sequencer   a controller in a PLC which operates an application through a sequence of events.

servo motor   a specialty motor that uses a closed-loop feedback signal to control its position and speed.

servo valve   a transducer which converts a low-energy electrical signal into a proportional, high-energy hydraulic output.

servomechanism   a closed-loop motion device which automatically controls velocity and position.

set point   the input applied to the elements of a control system that represents the desired value of the controlled variable.

shift register   a memory device in a PLC that is used to store or move binary words within itself.

sight glass   a transparent tube connected to the side of a vessel to measure level.

Silicon-Controlled Rectifier (SCR)   a three-terminal unidirectional solid-state device that passes current in one direction when it is triggered.

single axis positioning   a motion control method that causes a load to make linear movements in both directions.

single-phased   a three-phase motor condition in which one of the power lines is open.

sinking   the type of sensor connection that draws conventional current into its output terminal through the load device to which it is connected.

slip   the difference between rotor speed and synchronous speed in an AC motor.

soft start   when a DC drive accelerates the motor slowly instead of abruptly.

sourcing   the type of sensor connection that supplies conventional current from its output to the load device to which it is connected.

specific gravity   the relative weight of any liquid when compared to water at a 60-degree F temperature.

specular sensing method   a photoelectric sensor in which the transmitter and receiver are placed at equal angles from an object. The object to be detected is used to reflect light at one set distance. This method is used to differentiate between shiny and dull surfaces.

speed regulation   the amount the motor speed will vary with a change in the physical load it is driving.

stability   the characteristic of good speed regulation in a closed-loop velocity control system.

static   a state in which the controlled variable does not move or change appreciably within an arbitrary time interval.

static head   the force developed at the bottom of a tank which results from the weight of fluid placed above it.

stator   a term that refers to the main field pole assembly.

steady-state error   the error that remains between the set point and the desired output condition after a transient response by the controller element is completed. Also known as offset.

step angle   the number of degrees per arc the rotor of a stepper motor moves per step.

stepping rate   the maximum number of step movements the stepper motor can make in one second.

stiffness   the reluctance of an actuator to deviate from the desired position specified by the command signal.

synchronous speed   the rate at which the magnetic field in the stator rotates.

T _____

tachometer   a device which measures the angular velocity of a rotary shaft.

target   the object detected by a sensor.

thermal energy   molecular movement that creates heat.

thermistor a temperature sensor which exhibits a large change in resistance when subjected to a small change in temperature.

thermocouple a temperature sensor which converts heat (to which it is exposed) into a voltage.

three-phase power AC power consisting of three alternating currents of equal frequency and amplitude, each differing in phase from the others by one-third of a period.

thyristor a four-layer semiconductor device.

torque the measure of a motor's rotary force.

tracking the movement of the load by the actuator as it attempts to follow a changing command signal in a motion control position system.

transient response the response time of the actuator to sudden changes of speed or position signals.

transistor a solid-state device that amplifies and switches on and off.

traverse rate the fast rate of speed at which a load is moved from one position to another in a point-to-point servo system.

triac a three-terminal bidirectional solid-state device that passes AC current when it is triggered.

U

underdamped a system that is tuned with too much gain and responds to a changed command signal by causing the controlled variable to react too quickly and overshoot its end position or desired state.

Unijunction Transistor (UJT) a three-terminal solid-state device which is used primarily to trigger an SCR.

V

vacuum the absence of a gas inside a container. Any reduction of pressure compared to atmospheric pressure is called a partial vacuum.

velocity the speed at which an object or a material moves.

viscosity 1. the ability of a liquid to flow and take the shape of a container. 2. the ease with which a liquid flows.

Voltage-to-Frequency (V/Hz) ratio a method in which the bus line voltage of an AC drive is varied in direct proportion to the inverter output frequency to maintain constant current and motor torque.

volumetric flow rate the measurement of flow which is determined by the volume of material that flows during a given time period.

W

wound rotor a rotor that has coils wound around an iron core in an AC motor.

INDEX

self-diagnostics, 348, 356
special purpose modules, 409–13
troubleshooting, 413–15
Programmable logic controller
(PLC), 329. *See also*
Programmable controllers
Programmers, hand-held, 342
Programming languages, 357–58,
387, 388
Programming units, 341–42
Programs, computer. *See* Computer
programs
Proportional band, 72–73
Proportional control, 70–76, 244,
286, 477
Proportional gain, 476
Proportional-integral control, 76–79
Proportional-integral-derivative
(PID) control, 79–83, 412
Proportional-integral-derivative
(PID) mode, 477
Proportional op amp, 75, 79
Proportional-plus-integral (PI)
regulator, 476
Proportional position servos,
466–70
Proximity detectors, 301–6
psi (pounds per square inch), 222
psia (pounds per square inch,
absolute), 226
psid (pounds per square inch,
differential), 227
psig (pounds per square inch,
gage), 226
P-type (positive) material, 18
PTC. *See* Positive temperature
coefficient
Pull-in torque, 153
Pull-out torque, 149, 153
Pulleys, 456–57
Pulse width modulation (PWM)
drives, 206–12
Pulses, 55
duration, 60, 162
optical encoders, 422–24
in PWM drive, 211
smoothing, 202
for stepper motors, 413
strain, 439, 440

Pumps, 287, 489–91
Purge method level measurement,
292–93
Push-roller, 300
PUT, 374, 375, 377, 378
PWM. *See* Pulse width modulation
drives
Pyrometer, 259

Q

QPO. *See* Quadrature phase output
Quadrature phase output (QPO),
433

R

Rack, 455
Rack-and-pinion transmission,
455–56
Rack assembly, 340
Radial approach, 302
Radiation, 245
Radiation pyrometers, 259, 261–64
Radiation thermometry, 259–64
Radiator, 248
Radio frequency (RF) modules,
411
Rails, 332
RAM, 348
Rate control. *See* Derivative control
Rated load torque, 113
Ratio pyrometers, 262–63
Ratiometric tracking, 430–32
RC network, 42, 180
in SCR, 37, 38
in sensors, 321
R/D. *See* Resolver-to-digital
converter
Reactive power, 154, 155
Reactive voltages, 146
Rebounding, 300
Receiver, 310, 311
Rectification, 37, 179
Rectifiers, 179–80
chopper control, 204

converters, 194–95
in inductive proximity switch, 303
in LVDTs, 437–38
phase control, 201–2
three-phase, 206–9
Reed switch, 290
Reference. *See* Set point
Reference column, 229, 230
Reference phase output (RPO), 432
Reference point, 299
Reference voltage, 50, 202
Reflex scanning, 314
Refrigeration, 246–48
Regenerative braking, 187–88
Regenerative feedback. *See*
Positive feedback
Regions, 18
Registers, 349–50, 513
Relaxation oscillator, 40
Relay diagrams, 358
Relay ladder logic, 330–34, 358
arithmetic operations, 380–83
circuits, 334–39
counter instructions, 370–74
data manipulation instructions,
374–80
relay-type instructions, 359–65
timer instructions, 365–70
Relay-type instructions, 359–65
Relays, 331
Release point, 308
Remote devices, 178
Remote sensing, 317
Repeatability, 470, 474, 476
Reservoir, 9–10
Reset control. *See* Integral control
Resistance, 12, 20
of metals, 253
in photodiodes, 310
in potentiometers, 420–21
of rheostat, 202, 206
in strain gauges, 233–34
in thermistors, 252–54
of transistor, 180
Resistance furnaces, 246
Resistance-start induction-run
motors, 139–41, 145
Resistance temperature detectors
(RTDs), 254